Management Systems
Conceptual Considerations

Management Systems
Conceptual Considerations

Fourth Edition

Peter P. Schoderbek, Ph.D.
The University of Iowa

Charles G. Schoderbek, Ph.D.
Rome, Italy

Asterios G. Kefalas, Ph.D.
The University of Georgia

Homewood, IL 60430
Boston, MA 02116

Sponsoring editor: Craig S. Beytien
Project editor: Rita McMullen
Production manager: Bette Ittersagen
Cover designer: Robin Basquin
Compositor: Graphic World Inc.
Typeface: 10/12 Times Roman
Printer: R. R. Donnelley & Sons Company

Library of Congress Cataloging-in-Publication Data

Schoderbek, Peter P.

Management systems : conceptual considerations / Peter P.

Schoderbek, Charles G. Schoderbek, Asterios G. Kefalas. — 4th ed.

p.   cm.

Includes indexes.

ISBN 0-256-07897-1

1. Management.   2. System analysis.   I. Schoderbek, Charles G.

II. Kefalas, Asterios G.   III. Title.

HD38.S36515   1990

658.4′032 — dc20   89–15246
                                                                                               CIP

Printed in the United States of America
1 2 3 4 5 6 7 8 9 0 D O 6 5 4 3 2 1 0 9

Let no man say that I have said nothing new; the arrangement of the material is new. Just as the same words differently arranged form different thoughts.

Blaise Pascal

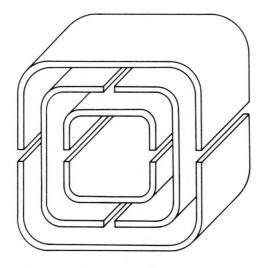

Systems are components in hierarchies. In another sense, hierarchies are systems and each system is itself a hierarchy.

Ervin Laszlo

Foreword

A NOTE ON SYSTEMS SCIENCE*

World War II marked the end of an era of Western culture that began with the Renaissance, the Machine Age, and the beginning of a new era, the Systems Age.

In the Machine Age man sought to take the world apart, to analyze its contents and our experiences of them down to ultimate indivisible parts: atoms, chemical elements, cells, instincts, elementary perceptions, and so on. These elements were taken to be related by causal laws, laws which made the world behave like a machine. This mechanistic concept of the world left no place in science for the study of free will, goal seeking, and purposes. Such concepts were either taken to be meaningless or were relegated to the realm of pure speculation, metaphysics.

It was natural for men who believed (1) the world to be a machine that God had created to serve his purposes, and (2) that man was created in His image, to seek to develop machines that would do man's work. Man succeeded and brought about mechanization, the replacement of man by machine as a source of physical work.

Work itself was broken down into its smallest elements. These were assigned to machines and men, and assembled into the modern production line. Productivity increased and work was dehumanized. The process which replaced man by machine reduced man to behaving like a machine — to performing simple, dull, repetitive tasks.

With World War II we began to shift into the Systems Age. A system is a whole that cannot be taken apart without loss of its essential characteristics, and hence it must be studied as a whole. Now, instead of explaining a whole in terms of its parts, parts began to be explained in terms of the whole. Therefore, things to be explained are viewed as parts of larger wholes rather than as wholes to be taken apart. Furthermore, nonmechanistic ways of viewing the world were developed which were compatible with the older mechanistic view and which

---

*By Russell L. Ackoff who is Professor of Systems Sciences in The Wharton School of Finance and Commerce, University of Pennsylvania. He was Editor of *Management Science* from 1965 to 1970, and is now on the Advisory Board of the *Mathematical Spectrum,* on the Editorial Board of *Management Decision* and is Advisory Editor in Management Sciences for John Wiley & Sons. He is coauthor of more than 10 books and is the author of more than 100 articles in a variety of journals and books. Copyright © 1972, The Institute of Management Sciences.

made it possible to deal with free will, goal seeking, and purposes within the framework of science. Instead of thinking of men in machine-like terms we began to think of machines in man-like terms.

The Systems Age brought with it the Post-Industrial Revolution. This very young revolution is based on machines that can observe (generate data), communicate it, and manipulate it logically. Such machines make it possible to mechanize mental work, to automate.

In the Machine Age science not only took the world apart, but it took itself apart, dividing itself into narrower and narrower disciplines. Each discipline represented a different way of looking at the same world. Shortly before World War II science began to put itself back together again so that it could study phenomena as a whole, from all points of view. As a result, a host of new interdisciplines emerged such as Operations Research, Cybernetics, Systems Engineering, Communications Sciences, and Environmental Sciences. Unlike earlier scientific disciplines which sought to separate themselves from each other and to subdivide, the new interdisciplines seek to enlarge themselves, to combine to take into account more and more aspects of reality. Systems Science is the limit of this process, an amalgamation of all the parts of science into an integrated whole. Thus, Systems Science is not a science, but is science taken as a whole and applied to the study of wholes.

Systems Science goes even one step further; it denies the value of the separation of science and the humanities. It views these as two sides of the same coin; they can be viewed and discussed separately, but cannot be separated. Science is conceived as the search for similarities among things that appear to be different; the humanities as the search for differences among things that appear to be the same. Both are necessary. For example, to solve a problem we need to know both (1) in what respects it is similar to problems already solved so that we can use what we have already learned; and (2) in what respects it differs from any problem yet solved so that we can determine what we must yet learn. Thus the humanities have the function of identifying problems to be solved, and science has the function of solving them.

The emergence of Systems Science does not constitute a rejection of traditional scientific and humanistic disciplines. It supplements them with a new way of thinking that is better suited than they to deal with large-scale societal problems. It offers us some hope of dealing successfully with such problems as poverty, racial and other types of discrimination, crime, environmental deterioration, and underdevelopment of countries. Systems Sciences may not only be able to assure man of a future, but it may also enable him to gain control of it.

Preface

While probing the historical antecedents of systems thinking, future scholars will, no doubt, uncover its roots in the fertile soil of the present. Only the latter half of this century, however, will they characterize as the Age of Systems. Probably most often to be cited in their studies will be the treatment of organizations and similar social phenomena not as detached parts but as integral wholes. Underlying this shift in emphasis would be the belief that only in such a holistic framework could a system's essential elements be realistically understood.

Unlike other intellectual movements, sprung from a specific discipline and nurtured within restrictive and narrow confines, systems thinking was born free of particularized scientific fetters and reared in an interdisciplinary environment. Because it deals with wholes in general and not with specific parts, it transcends the usual strictly defined disciplinary boundaries of the traditional sciences. It has indeed become an interdisciplinary movement.

This text is designed for an introductory course that could well be entitled Management Systems, or Introduction to Systems Concepts. Since it is introductory, it spans many diverse areas, all of which have something to contribute to an understanding of systems fundamentals. The overriding objective has been to provide the wherewithal for a clear understanding of systems postulates that underlie all applications. While it is indeed introductory, this in no way eliminates the need for the sustained serious concentration that characterizes the student of the sciences or of the humanities. Grappling with concepts, like engaging one's fellow in a game of intellectual wizardry, can be a highly satisfying but demanding pastime. We are convinced that the efforts exerted to master the systems concepts will be richly rewarding.

Because of its diversity and scope, this text can serve as an excellent point of departure in a systems curriculum. The material presented in these chapters should provide the springboard in advanced courses for class discussion on loftier, more discerning and recondite levels.

This book is organized around four logically related conceptual areas.

In Part One, The Basics of Systems, the reader is given a brief overview of systems thinking, its origins, and its development from the seminal ideas of biologists to the developed and developing concepts of its present-day general practitioners. The inquisitive student, while not encumbered with an extensively detailed account of the evolutionary history of systems thinking, will find here enough information to begin his or her own digging in this new and exciting field of investigation.

To appreciate what he or she is digging for, the reader will find useful the presentation of the ABCs of systems, the various concepts needed to understand what a system really is—its inputs, outputs, and environment, and its behavioral characteristics.

Part Two is concerned with the presentation of several of the approaches purporting to satisfy the tenets of systems thinking. It begins with two chapters on cybernetics, the science of communication and control. Cybernetics not only appears to have much promise in pointing the way to controlling any type of system, but it has lived up to its promises. These chapters are critical for the reader because of the substantive nature of the material. The subject matter, it should be pointed out, goes far beyond the basic notion of feedback control. It has implications for goal formulation as well as for systems effectiveness—topics that will be given detailed treatment in a later chapter. That cybernetics is not the sum and substance of systems thinking will be evident when one considers the Operations Research (OR) approach, the Systems Engineering approach, and the Systems Analysis approach. While each of these approaches employ elements of cybernetic systems, still they merit consideration on their own.

Organizations scan the environment for information that is needed for decision making—the prime determinant of survival in a highly competitive world. The main thrust of Part Three, therefore, is on managerial decision making, based on information scanned from the firm's external environment. First the concept of information is explored and its relevance to the organization is discussed. Since organizations acquire information from their respective environments, one would expect that different environments would call for different information-acquisition behaviors. This is precisely the thrust of these chapters. After exploring the various types of organizational environments, the discussion shifts to how organizations go about acquiring information and how they alter their structures and processes in doing so. The last chapter in this section is entitled "Organizational Effectiveness." Different organizations measure their effectiveness by different criteria. The major theoretical approaches to this question are addressed.

Part Four begins with a brief review of some of the major concepts of systems thinking. It then moves on from the micro approach to the macro approach in its illustration of the systems approach to problems. It first treats the organization as a system and then only does it delve into the present efforts to treat the universe from a systems perspective. It then examines forecasting and the use of sophisticated information technology to assist executives in decision making. The field of artificial intelligence is explored and its development from the joint parentage of cybernetics and information technology. Part Four concludes with a description of an Executive Support System whose aim is to provide environmental knowledge to the decision maker.

ACKNOWLEDGMENTS

We are particularly grateful to those professors who took the time to critically review our past editions and to suggest concrete improvements thereto. A special debt of thanks is owed to Samuel J. Mantel, Jr., The University of Cincinnati; Martin E. Rosenfeldt, University of North Texas; Charles Bartfeld, American University; Ed Baylin, John Abbot College; William J. Murin, University of Wisconsin-Parkside; and Helen W. Wolfe, Post College.

Peter P. Schoderbek
Charles G. Schoderbek
Asterios G. Kefalas

Contents

PART ONE

The Basics of Systems

Here and elsewhere we shall not obtain the best insight into things until we actually see them growing from the beginning.

Aristotle

PART OUTLINE

CONNECTIVE SUMMARY

The purpose of this part is not to present a complete account of the history of systems thinking in the so-called hard sciences or even to display for the general reader the most important attempts by social scientists to transplant systems thinking into their own fields of interest. Rather, the intention is to explore aspects of systems thinking essential for an understanding of its nature: to elucidate the role of systems thinking in the development of systems theories and disciplines within the social sciences. Unfortunately, the social scientist's eagerness to "get right to the point," a technique widely recommended and practiced in academia, has almost universally forced writers to skip this important and indispensable task.

For this reason Part One examines the origins of systems thinking and identifies certain common elements — theories, principles, postulates, etc., developed in dealing with physical phenomena in the study of man-made systems.

Since the term *system* appears in the first chapter, it was necessary to develop a language that could serve as the medium over which the messages of the rest of the book could be transmitted. To this end, Chapters One and Two present a basic vocabulary of systems definitions, parameters, properties, and classifications. The authors strongly believe that the material presented in these first two chapters is the minimum amount of "homework" that students of management must do before tackling systems theory.

A careful study of this text will not immediately make everything perfectly clear and fully understandable. It will, however, bring together ideas garnered from many disciplines and weld them into a meaningful whole, into a system. Perhaps for the first time students of management will get a glimpse of the "big picture" that has so often eluded them in the past.

As Part One unmistakably shows, the book is not meant to be a cookbook type of text on management systems. The role of theory is given prime consideration. Here the authors have built upon the foundations laid by systems theorists, researchers, philosophers, and social scientists. They have tried to integrate their contributions into what they hope is a consistent and coherent whole. At times they have taken the liberty to transcribe the insights and wisdom of the pathfinders into language that students can more readily understand.

None of us in contact with the real world can long remain a pure theoretician. Because knowledge is valued not only in itself but in what we can do with it, we are by necessity practitioners as well. We need to apply the concepts of management systems to the world about us — to the worlds of business, government, medicine, and sports, even to our own personal lives. When referring the insights of others to organizational situations, we not only guarantee that what we learn will not soon evaporate, but we keep extending the limits of our knowledge, thus increasing our power. If knowledge is power, then knowledge of systems ought to be **superpower.**

CHAPTER 1

The Systems View

I find as impossible to know the parts without knowing the whole, as to know the whole without specifically knowing the parts.

Blaise Pascal

CHAPTER OUTLINE

INTRODUCTION

SYSTEMS APPROACHES

SYSTEMS CHARACTERISTICS
1. Objectives
2. Environment
3. Resources
4. Components
5. Management

WHAT IS A SYSTEM?
1. Set
2. Objects
    Inputs
    Process
    Outputs
3. Relationships
4. Attributes
5. Environment
6. Whole

DIAGRAM OF A SYSTEM

BOUNDARIES OF A SYSTEM

SUMMARY

KEY WORDS

REVIEW QUESTIONS

CASE

ADDITIONAL REFERENCES

INTRODUCTION

Today we live in a world of organized complexity — complexity being defined by the number of elements in the system, their attributes, the interactions among the elements, and the degree of organization inherent in the system.

Systems may be natural, such as living organisms; systems may be contrived, such as social organizations. Systems grow, as do government bureaucracies; systems die, as do individual families. There are public systems such as the federal and state governments; there are private systems such as family-owned businesses and personal computer systems. There are systems that operate in *relative* isolation, such as the one-room schoolhouse, or a well-water system in the country; there are systems that transcend several domains, such as the air transport system with its airplanes, airports, baggage-handling facilities, air traffic controllers with their radar equipment, communications subsystems, maintenance crews, ticket agents, food preparation subsystems, and training schools for flight attendants. The air transport system also had to contend with regulatory agencies for a long while and still has to contend with a multiplicity of unions, each of which has the power to significantly alter the performance of the entire system.

One could similarly spell out the components of the rail transport system, the water transport system, the educational system, the judicial system, the economic system, and the political system. Common to each of these are the primary phenomena of complexity and interrelatedness. Each of these systems is quite complex, and each has many interacting elements, all organized to accomplish certain objectives.

Problem solving today likewise necessitates a broad look at a system rather than an overly obsessive scrutiny of the particular problem in question. Examples of the need for a systems approach to problems abound. The transportation system can well illustrate this point. It does little good to design massive highway systems if motor fuel is unavailable for cars, buses, and trucks. It is sheer folly to design aircraft to carry hundreds of passengers if airports lack facilities to accommodate the passengers and motor vehicles bringing people to the airport. In these instances it is necessary to view the problem from a broader perspective, from a systems viewpoint, from a holistic viewpoint.

This viewing of the problem as a whole is termed the *systems view* or *systems approach.* It is the conviction of the authors that this view is indispensable for the solution of present-day problems.

The result of not viewing a problem from the systems approach can at times be catastrophic. The mass unemployment incurred in the auto industry in the early 1980s is clear evidence of the inability and/or the unwillingness of American car manufacturers to face up to changes concerning a basic environmental element — gasoline. As long as gasoline was plentiful and its cost low, automakers felt no constraint to consider in the design of their cars the amount of gasoline consumed per mile. However, the tripling and quadrupling of the price of gasoline became an important factor that American automakers could no longer afford to ignore when designing their cars. Because the Japanese incorporated this element into their system, they were able to make significant inroads into the American automobile market. Their cars were smaller, lighter, and more efficient.

A perennial problem in the U.S. economy has been farming. Each administration, in turn, has generated particularized solutions to the farming problem. Some, like President Carter, who refused to sell U.S. grain crops to the Russians in retaliation for their invasion of Afghanistan, have used grain as a political weapon. This certainly extends the boundaries of the farming system beyond traditionally accepted ones. Other administrations have tried to solve the farm problem by giving subsidies to farmers either in cash or in kind. The point made here is that the farming problem transcends such traditional factors as the amount of rainfall, the quality of the soil, national and regional demands for farm products, and tax breaks.

The systems approach contrasts with the analytical method. When an entity is examined primarily from the viewpoint of its constituent elements or components, the *analytical* viewpoint is said to be employed; analyzing is the process of segmenting the whole into smaller parts to better understand the functioning of the whole. Throughout history, man has used this method to unravel the mysteries of the world surrounding him. Indeed, many of the laws of nature have been discovered through the use of this time-proven methodology. In fact, the analytical method has been traced back to René Descartes (1596-1650), who in his *Discours de la Méthode* speaks of breaking down every problem into as many separate simple elements as possible. The analytical method has since been identified with the scientific method, the conceptual paradigm used by scientists from the early days of the scientific revolution to our present-day researchers into carcinogens, drug addiction, and a host of other problems clamoring for solution.

The underlying reason for the use of the analytical method may be that the human mind is a finite one, capable of grasping only so many concepts at one time. In order to exhaust a subject and thus to understand it, the mind must, therefore, attend to ideas in sequence. By breaking down the whole into smaller parts and then examining each of these in detail, one can attain, it is believed, a complete and accurate understanding of the individual aspects of a subject. Once having mentally broken down the subject into manageable components, the analyst then proceeds to put together (to synthesize) the various pieces previously broken down (analyzed). In this way, the investigator hopes to understand the whole thing.

Biologists were among the first scientists to become disenchanted with the analytical approach. The whole question of organized complexity found in all living organisms, and the question of goal directedness (entelechy), were either denied or ignored by this approach. Researchers increasingly felt that an organism must be studied as a system, as a whole. Because of the mutual interaction of the parts, the whole takes on distinctive properties that would be lacking were one to remove a part. Social scientists too became dissatisfied with the analytical approach. When a member of the family leaves home, the family no longer has the same interaction patterns as before, and when studying a child removed from the family environment, one observes characteristics typically different from those encountered when studying the child in the environment of the family. Because of this, some researchers believe that organizations cannot be studied without studying the people in them. On the other hand, the point can also be made that human beings cannot really be understood when studied apart from their real-world, real-time organizational ties.

Perhaps one can get a better idea of how the analytical approach differs from the systems approach by contrasting them in the following manner:

	Analytical Approach	*Systems Approach*
Emphasis	On the parts	On the whole
Type	Relatively closed	Open
Environment	None explicitly identified	One or more
Entropy	Tends toward entropy since system is closed	Not applicable since the system interacts with environment
Concept of goals	Maintenance	Changing and learning
Hierarchy	Few	Possibly many
State	Stable	Adaptive, seeks new equilibria

The phenomenon of complex organizations also contributed to the rise of the systems approach. A half century ago organizations were not as complex as they are today. Then the traditional methods still served well. Today the complexities include global and multinational firms, vertical integration, intense competition for limited resources, rapid technological change, extensive government regulation, and increased impact of governmental decisions that complicate the structures. Dealing with such complexity called for new approaches.

From the above, one must not get the notion that the systems approach is in stark *contradiction* to the analytical approach. Systems thinking does not do away with analytical thinking. Systems thinking *supplements* rather than *replaces* the latter. It would be sheer folly to expect to understand the whole without specifically knowing the parts, as Blaise Pascal noted in his *Pensées* some 300 years ago. For to isolate the system in question from all other systems and from its environment, one must define the parts and their interrelationships. And here is the rub: when one isolates variables for study, one employs the analytical method. One must be very, very careful, therefore, that in the process of isolating the system for study, one does not ignore or cut out the essential interrelationships existing among the various components. The systems analyst, instead of "microanalyzing" the parts, prefers to focus on the processes that link the parts together.

As regards terminology, some systems scientists make a distinction between systems philosophy, systems methodology, and systems techniques. This distinction is both theoretically well founded and eminently practical. By *philosophy* they understand the broad nonspecific guidelines for action that such a way of thinking embodies. Often this is equivalent to what the present authors mean by systems thinking, systems view, or systems approach. By *methodology* they understand the more specific (less general) and firmer guidelines to action than those provided by a philosophy, and by *techniques* they mean the precise and specific programs of action undertaken to produce a result. These techniques, of course, can be applied to diverse areas of concern — to engineering, information and decision sciences, ecology, management, medicine, the social sciences, etc.[1]

SYSTEMS APPROACHES

The systems approach, therefore, implies some form of departure from the traditional analytical method so successfully employed with simpler problems. The increasing complexities of various modern-day projects make it impossible to look for isolated solutions to problems.

For the systems approach, various specialized frameworks exist, all discussed in the systems literature. Among the more popular are general systems theory (GST) and various specialized systems theories like cybernetics, systems analysis, systems engineering, etc. These too can subsume other even more specialized systems approaches, e.g., information theory, under cybernetics. Figure 1–1 illustrates a possible ordering of these various systems approaches.

The systems approach calls for more than talk and discussion of differences: one must develop a methodology for conceptualizing and operationalizing the systems approach. Although there are other techniques available for treating systems, the authors believe that the best place to begin is with the identification of systems characteristics.

SYSTEMS CHARACTERISTICS

Of all the proponents of systems, C. West Churchman has given us perhaps one of the more logical expositions.[2] He outlines five basic considerations concerning systems thinking.

1. Objectives of the total system together with performance measures.
2. The system's environment.
3. The resources of the system.
4. The components of the system.
5. The management of the system.

These five basic considerations are not meant to be all-inclusive, but even a cursory comparison with the basic system properties outlined by other systems

---

[1] Robert L. Flood, "Introductory Statement," *Systems Practice* 1, no. 1 (1988), pp. 1–2.
[2] See C. West Churchman, *The Systems Approach* (New York: Delacorte Press, 1968), chap. 3.

FIGURE 1–1   Ordering of Various Systems Approaches

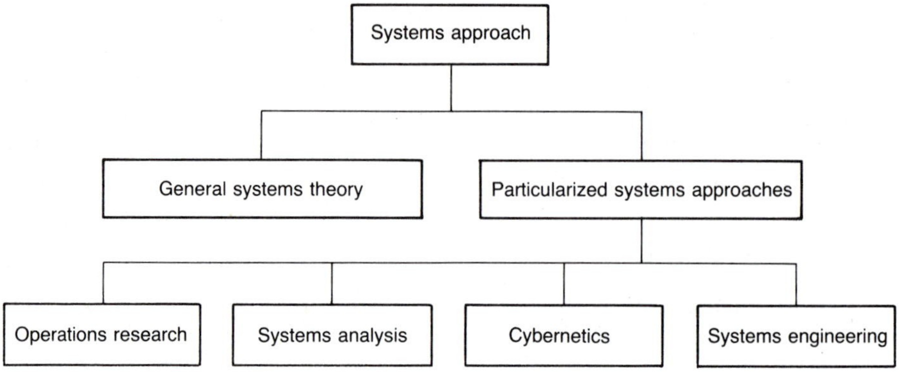

```
                        ┌──────────────────┐
                        │ Systems approach │
                        └──────────────────┘
                                 │
              ┌──────────────────┴──────────────────┐
    ┌─────────────────────────┐      ┌──────────────────────────────────┐
    │ General systems theory  │      │ Particularized systems approaches │
    └─────────────────────────┘      └──────────────────────────────────┘
         │               │                  │                  │
┌──────────────────┐ ┌──────────────┐ ┌──────────────┐ ┌──────────────────┐
│Operations research│ │Systems analysis│ │ Cybernetics │ │Systems engineering│
└──────────────────┘ └──────────────┘ └──────────────┘ └──────────────────┘
```

thinkers will reveal that most of these properties are either included or implied in Churchman's delineation.

The above outline merits further consideration. Here a brief explanation will be given for each of the five points, even though several of them will be treated at greater length in this and in subsequent chapters.

1. Objectives.   By objectives of the system Churchman means those goals or ends toward which the system tends. Hence, goal seeking, or teleology, is a characteristic of systems. With mechanical systems the determination of objectives is not really difficult, since the objectives had been determined even before the mechanical system took shape. A watch is made to tell time, either in hours, minutes, seconds, or in days; it isn't supposed to cut grass or to slice tomatoes. The determination of objectives of human systems, however, can be a very formidable task. One must beware of the distinction (often a real one) between the *stated* objectives and the *real* objectives of the system. Students, to use Churchman's telling illustration, may state as their objective the attainment of knowledge, while in fact the real objective may be the attainment of high scholastic grades. The test that Churchman proposes for distinguishing the real from the stated objective may be called the principle of primacy: Will the system knowingly sacrifice other goals to obtain the stated objective? If the answer is yes, then the stated and real objectives are identical.

Objectives, however real, need to be operationalized—to be defined in terms of identifiable and repeatable operations. Unless they are made quantifiable in some manner, it will be impossible to measure the performance of the total system. In other words, one cannot state to what precise degree the system's objectives are being realized without having on hand some objective measure of the performance of the overall system. Objectives are realized only through the medium of activity. In evaluating objectives one should therefore examine both the activities and their consequences.

Since conceptualization of any system must start with its purpose, objective, or goal, this first characteristic will be dealt with more thoroughly in a separate section under the heading of "Goals." Organization theorists generally agree that goal formation and systems effectiveness are intrinsically related. Because of the importance of this relationship, separate treatment will be accorded to this topic in the chapter entitled Organizational Effectiveness.

2. Environment.   The environment constitutes all that is "outside" the system. This concept, though obvious on the surface, needs and does receive at Churchman's hands further clarification. Two features characterize the environment.

First, the environment includes all that lies outside the system's *control.* The system can do relatively little or nothing about the characteristics or behavior of the environment. Because of this, the environment is often considered to be "fixed"—the "given" to be incorporated into any system's problem. Second, the environment must also include all that *determines,* in part at least, how the system performs. Both features must be present simultaneously: the environment must be beyond the system's control and must also exert some determination on the system's performance. Implied here in the concept of environment are the notions of interdependence and interaction, used so frequently by other proponents of systems.

3. Resources.   Resources are all the means available to the system for the execution of the activities necessary for goal realization. Resources are inside the system; they include all the things that the system can change and use to its own advantage. The real resources of human systems are not only people, money and equipment, but also the opportunities (used or neglected) for the aggrandizement of the human and nonhuman resources of the system.

Management scholars like Peter Drucker are convinced that people are our most valuable resource. People can be underutilized on jobs that tap little of their potential, thus making for dissatisfaction, absenteeism, or turn-over. People can be overutilized, driven past their ability to cope successfully with their job, thus making for burnout. The idea of *lost opportunities* as a consequence of unutilized resources is an ancient though very useful one even today. The adoption of one option by an organization generally implies the nonadoption—the neglect or disregard of others. For instance, universities in principle stress quality in both research and teaching as integral parts of their mission. Yet few, if any, have the necessary financial and personnel resources to excel in both.

Unfortunately, typical income statements or corporation reports reveal little of opportunities spurned or unrecognized; they only report on what occurred, not what could or should have occurred.

4. Components.   By components Churchman means the "mission," "jobs," or "activities" the system must perform to realize its objectives. (In this context these terms can be used interchangeably.) Components are all those activities that contribute to the realization of the system's objectives. Traditionally, organizations have been departmentalized. There is the production department, marketing department, accounting department, and so on. Churchman rejects this traditional partitioning because of the transcendental nature of these functions. For example, production has many functions that affect sales, such as not having products available, not getting products out on time (getting a daily newspaper out one day late will reduce sales drastically), and having defective products. The amount of sales likewise affects production.

Churchman's thrust here is to look at mission and not at what people *actually do.* Mission is *what people should be doing* to achieve the desired objectives.

It seems that Churchman is suggesting that organizations should be designed, not according to the similarity of functions, but according to the similarity of missions. This obviously flies in the face of most management writers who treat organizational design from the primacy of similarly grouped functions. Churchman's emphasis on mission has not yet received wide acceptance from those who design organizations.

The authors suspect that Churchman's ideal organizational structure would be one with the least structure possible, where decisions would be made by those people having the requisite information, regardless of their position.

Churchman's concern is a legitimate one because in many instances departments, while accomplishing their specific goals (optimizing), fail to achieve the goals of the total system. A marketing department may set for itself a sales goal that exceeds the organization's productive capacity or the market's demand. By analyzing activities, one can partially estimate the worth of the activity for the entire system. There seems to be no feasible way to estimate the worth of a department's performance to the whole system—at least, not yet.

The rationale behind this line of thinking is the identification of those activities whose measures of performance are, in fact, related to the measures of performance of the system itself. If all other elements are controlled, then, in the ideal case, as the measure of performance of an activity increases, so should the measure of performance of the total system.

5. Management.   By systems management Churchman means to include two basic functions: planning the system and controlling the system. Planning the system involves all the aspects of the system previously encountered, viz., its goals or objectives, its environment, its utilization of resources, and its components or activities.

Controlling the system involves both examination of the execution of the plans and planning for change. Managers must make sure that the plans as originally conceived and decided on are being executed; if not, then it must be discovered why not. This constitutes control in the most primary sense. In a secondary sense, control also concerns planning for change.

In any open system, either substantial or partial change is inevitable. Hence, in any ongoing system, plans must be subject to periodic review and reevaluation. Essential to all realistic planning, therefore, is the planning for a change of plans, since no manager or managers can possibly set down the system objectives that are valid for all times and under all conditions; or once and for all define the organizational environment so subject to constant change; or permanently delineate all the relevant resources available to the organization; or outline measures of performance that would never need improvement or updating.

Associated with the planning and control function of systems is the notion of information flow or *feedback,* so characteristic of cybernetic systems. Without adequate feedback, the planning and control functions would be almost totally inadequate.

These five basic characteristics of systems as proposed by Churchman merit more consideration from students of systems. One needs no special course in logic to deduce from these basic premises other characteristics associated with systems, such as wholeness, order, and the like.

WHAT IS A SYSTEM?

The "system" concept has been borrowed by the social scientist from the exact sciences, specifically from physics, which deals with matter, energy, motion, and force. All of these concepts lend themselves to exact measurement and obey certain laws. There a system is defined in very precise terms and in a mathematical equation that describes certain relationships among the variables. This kind of definition, however, is of little use to the social scientist, whose variables are very complex and often multidimensional.

The definition given here is a verbal, operational one which, though nonmathematical, is quite precise and as inclusive as that of the exact sciences. **A system is here defined as a *set* of *objects* together with *relationships* between the objects and between their *attributes* related to each other and to their *environment* so as to form a *whole*.**[3]

This definition has a dual property: it is extensive enough to allow for wide applicability, and at the same time it is intensive enough to include all the elements necessary for the detection and identification of a system. To further reduce the vagueness inherent in the terms, the key concepts — namely, set, objects, relationships, attributes, environment, and whole — will be explained.[4]

But before we begin, it should be pointed out that the above definition of systems, like that in classical physics, prescinds from the interaction between the observer and the system being observed. In other words the observer is not coupled to the system and does not affect its functioning. The systems approach of Guy M. Jumarie uses internal variables that define the state of the system and external variables that depend explicitly upon the observer who examines the system. The observer is an integral component of the system.[5]

---

[3] This is a commonly accepted definition. See, for example, A. D. Hall and R. E. Fagen, "Definition of System" in W. Buckley, *Modern Systems Research for the Behavioral Scientist* (Chicago: Aldine Publishing, 1968), pp. 81 ff; S. Optner, *Systems Analysis for Industrial and Business Problem Solving* (Englewood Cliffs, N.J.: Prentice-Hall, 1965).

[4] The following analysis and elaboration draw heavily upon Optner, *Systems Analysis,* chap. 2.

[5] See Guy M. Jumarie, *Subjectivity, Information, Systems: Introduction to a Theory of Relativistic Cybernetics* (New York: Gordon and Breach, 1986), pp. 3 and 19.

Set

The concept of set is not a difficult one to grasp. By *set* one understands any well-defined collection of elements or objects within some frame of discourse. Note that a set is not only a collection of objects like dishes or stamps, not only a collection of symbols such as letters of the alphabet or numerals of a number system; it is also a "well-defined" one. This means that it must be possible to tell beyond doubt whether or not a given object or symbol belongs to the collection under consideration. In systems analysis one must be able to state whether element X belongs to the system or not. The technical connotation of set is thus seen to be the same as its nontechnical, everyday one.

Objects

Objects are the elements of a system. From the static viewpoint, the objects of a system would be the parts of which the system consists. From the functional viewpoint, however, a system's objects are the basic functions performed by the system's parts. What we are particularly interested in are not the parts in themselves but the parts as they function, as they do things. There are three kinds of objects: inputs, processes, and outputs.

Inputs.   Inputs to a system may be matter, energy, humans, or simply information. Inputs are the start-up force that provides the system with its operating necessities. Inputs may vary from raw materials which are used in the manufacturing process to specific tasks performed by people, such as the typing of this manuscript or discussion used in the educational setting. There may be financial inputs or inputs consisting of services of other organizations, internal records (information), or the like. Systems may have numerous inputs which are the outputs of other systems. It is sometimes desirable to classify inputs into two basic categories: serial inputs and feedback inputs. The basis of this classification is how or where the input comes into the system.

A *serial* input is the result of a previous system with which the focal system (the one in question) is directly related. These kinds of inputs are easy to identify and study. They present little problem to the researcher because their absence would be felt immediately as the lack of movement in the system. Serial, or in-line, inputs are sometimes referred to as "direct-coupling" or "hooked-in" inputs.

Let us identify some of the most common serial inputs of a manufacturing firm. A manufacturing organization can be thought of as a transformation system which converts the three basic factors of production (human activity, materials, and capital equipment) into marketable products. This transformation process is accomplished through the interaction of a vast number of subsystems, each performing its own transformation process. The output of each of these subsystems becomes the input to other subsystems.

In Figure 1–2 the interaction of two such subsystems is shown. The production subsystem is concerned with the actual physical conversion or transformation process. In order to perform this task, however, this subsystem needs several

FIGURE 1–2   Serial or In-Line Input

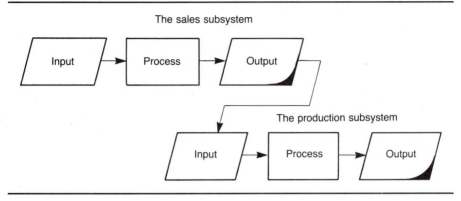

inputs, one of which is information on the volume of production, that is to say, the number of units (a quantitative attribute of this system's output) as well as certain qualitative characteristics of the output. These inputs are supplied to the production subsystem by the sales subsystem. The output of the sales subsystem that becomes the input to the production subsystem is a serial input because the two subsystems are directly related. In other words, the sales subsystem's output is produced for the specific purpose of providing the energizing or start-up function to the production subsystem without which it cannot function rationally.

Serial inputs may come to the manufacturing firm from the outside environment as well. For example, most of the energy resources needed for the production process (e.g., electricity, water) will be supplied by local subsystems to which the firm is "hooked up." Most of the laborers will also be supplied by the labor force of the community.

One can easily find examples for nonmanufacturing firms, such as a bank or a hospital. Energy inputs to the bank or hospital system, for example, will be provided by the municipal power plant and water reservoir. In all these instances the focal system (the system whose behavior is under study) is linked directly to a specific system upon which it depends for one or more inputs.

The second kind of input represents a reintroduction of a portion of the output of a system as an input to the *same* system. This kind of input has the very descriptive name *feedback.* A feedback input differs from other inputs only insofar as it is an output of the focal system itself and not of some other system. This output reenters the focal system as an input, as a *feedback* input. The subject will be treated extensively in the chapter on cybernetics.

Process.   A process is that which transforms the input into an output. As such it may be a machine, an individual, a computer, a chemical, or equipment, tasks performed by members of the organization, and so on. We generally know how inputs are transformed into outputs. However, in some situations, the process is not known in detail because this transformation is too complex. Different combinations of inputs, or their combination in different sequential order, may result in different output states.

A process may represent an assembly whereby an array of inputs is transformed into one output (e.g., a car assembly line) or it may be a disassembly (e.g., meat packing) where one input is converted into many outputs.

Many managers in large organizations cannot determine the interrelationships of the many components of the systems and therefore cannot understand what factors contribute to the attainment of an objective. For example, if the system objective is profit and indeed it is attained, one should be able to determine the factors of that result. However, executives often simply cannot tell you whether it was due to the packaging of the product, the quality, the channels of distribution, service, reputation, advertising, price, design and styling, or some other factor. This is precisely what management science is constantly trying to get at — some input contribution to profitability.

Outputs.   Outputs, like inputs, may take the form of products, services, information such as a computer printout, or energy such as the output of a hydroelectric plant. Outputs are the results of the operation of the process, or alternatively, the purpose for which the system exists.

As mentioned above, the output of one system becomes the input to another system, which in turn is processed to become another output, and the cycle repeats itself indefinitely. This is true for all living systems, from what biologists call the *food chain* to contemporary product and service enterprises.

All transformation processes may lead to more than one type of output. It is convenient to classify the output of a system into three main categories. One category includes outputs which are directly consumed by other systems. The main output of a business manufacturing firm, for instance, is sold to the customers for either consumption or further processing. A hospital or an educational institution renders services directly to the clients. The system's objective is to maximize this type of output. The percentage ratio of this output to input is usually termed efficiency.

A second category of outputs is the portion of the output which is consumed by the same system in the next production cycle. Defective products of a manufacturing process, for example, are usually reintroduced into the same production process. The output of the accounting subsystem of a bank or a hospital, in addition to being used for satisfying stockholder or taxpayer demand, is used to improve the performance of the system itself.

Finally, a third category of outputs consists of the portion of the total output which is consumed neither by other systems nor by the system itself but rather is disposed of as waste which enters the ecological system as an input. The focal system's objective or goal is to attempt to minimize that kind of output. Failure to do so can result in expensive lawsuits and punitive damages. Recently the disposal of nuclear wastes and of toxic chemicals has loomed large in the public's mind as a critical problem clamoring for a solution.

Relationships

Relationships are the bonds that link the objects together. In complex systems in which each object or parameter is a subsystem, relationships are the bonds that link these subsystems together. Although each relationship is unique and should therefore be considered in the context of a given set of objects, still the relationships most likely to be found in the empirical world belong to one of the three following categories: symbiotic, synergistic, and redundant.

A *symbiotic* relationship is one in which the connected systems cannot continue to function alone. Examples of this kind of relationship abound. In certain cases the symbiotic relationship is unipolar, running in one direction; in other situations the relationship is bipolar. For example, the symbiotic relationship between a parasite and a plant is unipolar to the extent that the parasite cannot live without the plant while the latter can — parasitic symbiosis. However, the symbiotic relationship between the production and sales subsystems of a manufacturing system is bipolar: no production — no sales, no sales — no production — mutualistic symbiosis. Despite the tremendous importance of symbiotic relationships, they are the least interesting from the researcher's point of view because they are relatively easy to identify and explain.

A *synergistic* relationship, though not functionally necessary, is nevertheless useful because its presence adds substantially to the system's performance. Synergy means "combined action." In systems nomenclature, however, the term means more than just cooperative effort. Synergistic relationships are those in which the cooperative action of semi-independent subsystems taken together produces a total output greater than the sum of their outputs taken independently. A colloquial and convenient expression of synergy is to say that $2 + 2 = 5$ or $1 + 1 > 2$.[6]

Numerous examples of synergistic relationships can be found in nature as well as in the sciences, especially in chemistry. Fuller states, "Synergy is the essence of chemistry. The tensile strength of chrome-nickel steel, which is approximately 350,000 pounds per square inch, is 100,000 P.S.I. greater than the sum of the tensile strengths of each of all its component, metallic elements. Here is a 'chain' that is 50 percent stronger than the sum of the strengths of all its links." [7]

A simple example of a synergistic relationship from the business world would be the following: Suppose a firm aspires to increase its sales by, let us say, 10 percent. The firm has two strategies available: (1) a $100,000 expenditure for advertisement which, according to the advertising agency, is supposed to increase sales by 5 percent; (2) a $100,000 expenditure for increasing the sales force by 20 percent, which is supposed to increase sales by another 5 percent. The two

---

[6] For a more detailed explanation of the concept of synergism as it applies to business enterprises, see "Business Synergism: When $1 + 1 > 2$," *Innovation* no. 31 (May 1972).

[7] R. Buckminster Fuller, *Operating Manual for Spaceship Earth* (New York: Simon and Schuster, 1969), p. 71.

strategies are scheduled to be put into effect sequentially: first advertise, then hit the market with sales people. Suppose that both strategies are effective; i.e., total increase in sales equals 10 percent.

A synergistically oriented sales promotion manager would have launched both strategies at the same time. Suppose the increase in sales was 12 percent; the 2 percent difference in increase would be the synergistic effect.

Redundant relationships are those that duplicate other relationships. The reason for having redundancy is reliability. Redundant relationships increase the probability that a system will operate as it was designed to. The greater the redundancy, the greater the system's reliablity but the greater the expense. Redundant or backup relationships are abundant in the man-made world, and satellites, spaceships, space shuttles, and airplanes have systems with redundant relationships designed to secure operation of the system under virtually any condition.

Attributes

Attributes are properties of both objects and relationships. They manifest the way something is known, observed, or introduced in a process. A machine, for instance, may have as its attributes the following characteristics: a machine number, a machine capacity (output per time), a required electrical current, x years of technical life, y years of economic life, and so on.

Besides the usual quantitative and qualitative properties of systems, attributes may also be of two general kinds: defining or accompanying. *Defining* characteristics are those without which an entity would not be designated or defined as it is. *Accompanying* characteristics or attributes are those whose presence or absence would not make any difference with respect to the use of the term describing it.

This division of the attributes of a system's objects into defining and accompanying characteristics has some very useful implications for the manager who desires either to design or to use a system. Consider, for example, a company which transports perishable items. The products are transported in refrigerated trucks to various destinations. The company is considering the acquisition of five new refrigerator trucks to replace some old trucks, as well as to increase the size of the fleet. The manager in charge of this acquisition would be interested in certain characteristics of each truck. For instance, he or she would want to know the maximum load capacity of each truck, its speed, frequency of maintenance, fuel consumption, and several other technical and economic characteristics all of which are necessary for an accurate description of the equipment. These would be the defining characteristics of a truck.

A truck, however, is characterized by certain other features which in a particular time span do not appear to be necessary for its definition, but nevertheless are attributes of that system. One of these characteristics is, for instance, the amount of pollution created by the engine of the vehicle. If the decision approving the acquisition of the trucks had been made 15 or more years ago, this attribute of the truck's engine would have been of no significance. That is to say, it would have been only an accompanying characteristic. Today, however, this attribute of pollution creation is one of the most significant characteristics for the description of the truck; it is a defining characteristic which must be taken into consideration along with the other defining characteristics of capacity, speed, fuel consumption, and so on.

The reverse situation is also conceivable. Certain characteristics of a system's objects which were at one time considered to be defining may at some other time, or in different circumstances, turn out to be accompanying characteristics. For example, the sex or race of an individual applying for a position within an organization 10 or more years ago may then have been considered a defining characteristic. Today, however, with the passage of the Equal Employment Opportunity Act, sex and race are expected to be of no significance for evaluating the applicant's suitability for the particular position. They are merely accompanying characteristics. However, for many firms attempting to comply with affirmative-action programs, both sex and race are once again defining characteristics.

Environment

Each system has something internal and something external to it. What is external to the system can pertain but to its environment and not to the system itself. However, the environment of a system includes not only that which lies outside the system's complete control but that which at the same time also determines in some way the system's performance. Because the environment lies outside the system, there is little if anything that the system can do to directly control its behavior. Because of this, the environment can be considered to be fixed or a "given," to be incorporated into the system's problems. The environment, besides being external, must also exert considerable or significant influence on the system's behavior. Otherwise, everything in the universe external to the system would constitute the system's environment, something to be programmed into the system's problem solving framework. Both features must be present together: the environment must be beyond the system's control and must also exert significant determination on the system's performance.

Some writers speak of internal environment and external environment. The present authors eschew this distinction, since for them environment is necessarily something *external* to the system. Anything *internal* belongs, therefore, not to the environment but to the system itself.

There can be little doubt that the environment affects the performance of a system. Firm X's profits are obviously affected by the number and aggressiveness of its competitors; the quality and price of the competitor's products; the purchasing power of the dollar; current federal, state, and local tax structures; pending congressional legislation and lawsuits; the political climate; and a host of other uncontrollable factors.

While external to the system's control, the environment is not impervious to the system's behavior. It is perhaps for this reason that some systems analysts also include in their definition the notion that the environment embraces also those objects whose attributes are changed by the behavior of the system.[8] This makes even more explicit the concept of interaction between systems and environment: environment affects systems, and systems in turn affect the environment. Thus, Company X and Company Y, who are competing with one another, must each include the other in its environment. Company Y is in Company X's environment, and Company X is in Company Y's environment.

Of course, one must take into consideration the state of the environment at any given point in time. At one time a firm's environment may be very stable, with little change occurring; at another time the environment may be very dynamic. It is the dynamic environment that presents problems in scanning. A company in a dynamic environment is thoroughly enveloped in change, and its very survival hinges on how well it can respond to these changes. In the early 1980s the personal computer industry was immersed in a very dynamic environment, with scores of competitors for every computer hardware and software product. The railroads from the 1860s to the end of World War II also operated in a very dynamic environment. Since then the state of the railroad industry has stabilized: most railroads are dying out; only a few are profitable without subsidy.

Perhaps one way to reduce the apparent arbitrariness of what constitutes the environment is to pose certain questions proposed by Churchman.[9] First, Is the factor in question related to the objective of the system? Secondly, Can I do anything about it? If the answer to the first question is yes and the answer to the second question is no, then the factor is in the environment. If the answer to both questions is yes, then the factor is in the system itself. If the answer to the first question is in the negative, then the factor is neither in the system nor in the environment (Figure 1–3).

From Figure 1–3 it should be apparent that relatedness is linked with relevance. What we are concerned with in any system is *relevant* relatedness. When dealing with systems, one must be careful to acknowledge relatedness only when one is ready to declare relevancy. One can easily relate something in this world to almost anything else by reason of color, size, shape, density, distance, and so on. Many of these relationships may be spurious; they lack relevance. Perhaps this is why Beer states that there seem to be three stages in the recognition of a system. First, "we acknowledge particular relationships which are obtrusive: this turns a

---

[8] A. D. Hall and R. E. Fagen, "Definition of System," *General Systems Yearbook* 1 (1956), pp. 18–28; reprinted in Buckley, *Modern Systems Research,* p. 83.

[9] Churchman, *Systems Approach,* chap. 3.

FIGURE 1–3   Environmental Determination

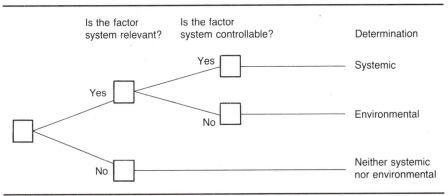

mere collection into something that may be called assemblage. Secondly, we detect a pattern in the set of relationships concerned: this turns an assemblage into a systematically arranged assemblage. Thirdly, we perceive a purpose served by this arrangement: and there is a system."[10]

Figure 1–4 attempts to further clarify the relationship between a system and its environment by using as a criterion of differentiation the relative degree of control which can be exercised by the organization over the factors surrounding it. Ten external factors have been chosen as indicative of the multiplicity of factors usually referred to as "the environment." External factors over which the organization has a high degree of control can be considered the resources of the organization. On the other hand, external factors over which the organization has a relatively low degree of control can be defined as the environment of the organization. The relative degree of control has been depicted in Figure 1–4 as shaded.

As can be seen from this figure, the four major inputs of the organization (land, capital, material/equipment, and labor—we prescind in our discussion from the all-important entrepreneurial ability) are relatively highly controllable by the organization. These are, therefore, the organization's major resources. On the other hand, the degree of control of the four major external factors depicted in the right-hand side of Figure 1–4 (ecology, government, general public, and competitors) is very low. These are, therefore, the organization's major environmental factors. Between these two extremes of the largely controllable factors (resources) and the largely uncontrollable variables (environment) lie two additional sets of factors which are relatively less controllable than resources but relatively more controllable than the environment. These factors are consumers and technology.

The degree of controllability reflects the organization's ability to use its resources to influence the external factors. This ability is, in turn, a function of the

---

[10]Stafford Beer, *Decision and Control* (London: John Wiley & Sons, 1966), p. 242.

FIGURE 1–4 The Organization. Its Resources and Its Environment

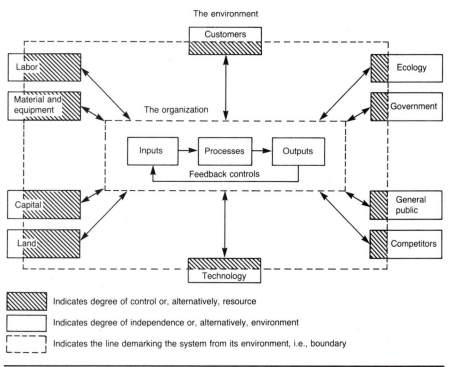

existence of resources, managerial talent, and the availability of organizational intelligence. Organizational intelligence refers to the organization's ability to recognize the need for control of an external factor, as well as the ability to devise the appropriate influencing strategy.

A few examples should suffice to demonstrate the differences between an organization's resources and its environment. Labor, material, money, and land have always been the exclusive concern of management, primarily because of the necessity of these factors for performing the basic functions of an enterprise and also as a result of the early developments in the discipline of economics. Knowledge of the basic principles of the economics of the firm, or what is usually referred to as microeconomics, enables management to recognize the need of influencing the behavior of these basic subsystems, as well as to develop sophisticated techniques in dealing with them. Thus, labor economics, material- and equipment-handling techniques, money management (finance), and land acquisition and utilization procedures are some of the most highly developed managerial tools.

At the other end of the continuum, knowledge of ecology, government regulation, social or public responsibility, and competitive strategy development has not advanced enough to enable the construction of a framework for recognizing the need for influence, as well as for enabling management to devise effective techniques for dealing with these external factors. For this reason the degree of controllability available to the organizations is relatively small but not negligible.

The relative degree of control that a particular organization can exert upon these four environmental factors will depend on the organization's ability to employ some conventional techniques, as well as to devise some new effective means for dealing with them. Concerning the environmental factor labeled "ecology," the organization can employ sound manufacturing processes to minimize the amount of waste created by its operations. The fact that during the last dozen or so years an environmental crisis has arisen that has forced governments all over the world to adopt policies of the type known in the United States as the National Environmental Policy Act indicates that organizations have not been very successful in employing conventional management techniques to this end. Since 1970 companies have been increasingly compelled to devise new techniques for pollution minimization.

Traditionally, organizations attempted to increase their degree of influence over the external factor labeled "government" via conventional lobbying and financial contribution to political parties in an effort to influence favorable legislation or to prevent excessive governmental surveillance of their activities. These techniques are increasingly proving to be ineffective. In the future, organizations will have to devise more sophisticated techniques in dealing with that sector of the external environment in particular, because the degree of government interference with free enterprise is estimated as likely to increase, thereby curtailing the organization's control even further.

Participation in community programs and heavy advertisement are two of the most common and conventional techniques employed by organizations in dealing with the public sector. Consumerism, affirmative-action programs, social responsibility demands, and other new signs of the public's desire to intervene in the day-to-day activities of an organization have contributed considerably toward a decline in the organization's control of this sector of the external environment. A positive reaction created as a result of XYZ Oil Company's contribution toward a clean environment was nullified by its announcement of a 400 percent increase in its corporate profits.

Because of the passivity of the government during the Reagan administration regarding the enforcement of antitrust legislation, there has been a plethora of megamergers and acquisitions. The effect of all this has been to lessen the extent of competition, particularly in the airlines.

Customers and technology are two sectors of the external environment which are somewhere between the two extremes of relatively high controllability (resources) and relatively high uncontrollability (environment). Marketing, which started as a managerial function primarily concerned with sales effectiveness, has grown into a full discipline employing sophisticated quantitative and behavioral techniques. Research and development, which started as a hit-or-miss type of engineering ingenuity, has developed into a sophisticated organizational entity employing the latest techniques of technological forecasting and the latest developments in quantitative methods and computer sciences. Thus, these two sectors or environmental variables, although not completely controllable by the organization to the extent that they do not represent its resources, are not beyond organizational influence.

This brief discussion of the system's environment and its resources was not intended to clarify the subject completely. Several subsequent chapters deal with these considerations in greater detail. It should suffice at this point to emphasize that the line separating the system from its environment is indeed not a wall insulating the system from external influences but more like a semipermeable osmotic membrane. Since organizations are open systems, their interaction and mutual influence with the environment are indeed a necessary condition for their survival.

Whole

The concept of whole was purposely left as an undefined term much as a point and a straight line are in geometry. Philosophically, wholeness is an attribute – a defining attribute – of a thing or being. Whenever one thinks of any object – one's ranch house, one's hi-fi set, one's hunting rifle, one's dog – the object is always conceived of as a unity, a whole to which belongs every datum within the unity. Thus, my dog, Snoopy, is thought of as a whole, and to Snoopy is ascribed the totality of data, whether of shape, color, breed, sex, sound or odor, or of movement. Unity, wholeness, entirety – these can best be left as undefined.

Throughout this text we will be emphasizing the study of wholes rather than of parts. A well-known synergistic principle, previously referred to, is that "the whole is greater than the sum of its parts." This tenet makes quite clear that "whole" and "the sum of its parts" are quite different entities. The whole is not of the same character and substance as the parts. When one aggregates the parts, then the parts are added; however, the whole is more than the aggregate of parts. The whole is an independent framework in which the parts play a distinctive role.

Were one to examine a system composed of six family members (two parents, three children, and a grandparent), one would soon see that the family as a system is more than the individual members together. Removing a member from the family through death, divorce, or separation causes the system to take on a quite different configuration. Its structure is changed.

DIAGRAM OF A SYSTEM

Usually a somewhat detailed schematic presentation of systems can be found in most introductory engineering books. Here a detailed but still incomplete explanation of the major symbols used in diagramming a system will be given.

The first thing that one should notice when looking at Figure 1–5 is that the

FIGURE 1–5   Diagram of a System's Parameters, Boundary, and Environment

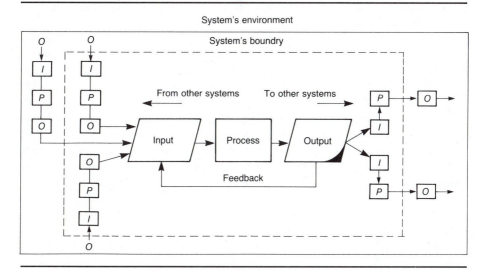

input to a system is the output of another system, and that the output of the system becomes the input to another system.

Second, one should notice that the line demarcating the system from its environment (i.e., the systems boundary) is not solid. There are two reasons for this: First, such a line indicates that there is a continuous interchange of energy and/or information between the *open* system and its environment. This kind of boundary serves the same purpose as a cell membrane: it connects the exterior to the interior. Second, the broken line indicates that the boundary's actual position is more or less arbitrarily determined by the designers, investigators, or observers of the system's structure. They tentatively assign a boundary, examine what is happening inside the system, and accordingly readjust the boundary.

Third, in this diagrammatical presentation of a system the component control positioned over the output or process box in conventional diagrams has been deleted. Instead, the control function has been incorporated into the feedback component for reasons that will become clear when the science of control and communication (i.e., cybernetics) is scrutinized. Finally, it should be noticed that the lines connecting the system's parameters to each other as well as the system to its environment represent the system relationships.

BOUNDARIES OF A SYSTEM

Closely allied to the question of environment is that of system boundaries. Chin gives as his operational definition of the boundary of a system "the line forming a closed circle around selected variables, where there is less interchange of energy (or communication, and so on) *across* the line of the circle than within the delimiting circle."[11] One can readily see that the boundary demarcates the system from its environment.

The boundary of a system is often arbitrarily drawn depending upon the particular variables under focus. One can adjust the boundary to determine whether certain variables are relevant or irrelevant, within or without the system. A system viewed from two different levels may have different boundaries. This arbitrariness is not necessarily undesirable, since researchers and organizational officials tend to view a particular system from their own intellectual perspectives much as managers tend to evaluate case study problems from the vantage point of their own specialties. If researchers from two different disciplines were to examine the same organization, no doubt they would view the organization from different perspectives; at the same time they would most probably identify different parameters and operate at different levels of analysis. The two researchers, while studying the same system, would be doing so at different resolution levels. The particular levels researchers choose are subject to such factors as the complexity of the system, their understanding of the system, the resources available to them, their comprehension of the problem, and so on. The resolution level of an experienced brain surgeon would probably be not only different from that of a medical student but also higher. Likewise, a behavioralist analyzing a firm would typically choose a resolution level different from that of a systems engineer with the same task. This does not mean that a higher level is always more useful or more desirable; it is merely different.

Different resolution levels call for different definitions of the system, different objectives of the investigation, different parameters, and different boundaries separating the system from its environment. It is immaterial that one's system is viewed by others differently or as a subsystem of a larger and different system. What is important in system analysis is that one clearly discriminates between what is in the system and what is in the environment.

The practical problem here is, How does one go about determining the boundaries of a system? or, to put it in a slightly different way, How does one determine what constitutes the focal system, the system under study, the system in focus? Unfortunately, no guidelines exist as to how big or how small a view one should take, what system or subsystems should be under focus. If the system is too narrow (as is generally the case), no meaningful solution may be forthcoming. Symptoms are treated: the real causes are left untouched. If the system is too expansive, no solution is even started. The only conclusion drawn is that no conclusion can be drawn until further research into the problem is undertaken. And this goes on ad infinitum. Perhaps the example noted by

---

[11] Robert Chin, "The Utility of System Models and Developmental Models for Practitioners," in Warren G. Bennis, et al., *The Planning of Change* (New York: Holt, Rinehart and Winston, 1961), pp. 201–14.

Churchman in a number of sources can illustrate the point of focal-system determination.

Suppose certain government officials in HHS have as one of their system goals a decrease in drug usage by teenagers. To add specificity to the goal formulation, suppose they even quantify this goal — a decrease in drug usage by teenagers from x percent currently to $x - y$ percent in five years' time. The first problem that the government officials will encounter will be the determining of the boundaries of the system they are studying. Who are the teenagers in the study? Will this be a pilot study or one undertaken on a national scale? How extensive and how intensive will be the effort? (Finances are a prime consideration here.)

Once the preliminary definition of the problem has been made, the officials can then turn to the next problem, namely, determining the activities whereby the problem can be solved, the goal attained. Depending on their previous training and biases, they can approach this problem in a variety of ways. Suppose they choose the preventive approach. The program(s) designed toward prevention might well include (1) an educational program for teenagers which would show the physical, social, and psychological consequences of drug addiction and drug abuse, (2) an all-out attempt to cut off the supply of drugs to teenagers either through laws and regulations that prohibit their importation or through the seizure of illegal drugs, (3) laws with stiff penalties which will make the possession of various drugs even less likely, and so on.

But is the solution to the problem one of prevention? Should officials strengthen the agency concerned with border patrol guards the better to prevent drugs from entering this country, or should we pay foreign countries to control more closely the cultivation of the plants from which these dangerous drugs are derived?

Perhaps underlying the problem of drug abuse are social factors such as the leniency of the judicial system itself that renders futile attempts made in other areas to solve the problem. For until current and potential offenders are swayed by the severity of the punishment, little progress will be made.

Several observations are in order here. First, that as one employs the systems approach, other subsystems are identified which may have conflicting goals. Second, the very discipline of the systems approach forces one to an examination and identification of the systems components. Third, problem identification often dictates the boundaries implicitly accepted by the researcher. Fourth, systems goals may change as one varies the boundaries of the system. Last, the actual setting of the systems boundaries may also be related to who determines the systems goals.

The last of these observations deserves further comment. A labor union official, for instance, could well set goals quite different from those set by top officials of a corporation. Stockholders may well set goals inconsistent with the long-term plans of the policymakers. While it may be true that setting the systems boundaries may be related to who determines the organizational goals, there is no relation between regulation of the system and how the goals are set. Whether the goals are company imposed or employee determined is immaterial for regulation.

(Goal acceptance, though, is another matter.) But system regulation is independent of goal determination, for in all cases systems are regulated in the same way, namely, a comparison of actual performance against the systems goals via some feedback mechanism.

If one were to devise a rule of thumb for determining the boundaries of a system, perhaps the following might do: Starting from a small, manageable system, the researcher should gradually enlarge its scope until the factors brought in no longer make any tangible difference in the results. Then carefully ascertain and define the interrelationships between the variables. The most critical aspect of all model building is to make sure that one has included all the important variables in the problem. Then and only then should one venture forth to study the interrelationships.

SUMMARY

This chapter was concerned with introducing the reader to a distinctively different way of thinking. The reader has been asked to go beyond the realm of the particular object or problem under consideration and to try to visualize other objects with which the original object or problem interrelates. In looking at a house, one is asked to consider not only the house and its inhabitants but also its surroundings; in visualizing a movie theater, one is asked to think of it along with the building, the projectionist, and the cinema community; in observing an automobile, one is asked to think also of oil imports, pollution, highways and speed limits, police officers and traffic tickets, accidents and hospitals; in thinking of employment, one is asked to think of supervisors, products or services provided, co-workers, customers, departments, etc.; in sitting down to eat, one is asked to think of farmers, truckers, dealers, imports, diets, precipitation, the Department of Agriculture, tractors, and fertilizer. If one can expand this conceptualization of the object or problem to include other components with which the object may interact, then one is traveling down the road of systems thinking.

One should not expect to grasp fully every concept presented in these chapters. What one is looking for is an awareness and appreciation of the whole of the discipline. An understanding may not occur until the text is completed in rigorous study. Students are urged to mull over the concepts, to turn them around so as to view them from various perspectives, and to apply them to what they already know and to what they are presently encountering in the whole area of living.

In this chapter, systems thinking was contrasted with the analytical method long enshrined in the halls of science. In the analytical method the whole is broken down (analyzed) into smaller and smaller segments the better to understand the functioning of the whole.

Initially investigated were the characteristics of systems as spelled out by Churchman, namely, objectives, environment, resources, components, and finally management. These goals have to be made quantifiable. To differentiate real from merely stated objectives, one uses the principle of primacy: will the system sacrifice other goals to attain this objective?

The *environment* includes all that lies outside the system's control and that determines, in part at least, how the system performs. *Resources* are the means the system uses to achieve its objectives. Resources include not only people, money, and equipment but also the available opportunities for making the most of the system's other resources. Lost opportunities were seen to be unutilized system resources. By *components* Churchman understands mission, or goal-directed activities. By *management* he means the planning and controlling of the system. Planning involves also planning for change, for a change of plans, while control involves cybernetic feedback.

A system was then defined as a set of objects together with relationships between the objects and between their attributes, related to each other and to their environment so as to form a whole.

A detailed explanation followed of all the main concepts comprising the definition: set, objects, relationships, attributes, environment, and whole. Under *objects* we distinguished among inputs (serial and feedback), process and outputs (system consumable, other consumable, and nonconsumable). *Relationships,* the "glue" that binds the parts together, can be viewed as symbiotic, synergistic, or redundant. *Attributes* are properties of both objects and relationships. Defining attributes differ from accompanying attributes in that the former must be present to make the entity what it is. The attribute of being a closed plane curve, all points of which are equidistant from a point within, would be a defining attribute of a circle.

Environment, it was stated, is sometimes used by writers to denote objects both within and without the system in question. In this text, environment is always external and is understood in the Churchmanian sense. To determine what constitutes the system, the environment, or neither, the criterion of differentiation can be employed. Is the factor in question related to the objective of the system? Is the factor controllable by the system? Linked to the concept of environment is that of boundaries. How we draw the boundaries of a system will depend on the level of resolution we want. The danger normally is that, without an understanding of system, we draw the boundaries too narrow, thus preventing any meaningful solution from emerging. By drawing the boundaries too wide, we eliminate from the very start any solution: we abort all possible solutions.

KEY WORDS

Objectives
Environment
Resources
Components
Management
System
Set
Inputs

Outputs
Process
Relationships
Attributes
Boundaries
Feedback
Control

REVIEW QUESTIONS

1. Contrast the analytical approach with the systems approach.

2. Identify other particularized approaches in addition to those mentioned in this chapter of the text.

3. Give a definition of the term *system* and illustrate the definition with some examples from your everyday life.

4. Identify the major parameters of a system and draw a diagram of this system. Explain your diagram.

5. Think of any kind of organization you are familiar with (e.g., a factory, a bank, a hospital, a school) and list its inputs, processes, and outputs.

6. In the system you described for question 5, classify its inputs into serial and feedback.

7. What are the main relationships in this system? Classify them into the three categories of symbiotic, synergistic, and redundant relationships.

8. Take a typical manufacturing company or service organization (e.g., a bank, a hospital, or an insurance company) and identify some of its resources and some of its environmental factors.

9. Choose several environmental variables you identified in your answer to question 8 and indicate the degree of controllability as well as the most successful strategies which the company employs or should employ.

10. Work through an example of an efficient or inefficient system (airline, railroad, telephone, educational, etc.) that you are familiar with.

Case: Pace Ltd.

Pace Limited was a medium-sized firm employing several thousand employees manufacturing component parts for an automobile company. As sales fluctuated, so did employment. Because the workers were generally only semiskilled, labor was seldom difficult to obtain. Wages were higher than those prevailing in the community, and the company did not have a union.

Employee turnover was a constant problem with the firm. Even when the work was steady employees seemed to leave. In an effort to solve this problem the plant manager called in the head of human resources and told him to start hiring some better people. The human resource person examined the firm's selection procedures and decided to implement some basic testing to screen out people thought unable to do the job. However, the testing did little to help the turnover problem. Then a training program was implemented; however, the problem still persisted. The firm discussed the possibility of increasing the starting wage in an effort to solve the problem.

In one of the monthly foremen meetings with the human resource department a new employee from the human resource department suggested that instead of examining the problem from the analytical approach maybe it should be looked at from the systems approach. "What are you talking about?" bellowed one of the oldtimers. "Well," said the young employee, "I once took a course that taught me to look at problems from a wider perspective rather than from a focused one. For example, maybe the turnover problem has nothing to do with the wages which are already one of the highest in the area. Maybe its the result of a number of things that are all interrelated." "Well, hotshot," said the general foreman, "the problem is yours, you work it out."

The new employee sat down with his manager and started to ask a lot of questions. "What's our layoff policy? How do people get ahead? Can they get ahead? Are there any career paths available for the rank and file? Does the company do job posting? Are the people promoted on merit? How long does the average employee work here? Has the company ever done exit interviews to see why employees leave and where they go? Do they leave for higher wages? Are our benefits comparable to other firms? How consistent is the production schedule? How much overtime is employed? How much backlog of orders is there? What subjects are discussed in our orientation training? Does this company have a culture?" At first the human resource manager was puzzled as to why her employee was asking questions about so many seemingly unrelated areas, but she soon saw the point. After many discussions with people from production control, inventory, scheduling, and new, present and past workers, the new employee was ready to offer some suggestions. "Well," said the general foreman, "let's hear them!"

The new employee felt that the problem was really the combination of the firm's policies and practices and had nothing to do with the present wage structure or training. He found that, while the starting wage was good, the job provided less security than jobs in other firms because the firm was constantly hiring and laying off people in response to changing demand for products. Because the firm had such high turnover it did not feel that the employees were even concerned with any "careers" with the company. Thus, while they were attracting good people they were soon losing them because of the fluctuating boom or bust work schedules. The workers in turn did not feel any loyalty to the firm and when other jobs came along that provided the same or even slightly less money, but more security, they took them. The company felt that all it had to do was provide good wages and everything else would take care of itself. Top management thought that high wages would show the workers that the company was interested in them.

"Well," said the plant manager, "where do we go from here?"

The young employee hesitated but finally stammered, "I . . . don't know how to solve the problem; the prof in the course only got up to the part on identifying problems."

Required: Contrast the analytical and the systems approach used in this case. What subsystems are involved in this exercise and how do they impact on the problem?

Additional References

Ackoff, Russell L. "Towards a System of Systems Concepts." *Management Science* 17, no. 11 (July 1971), 661–71; reprinted in John Beishon and Geoff Peters, *Systems Behavior,* pp. 83–90.

————. and F. E. Emery. *On Purposeful Systems.* Chicago: Aldine Publishing, 1972.

Berrien, F. Kenneth. *General and Social Systems.* New Brunswick, N.J.: Rutgers University Press, 1968. Ch. 2: "Basic Definitions and Assumptions."

Boguslaw, Robert. *The New Utopians: A Study of System Design and Social Change.* Englewood Cliffs, N.J.: Prentice-Hall, 1965. Chap. 2: "System Ideas—for Whom and for What."

Buckley, W. *Modern Systems Research for the Behavioral Scientist.* Chicago: Aldine Publishing, 1968.

Checkland, Peter. *Systems Thinking, Systems Practice.* New York: John Wiley & Sons, 1981.

Churchman, C. West. *The Systems Approach and Its Enemies.* New York: Basic Books, 1979.

Cummings, Thomas G., ed. *Systems Theory for Organizational Development.* New York: John Wiley & Sons, 1980.

DeGreene, Kenyon B. *The Adaptive Organization: Anticipation of Management Crisis.* New York: John Wiley, 1982.

Ellis, David O., and Fred J. Ludwig. *Systems Philosophy.* Englewood Cliffs, N.J.: Prentice-Hall, 1962.

Emery, F. E., ed. *Systems Thinking.* Harmondsworth, Middlesex, England: Penguin Books, 1970.

Flood, Robert L., and Ewart R. Carson. *Dealing with Complexity: An Introduction to the Theory and Application of Systems Science.* New York: Plenum Press, 1988.

Jenkins, Gwilyn M. "The Systems Approach." *Journal of Systems Engineering* 1, no. 1 (1969); reprinted in John Beishon and Geoff Peters, *Systems Behavior,* pp. 56–82.

Kuenzlen, Martin. *Playing Urban Games: The Systems Approach to Planning.* Boston: I Press, 1972. Chap. 4: "Introduction to Systems Thinking."

Leftwich, Robert H., and Ansel M. Sharp. *Economics of Social Issues.* 4th ed. Plano, Tex.: Business Publications, Inc., 1980.

Lilienfeld, Robert. *The Rise of Systems Theory: An Ideological Analysis.* New York: John Wiley & Sons, 1978.

Phillips, Denis C. *Holistic Thought in Social Science.* Stanford, Calif.: Stanford University Press, 1976.

Schoderbek, Peter, and Charles Schoderbek, eds. *Management Systems.* New York: John Wiley & Sons, 1971.

Stephaou, Stephen E., ed. *The Systems Approach to Societal Problems.* Malibu, Calif.: Daniel Spencer Publishers, 1982.

Von Foerster, Heinz et al. *Purposive Systems.* New York: Spartan Books, 1968.

Wilson, Brian. *Systems: Concepts, Methodologies, and Applications.* Chichester, England: John Wiley, 1984.

CHAPTER 2

General Systems Theory

Observe how system into system runs;
What other planets circle the suns.

Alexander Pope

CHAPTER OUTLINE

INTRODUCTION

In Chapter 1 it was noted that since the time of Galileo modern science has been dominated by the analytical approach, that is, by the reduction of complex problems to their smallest isolatable components. This approach yielded the causal relationships sought, the sum of which constituted a description of the phenomena themselves. However, with complex phenomena, the whole proved to be more than the simple sum of the properties of the parts taken separately. With these complex systems it was found that their behavior must be explained in terms of the relationships existing among the components themselves and with their environment. The shift from the analytical approach to problems to the systems approach to the study of problems as a whole can be viewed as a change in methodology.

No doubt researchers throughout history have here and there employed a method somewhat resembling this approach, but it remained for Ludwig von Bertalanffy to formalize and advocate this methodology in the 1930s — the treatment of organisms as open systems interacting with their environment. (Open systems, as we will see later in this chapter, are simply those in which resources can pass over from the environment into the system itself. These are renewable systems; these are open systems — open to receive additional resources.) The full-scale revolution of this approach came several decades later, in the 1960s. Today it is the one generally professed by scientists investigating modern complex problems.

To understand the systems approach, one should know something of its roots, its history, and its origin. The systems approach evolved out of a general systems theory (GST), formulated by an interdisciplinary team of scientists with common interests. All were groping for a universal science — one that would unite the many splintered disciplines with a law of laws applicable to all.

The prime mover of GST was the biologist Ludwig von Bertalanffy. Although he formulated his "general systems theory" in the early 1930s, it was his major publication in *Science* in 1950 that provided the impetus for further development. In this article he presented the idea that all living systems are open systems and as such interact with their environment. The open system became for von Bertalanffy the general system model.

As mentioned previously, Bertalanffy had become disenchanted with the analytical method. Steeped in the Aristotelian philosophy that viewed objects as wholes and as endowed with intrinsic goals (*telos*), he began to view his own discipline in this way and was impressed by how well this methodology could explain some of the life problems with which he wrestled. He further maintained that all living organisms were goal directed, were endowed with intrinsic goals to which they tended. To understand the organism's behavior, one must view the organism as a whole, with its goal directedness, with its organization of interrelated and interacting parts. When one did this, then the statement of Aristotle that the whole is more than the sum of its parts aptly defined the basic systems problem.

General systems theory is not only a methodology; it is also a valid framework for viewing the empirical world. Its ideal of integrating all scientific knowledge through the discovery of analogies or isomorphisms is still to be realized. Its assumption of the unity of nature sustains the search for the isomorphy of concepts, laws, and models in the various scientific disciplines.

General systems theory, even stripped of its substantive laws (though few), has made its mark in the scientific world by providing the framework for viewing complex phenomena as systems, as wholes, with all their interrelated and interacting parts. Herein lies one of its merits, and its justification.

GENERAL SYSTEMS THEORY

Its Origin and Logic

At the 1954 annual meeting of the American Association for the Advancement of Science (AAAS), a society was founded under the leadership of biologist Ludwig von Bertalanffy, economist Kenneth Boulding, biomathematician Anatol Rapoport, and physiologist Ralph Gerard. This society was called the Society for General Systems Theory, later renamed the Society for General Systems Research. Its original purpose and functions were as follows:

> The Society for General Systems Research was organized in 1954 to further the development of theoretical systems which are applicable to more than one of the traditional departments of knowledge. Major functions are to: (1) investigate the isomorphy of concepts, laws, and models in various fields, and to help in useful transfers from one field to another; (2) encourage the development of adequate theoretical models in the fields which lack them; (3) minimize the duplication of theoretical effort in different fields; (4) promote the unity of science through improving communication among specialists.[1]

Certainly one would have to acknowledge that the first aim of the society has been somewhat fulfilled and that the progress to date has not been insignificant. This search for generalized laws continues and some of those first articulated by Bertalanffy still continue to be examined. Laws such as those for growth, evolution, and equilibrium still continue to find application in various fields. The law of growth, for instance, is the same for cells in biology, for crystals in crystallography, for populations in demography, and for compound interest in finance.

The second aim of the society is realizable through the use of mathematical models. Not only have varied disciplines like biology, physics, chemistry, engineering, and medicine contributed to the development of general systems theory, but general systems theory itself can contribute to the development of

---

[1] Ludwig von Bertalanffy, *General Systems Theory* (New York: George Braziller, 1968), p. 15.

specialized disciplines through its encouragement of the use of mathematical models.

To determine if a particular law or concept of one discipline can be applied in another, one must be capable of testing that law or concept in the other discipline. The one requirement, then, is for a language common to both disciplines—a language possessing little or no distortion. This language is the language of mathematics.

Rapoport believes that the language of mathematics is eminently qualified to serve as the language of general systems theory, because "the mathematical approach to general systems theory is characterized by complete rigour of its descriptions and by unchallengeable validity of the analogies that are revealed by these descriptions."[2]

Mathematics allows the systems theorist, irrespective of discipline, to employ and test laws for their generalizability. Unfortunately, nearly all of the developments in the field of general systems theory have come from the so-called hard sciences, traditionally endowed with precise measurement tools. The quantification problems inherent in the social and behavioral sciences, coupled with the complexity of the behavioral phenomena, have rendered their concepts less amenable to testing.

Its Postulates

In delving into the literature on general systems, one is soon overwhelmed by the many and diverse characteristics that it seems to manifest. Obviously this is due to the many specialized systems that the systems theorists have investigated, whose traits are extrapolated to and predicated of general systems theory. Perhaps what one ought to search for first is some kind of synthesis of the underlying premises or assumptions of general systems theory; then and only then ought one to investigate the alleged characteristics of general systems, compiled from particularistic investigations. Such a synthesis was attempted by Kenneth Boulding, and the result is both informative and fascinating—informative because he does manage in a way to "get under" the system and see it at its grass roots; and fascinating because he does this detective work with good grace and humor.

According to Boulding, there are five basic premises that any general systems theorist would most probably subscribe to.[3] These premises could just as well have been labeled postulates (P), presuppositions, or value judgments. As such they are statements that one accepts without further proof, for no proof is needed or even, at times, possible. Whether they are all independent assumptions or whether some are corollaries themselves derived from one or more of the basic premises is left

---

[2] Anatol Rapoport, *General System Theory: Essential Concepts and Applications* (Turnbridge, Wells, Kent: Abacus Press, 1986), p. 26.

[3] Kenneth Boulding, "General Systems as a Point of View," in Mihajlo D. Mesarovic, ed., *Views on General Systems Theory* (New York: John Wiley & Sons, 1964), pp. 25–38.

for the reader to decide. Even if the reader does not prefer to engage in the minute examination of the nature of each postulate, he will not fail to note that order is the order of the experiment.

P1.   Order, regularity, and nonrandomness are preferable to lack of order or to irregularity (= chaos) and to randomness.

The general systems theorist has a "rage for order." The theorist will be fond of all those things that foster or manifest order.

P2.   Orderliness in the empirical world makes the world good, interesting, and attractive to the systems theorist. "He loves regularity, his delight is in the law, and a law to him is a path through the jungle."

P3.   There is order in the orderliness of the external or empirical world (order to the second degree) — a law about laws.

The general systems theorist is not only in search of order and law in the empirical world; the theorist is in search of order in order, and of laws about laws.

P4.   To establish order, quantification and mathematization are highly valuable aids.

Because these will enable the general systems theorist to pursue the unrelenting quest for order and law, he or she will use them "in season and out of season," always mindful that there may be (and are) empirical elements displaying order but still not amenable to quantification and mathematization.

P5.   The search for order and law necessarily involves the quest for those realities that embody these abstract laws and order — their *empirical referents.* It is this ability to see the infinitely varied particularity of the world about us that gives this world its oneness, its "goodness."

In brief, the general systems researcher is not only a searcher after order in order, after law about laws; he or she is in quest of the concrete and particularistic embodiments of the abstract order and formal law that are discovered.

The search for the empirical referents of abstract order and formal law can begin from either one or the other of two starting points: the theoretical or empirical origin. The systems theorist can begin with some elegant mathematical relationship and then look around in the empirical world to see whether he or she can find something to match it, or the theorist may begin with some carefully and patiently constructed empirical order in the world of experience and then look around in the abstract world of mathematics to discover some relationship that will help to simplify it or to relate it to other laws that are already familiar.

General systems theory, then, like all of the true sciences, is grounded in a systematic search for law and order in the universe. Unlike the other sciences, *it tends to extend its search to a search for an order of order, a law of laws.* For this reason it can be called a *general* systems theory.

Its Properties

Having briefly considered some of the fundamental assumptions underlying general systems theory, one can now turn one's attention to the many and varied characteristics that systems theorists have attributed to general systems theory.

Since GST has as yet no definite body of doctrine (if it ever will),[4] one should be prepared to find little law or order in the characteristics of the systems theory that aims to search out order in order and to formulate a law of laws.

The following points do not comprise an all-inclusive list, nor do they constitute separate and distinct qualities. They do, however, reveal what theorists conceive as being the fundamental hallmarks of GST.[5]

1. Interrelationship and interdependence of objects and their attributes. Every systems theory must take cognizance of the elements or objects in the system, of the interrelationship existing between the various elements and of their interdependence. Unrelated and independent elements can never constitute a system.

2. Holism. The systems approach is not an analytical one where the whole is broken down into its constituent parts and then each of the decomposed elements is studied in isolation; rather, it is a *gestalt* type of approach, attempting to view the whole with all its interrelated and interdependent parts in interaction. The system is not a reconstituted (resynthesized) one; it is an undivided one – never having been broken down (analyzed).

3. Goal seeking. One of the major tenets of Bertalanffy's philosophy was the identification of a system's intrinsic goals (goal seeking, or teleology). Interestingly, this concept has recently been rediscovered by business organizations, inasmuch as management techniques are now being employed that stress the importance of goal formulation. All systems embody components that interact, and interaction results in some goal or final state being reached or some equilibrium point being approached.

4. Inputs and Outputs. All systems are dependent on some inputs that, when transformed into outputs, will enable the system to reach its ultimate goal. All ·systems produce some outputs needed by other systems. In closed systems the inputs are determined once and for all; in open systems additional inputs are admitted into the system from its environment.

5. Transformation Process. All systems are transformers of inputs into outputs. Among inputs that undergo transformation, one can include raw materials, power sources, energy, information, lectures, readings, examinations, etc. Through the transformation process, the form of the output is made different from that of the input. Whether one considers the production process, distribution process, educational process, religious conversion process, or other processes, the input is transformed from what it was initially to something else ultimately. All systems, if they are to attain their goal, must transform inputs into outputs.

6. Entropy Originally defined negatively in thermodynamics in terms of heat and temperature, entropy denotes the availability of the thermal energy of a

---

[4] The more general a science, the less content (body of doctrine) it will encompass. Mathematics owes its almost universal applicability to its amazingly contentless nature. Since GST aims to uncover the laws and order inherent in all systems, it ought to be the most contentless of all systems theories.

[5] Joseph A. Litterer, *Organizations: Systems, Control and Adaptation,* vol 2, 2d ed. (New York: John Wiley & Sons, 1969), pp. 3–6.

system for doing useful work. The more available the energy, the less the entropy; and conversely, the less available the energy, the greater the entropy. In thermodynamics, energy can be extracted for performing work only when a system changes from a more ordered state to a less ordered one. Thus in simple heat-energy systems, if useful work is to be done, there must be a difference in heat potential between two parts of the system. When two bodies are placed in contact with one another, heat will flow from the hot body to the cold body, and if some suitable device is hooked up to the system, the heat flow can be utilized to do work. But once the two bodies have reached the same temperature, no further work can be done.

The principle can be extended to fields other than heat energy. Water on top of a hill has a large amount of energy for doing work because of its position in the earth's gravitational field. As the water flows downhill its potential energy can be used to do work by turning a windmill, a turbine, etc. But once it reaches the bottom of the hill, the difference in potential becomes zero, and its entropy is at a maximum.

Because of the law of conservation of energy, energy is never lost; it tends to be degraded from useful forms to useless ones. This is the essence of the second law of thermodynamics: as work is performed, entropy increases — the availability of the system for doing more work declines.

Entropy thus is the amount of disorder or randomness present in any system. It has to do with the natural tendency of objects to fall into a state of disorder. All nonliving systems tend toward disorder; if left alone, they will eventually lose all motion and degenerate into an inert mass. When a permanent state is reached in which no observable events occur, the object is said to have attained maximum entropy. The process of degradation can take hours, years, even centuries.

Living systems, if isolated, would also follow suit and attain maximum entropy (death). However, by accepting inputs into the system from the external environment, living systems can, at least for a finite time, forestall this movement toward maximum entropy. Organizations too are constantly importing energy (people, raw material, capital, etc.) from their environment. By the use of "retained earnings" and diversification attempts, organizations can combat the tendency toward entropy. Because entropy presupposes a closed-system state, it will be of limited importance to our consideration of management systems.

7. Regulation.   If systems are sets of interrelated and interdependent objects in interaction, the interacting objects must be regulated (managed) in some fashion so that the systems objectives (goals) can ultimately be realized. In human organizations this implies the setting up of objectives and the determining of activities that will result in goal achievement. This setting up of objectives is commonly called planning. Regulation (control) implies that the original design for action will be adhered to and that deviations from the plan will be noted and corrected. Feedback is a requisite of effective control. The fundamental theme of cybernetics is always regulation and control. Because of its importance for the management of systems, the subject of cybernetics will be treated extensively in two subsequent chapters.

8. Hierarchy.   Systems are generally complex wholes made up of smaller subsystems. The nesting of systems within other systems is what is implied by hierarchy. The structure of any system has implications for its regulation. Simplistic structures (with few components) are typically more easily managed than more complex ones (with many interacting components). Methods of how to simplify complex systems will be presented later.

9. Differentiation.   In complex systems, specialized units perform specialized functions. This differentiation of functions by components is a characteristic of all systems and enables the focal system to adapt to its environment. Differentiation, specialization, and division of labor are virtually identical concepts.

10. Equifinality.   A direct cause-and-effect relationship can be found between the initial conditions and the final state of closed systems. In open systems, however, that is, in biological and social systems, the final states are not determined solely by their initial conditions but can be reached from different starting points and in different ways. Social systems are thus not constrained by the simple cause-and-effect relationships found in physical systems; they can attain their objectives with varying inputs and with varying transformational processes. Equifinality simply means that open systems have equally valid alternate ways to reach the same objectives.

Equifinality is a very useful concept since it has direct bearing on how social systems operate. The inflexible cause-and-effect relationships found in the physical sciences lead us to conclude that there is but one best (optimal) way to achieve a given objective. In the social sciences, however, we are led to believe that there is no one best solution to managerial problems but that there may be many *satisfactory* solutions to the same decision problem.

The 10 properties of open systems given above may seem at first glance a bit complicated, if not confusing, to the beginning student. Some of these properties, however, have already been touched upon in our treatment of the definition of system in Chapter One and so should not present an insurmountable problem. Some of the residual difficulty encountered by students may be due to their overexposure to closed systems in their formal study of the exact sciences, especially physics and mathematics. As the student is initiated more and more into the "secrets" of open systems as evidenced in the soft sciences — especially biology, sociology, and psychology — much of the haze should be lifted and a clear view of the entire scene eventually provided. As mentioned in the previous chapter, one should not expect to grasp fully every concept the first time it is presented. It is in the interplay of these concepts, their interrelationships, that a fuller understanding will arise.

One can easily envision a firm, a hospital, or a university as a system and apply the above tenets to that entity. Organizations, for example, obviously have many components which *interact* — production, marketing, accounting, research and development, all of which are dependent upon each other. This interaction can easily be seen by looking at the effects of a decision to increase production of a particular unit. Increased production can affect inventory levels, working capital, purchasing, number of employees required in production, quality control, maintenance, scheduling, transportation, sales levels, equipment utilization, overtime costs, delivery schedules, and a host of other factors. It is precisely this aspect of interrelatedness which is often neglected in the study of organizational problems which produces suboptimal solutions.

In attempting to understand the organization one must view it in its entire complexity rather than simply through one functional area or one component. A study of the production system would not yield satisfactory analysis if one ignored the marketing system or the personnel system. While students in their learning process do indeed study functional areas, which is necessary to determine the interactions, it is also necessary to study the entire organization as a system. It is for this precise reason that many schools offer a capstone course in business policies which attempts to view the organization as a *whole* and to integrate its composite dimensions. The use of management games also serves as a vehicle both to see the system in its entirety and to stress the interdependencies of the variables.

All living systems are *goal oriented* and, indeed, this particular element is widely noted and discussed in the literature. In business these goals commonly take the form of profit, share of the market, sales volume, labor productivity, and so on.

The organization is obviously dependent upon *inputs* which are then transformed into *outputs.* This *transformation* function may be production oriented, service oriented, or task oriented. In all cases however, inputs are transformed into outputs.

In regard to entropy, open systems (natural systems, organizations, individuals) have a continuous flow of inputs which move against the tendency toward entropy. To survive, a living system must have an incoming flow of imported energy as great as or greater than the outgoing flow. Indeed, firms try to build a reserve both to improve their chances of survival and to provide for further growth. Since the concept of entropy is mainly concerned with closed systems (systems in which no energy is received from an outside source and in which no energy is released to the environment), little attention will be devoted to the concepts of closed systems and of entropy. Social organizations, of course, are not closed systems; in fact, no system is ever completely closed — only relatively so. One frequently employs artificial closure in order to study a system at a particular time, treating the system as if it were shut off from its environment. This procedure is employed in controlled experiments where all conditions but one are controlled — held constant — the one remaining condition or factor is systematically varied, and the results are noted.

Regulation is synonymous with control, and it is quite obvious that organizations employ a spectrum of control methods, each designed to measure actual performance with a desired goal. Quality control, cost control, budgetary control, production control, and worker control are all illustrative of vehicles of regulation.

The organization is a *hierarchy* of subsystems and the typical organization chart often depicts this hierarchy. The major goals of the organization are typically segmented into divisional goals, which in turn are fragmented further into plant goals, department goals, and individual goals.

Differentiation and specialization are evident in the structuring of tasks, the layout of plants, and the assignment of specialists.

Equifinality simply means that there are alternative ways to reach goals. In some instances this may be achieved through the introduction of new products; in others, it may be through acquisition or greater market penetration.

BOULDING'S CLASSIFICATION OF SYSTEMS

For thousands of years man has been occupied with classifying phenomena. While early man must have classified plants and animals as either harmful or not harmful, useful or not useful, one of the early formal attempts at the classification of the thousand or so then known plants and animals was undertaken by Aristotle. His simplified division of animals into those with red blood (animals with backbones) and those with no red blood (animals without backbones) and his division of plants (by size and appearance) into herbs, shrubs, and trees served man until the 18th century, when the idea of structure (arrangement of parts) was adopted by Linnaeus.

Every classification scheme, though arbitrary in design, is drawn up with some particular purpose in mind. Students can be classified by level or by proven ability; climate can be classified on the basis of temperature and rainfall; books can be categorized as either fiction or nonfiction (fact). One classification of planets is by size, another by relative position with respect to the sun. Football players can be classified into offense and defense, into linemen and backs. In all of the above there is some order in the classifying scheme. The scheme adopted evidently presupposes some knowledge of the objects being classified and aids in the study of all other such objects.

As for systems, their classification is necessary if a methodology for their study is to be developed. The *first* classification of systems in which we are interested is that which utilizes the criterion of *complexity* as its distinctive feature. It is within a hierarchy of complexity that Boulding arranges his theoretical "system of systems" (see Figure 2–1).[6] As one progresses from level 1 to level 9, one encounters an increase in system complexity.

Level 1. Frameworks.   This is the level of static structures. Before one can deal with the dynamic behavior of a system, one must first be able to describe accurately their static relationships. These can be described by function, position, structure, or relationship; e.g., anatomy of an individual, the location of the stars in the solar system.

---

[6]Kenneth Boulding, "General Systems Theory – the Skeleton of Science," in *Management Systems,* 2d ed., ed. Peter P. Schoderbek (New York: John Wiley & Sons, 1971), pp. 20–28.

FIGURE 2–1   An Ordering of Systems by Complexity

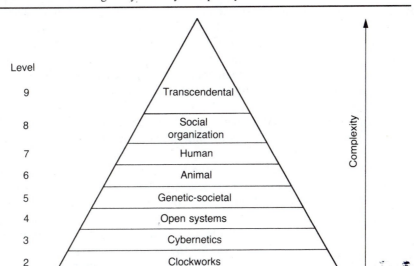

Level 2. Clockworks.   This is the level of simple dynamic systems with predetermined motions. The movements of the solar system, the theories of physics and chemistry fall into this category. Practically all systems which tend toward equilibrium, including machines, are included.

Level 3. Cybernetics.   This is the level of maintenance of a given equilibrium within certain limits. This level refers both to the engineering-type control (thermostat) as well as the physiological (the maintenance of body temperature). Here there is teleological behavior (goal seeking) but no automatic goal changing.

Level 4. Open System.   This level is concerned with self-maintenance of structure and therefore relies on throughput of material and energy. Closely connected with self-maintenance of structure is the property of self-reproduction. Self-maintaining and self-reproducing systems are definitively living systems. Hence this is the level of the cell.

Level 5. Genetic-Societal.   This level is typified by the plant and is characterized by a division of labor. Although there are many information receptors, they are not refined enough to accept information and to act upon it, but, rather, plant life or death is blueprinted by stages.

Level 6. Animal Level.   The notable characteristics of this level are increased mobility, teleological behavior (goal seeking), and self-awareness. Specialized information receptors are present which allow for a structuring of information and for the storage of information.

Level 7. Human.   In addition to all of the characteristics of animal systems, humans have self-consciousness—i.e., they are not only aware, but are aware that they are aware. Their capacity for storing information and for formulating goals and their facility for speech are all well developed. Humans can reflect upon life, plan for it, and develop tools and machines to enhance their control of life. In recent decades they have even been able to develop nearly humanlike machines, digital computers, to significantly alter the way they live and work.

Level 8. Social Organization.   People are not isolated but rather are the product of the many roles that they play as well as that of society in general, and therefore are molded by, affected by, and affect the entire gamut of history and society.

Level 9. Transcendental.   This is the level of the unknowables which escape us and for which we have no answers. Yet these systems exhibit structure and relationships. These indeed are the most complex of all since they are indescribable.

The first three levels are made up of physical and mechanical systems and have been of particular interest to the physical scientists. The next three levels all deal with biological systems and are of concern to the biologist, botanist, and zoologist. The remaining three levels, those of humans, social organizations, and transcendental systems are primarily of interest to the social scientists.

As stated above, systems are classified with certain purposes in mind, and to simply list Boulding's classification scheme without examining the purpose would be unpardonable. Realizing that all theoretical knowledge must have empirical referents, he proposed his categorizations with a view to assessing the gap between theoretical models and empirical knowledge. In this regard he states that, while our theoretical knowledge may extend adequately up to the fourth level, empirical knowledge is deficient at all levels. There is agreement that, even at the first level of static structures, inadequate descriptions of many complex phenomena exist. While adequate models exist at the level of clockworks, and to some extent at the third and fourth levels, these are only a modest beginning. While one could hardly doubt the achievements made in medicine and the systematic knowledge acquired in this area, these too are the mere rudiments of theoretical systems.

To some readers Boulding's assessment of the gap between the theoretical models and empirical knowledge may seem too pessimistic. One should remember that this assessment was made back in 1956—almost 35 years ago. Since then great strides have been made, not only in medicine and science, but also in technology. Here are some of the significant breakthroughs that had not yet transpired when Boulding first made his assessment:

In Medicine

Dr. Albert Sabin perfects his oral polio vaccine (1956).

Dr. Christiaan Barnard performs the world's first human heart transplant (1967).

First test-tube baby born in England (1978).

Gene splitting produces natural virus-fighting substance (1980).

Swiss scientist reports first cloning (asexual reproduction) of a mammal (1981).

First permanent artificial heart installed (1982).

In Science and Technology

USSR launches Sputniks in outer space (1956).

U.S. Explorer I circles the globe and discovers the Van Allen radiation belt (1958).

Laser perfected and patented by two U.S. scientists (1960).

Triton, the first U.S. atomic submarine, circumnavigates the globe underwater (1960).

First weather satellite, Tiros I, launched (1960).

Russian cosmonaut becomes first human to orbit the earth (1961).

Soviet cosmonaut and also American astronaut "walk" in space (1965).

American astronaut becomes first human to land on moon (1969).

USSR orbits a spacecraft around the planet Mars (1971).

Pioneer 10 lifts off to photograph Jupiter (1972).

Soviet Venus 8 soft-lands on Venus (1972).

Two U.S. Viking spacecrafts land on Mars to photograph the planet, to analyze soil samples and to search for life (1976).

Pioneer 11 gets first close-up pictures of Saturn (1979).

Two Soviet cosmonauts rendezvous with space station (1980).

Columbia space shuttle lifts off (1981).

Voyager II discovers more moons around Venus (1982).

U.S. launches 25th space shuttle (1988).

In the list above, nothing was said of the electronic digital computer, which has undergone several additional generations since the first (1951-58). Then too, besides supercomputers, mainframes, and minicomputers, there are microcomputers (personal computers) that have not only revolutionized the way corporations, large and small, carry on their business but have altered in innumerable ways the lifestyle of almost every social institution, ranging from the governmental, through our educational system, down to the core of the family. The tremendous social and economic changes brought about by the introduction of computers have affected not only those in the United States but have spread globally—to Japan, Korea, Taiwan, and other countries.

This omission of computers from the above list was not accidental. It was purposely done—with malice aforethought—since the rate of change in the computer industry (both in hardware and software) has been so rapid and dynamic that we have hardly anything else to compare it with.

In just a few years, for example, we have come from the 8086/8088 chips to the 80286, 80386, 80486, and the 80860 RISC microprocessor ("Cray-on-a-Chip"), which have boosted processing power significantly, each time making multiuser, multitasking networking systems possible. Desktop publishing, which in the early days of the PCs was only a pipe dream, has become a reality. LANs (local area networks), which allow users to work independently, have been fast gaining acceptance among both large and small companies since they have become ever less costly, easier to install, and relatively more user-friendly. Erasable optical disks are redefining the meaning of mass storage, especially for desktop and laptop computers. According to some experts, EISA (Extended Industry Standard Architecture) will eventually become an open, industry-standard bus architecture for the entire microcomputer industry.

If the evolution of hardware is difficult to track, that of software is even more formidable. The many commercial programs that appear each month are not only more powerful and more sophisticated, but they have also given rise to what is currently known as freeware and shareware — types that have enabled ordinary users to extend their productivity in untold ways, besides providing them with interesting entertainment.

Computer languages too have proliferated from the early days of assembly language, having evolved into FORTRAN and BASIC, giving rise to FORTH and PL/I, LISP and PROLOG, Pascal, and C. Many believe that future years will see the maturing of the C language, which has been steadily growing in acceptance among programmers. SQL (Structured Query Language), a language common to mainframes and minicomputers, is now being adapted to personal computers. Unfortunately, like BASIC and FORTRAN, with every major institutional user producing its own version, no one standard version of SQL has gained prominence.

Hundreds of professional and popular computer magazines track the projected and current changes in the computer industry. As we shall see in a later chapter, the computer industry provides an excellent example of companies operating in a highly complex and dynamically changing environment.

A second purpose of Boulding's classification scheme is to "prevent us from accepting as final, a level of theoretical analysis which is below the level of the empirical world which we are investigating."

And third, the scheme serves as a mild warning to the management scientist who, despite the fact that new and powerful tools have led to more sophisticated theoretical formulations, must never forget that he or she is little beyond the third or fourth level of analysis and therefore cannot expect that the simpler system will hold true for the more complex ones.

Thus, for Boulding's purposes, his classification scheme is quite logical and consistent although one may disagree with his "perception of developments" or even his ordering of complexity. For even here Boulding's concept may indeed have different connotations for different students.

OPEN AND CLOSED SYSTEMS

There are many possible bases for distinguishing different kinds of systems. Authors speak of abstract and concrete systems, mechanistic and organismic systems, simple and complex systems, decomposable and nondecomposable systems, open and closed systems and many others. With regard to these, there is never a question of any one type being right or wrong; it is simply a question of being useful or not useful. Does the distinction enable us to better understand the subject, or does it simply multiply words for concepts that do not enlighten?

Of all the distinctions noted, that of open and closed systems appears to be most fundamental and most useful. This classification rests upon the basis of *resource availability.* The resources of a system are all those things available to the system for carrying out the activities necessary for goal realization. They include personnel, money, equipment, technological processes, information, and even the *opportunities* for the aggrandizement of the human and nonhuman resources of the system.

In a closed system, all the system's resources are present within it. No additional resources enter the system from the environment. In a sense, the closed system has no environment: no other system impinges upon it or interacts with it. The closed system can be viewed as being self-contained. Open systems, on the other hand, import resources from the environment, transform them into some useful output, and export the output into the environment. This input-transformation-output cycle undergoes continuous iteration. The open system definitely has an environment with which it interacts. Because it obtains its resources from the environment, it is self-renewing. It is open, ready to receive.

Some theorists have entertained serious doubts whether a closed system really exists outside the mind. Environments seem to be all around us. We exert little or no control over them, while they exert significant influence on us and on nearly every system we are familiar with. But regardless of whether or not closed systems actually exist, the concept at least is still quite useful, especially for research purposes.

Often we attempt to set up experimental models that are virtually closed systems. We try to keep all variables constant (with no environmental influences affecting them) while we systematically vary the value of the independent variable. The lab technician testing a blood sample of a hospital patient assumes a system closed to environmental factors that might alter the data. In laboratory research on human behavior, we also try to set up a closed system, at least for a time. We try to bring under direct control the three features of the causal proposition: time order between variables, covariance, and the exclusion of rival causal factors. Problem solving by business executives is often done in a quasi-closed system with the aim of simplifying the situation enough so that a rough estimate can be arrived at.

The one danger that lurks in treating open systems *as if* they were closed systems is the tendency to assume that, just because they are conceived as closed systems, they *really are* such. This is the fallacy of *reification.*

ISOMORPHIC SYSTEMS

Instances abound in which the structural relationships of one system are similar to those of another system. Models in general attempt to represent a correspondence of their structure with the real elements being modeled. Good models correspond point for point with the object modeled. In this case, a one-to-one correspondence is said to exist between the elements of the model and the components of the system being modeled. Where a one-to-one correspondence exists of the elements of one system with those of another, the systems are said to be isomorphic ("of like form"). One can map the 26 letters of the English alphabet on the set of cardinal numbers 1–26, thus setting up a one-to-one correspondence between the individual numbers and the letters of the alphabet.

The isomorphy most commonly acclaimed is that between mechanical and electrical systems. The two systems exhibit a one-to-one relationship in their structures. The relationship between quantities of the mechanical system and those of the electrical system is expressed by equations of the same form. The corresponding quantities encountered in these equations are force and voltage, speed and current, mass and inductance, mechanical resistance and electrical resistance, elasticity and capacitance. (See Figure 2–2 on isomorphisms.)

The application of certain laws across a number of different branches of science is well known. The exponential law, for example, has application in biology, economics, and psychology as well as in physics, and in all instances the equations are identical. The mathematical equation for information is identical with that of negative entropy. The reason for these isomorphisms is that the structures are similar when considered in the abstract. Telephone calls, radioactive disintegrations, and impacts of particles can all be considered as random events in time. Because they have the same abstract nature, they can be studied by exactly the same mathematical model.

Of importance in isomorphic mappings is that there is also a correspondence, not only of structure, but also of operational characteristics. It is this feature of isomorphic systems that has enabled the researcher to investigate and to predict properties of other systems. Cybernetics itself arose out of the realization by Norbert Wiener that the structure and operation of machines which are to be controlled are quite similar to those of animals.

Analogy

Perhaps something should be said on analogies and isomorphisms and the difference between them. Analogy seems to be the more general term. An analogy is the assertion that things that resemble each other in some respects will resemble each other in some further respect. People analogize every day. They compare two or more events or things because they resemble one another on one or more points. We hear such expressions as "He runs like a deer," "He lives like a king," "He's another Einstein," "He's another Napoleon." All of these are analogies.

FIGURE 2–2   Selected Isomorphisms

Magnetic	*Electrical*
Rowland's Law $\Phi = F/R$	Ohm's Law $I = E/R$
where	where
Φ = Flux in maxwells F = Gilberts R = Reluctance	I = Current flow in amperes E = Volts R = Ohms
Mechanical	*Electrical*
Velocity $v = s/t$	Current flow $I = Q/t$
where	where
v = Velocity s = Distance t = Time	I = Current flow in amperes Q = Coulombs t = Time

The first two statements are, in fact, similes; the latter two, metaphors. In all of them we expect that the items being compared have at least one property in common.

General systems theorists employ analogies all the time in the hope that some meaningful relationships might be discovered. However, the use of analogies has also occasioned much criticism. Many of the analogies of systems theorists, it is argued, are meaningless. But analogies do enable the systems theorist to create new approaches to problem areas, to pick up "creative hunches." After all, symbolic thought is, for the most part, analogical thinking. The history of science abounds in examples of analogies that led to the formation of theoretical ideas.

The "father" of general systems theory, Ludwig von Bertalanffy, stressed the idea of analogies—not any kind of analogy but the kind whose corresponding abstractions and conceptual models could be applied to quite different phenomena. The eminent mathematician and astronomer Johannes Kepler once remarked, "Above all, I value analogies, my faithful instructors. They are in possession of all the secrets of nature and should, therefore, be the last to be ignored." Kepler was probably referring to mathematical analogies, of which there are many. Simultaneous equations, for example, serve many problem areas depending on the meaning associated with the constant coefficients. Samuel Butler may have been ahead of his time when he wrote, "Though analogy is often misleading, it is the least misleading thing we have." Analogies can well serve the electrical, acoustical, mechanical, and structural engineer as well as the production scheduler, the sociologist, and a host of other professionals.

Analogy, as mentioned above, is at the root of most of our ordinary reasoning from past experience to future contingencies. Argument by analogy is thus a form of inductive logic and differs from deductive logic in that none of its conclusions follow with *logical necessity* from its premises. Consequently the conclusion in analogies is not to be considered *valid* or *invalid;* rather it should be considered more or less *probable* or *improbable.*

In any analogy three things are involved: (1) the entities themselves, (2) the attributes (or respects) in which the entities are said to resemble each other, and (3) the probability that the entities will be similar to each other in some other attribute. Schematically this can be represented thus:

> Given entities A, B, C, and D
>   and attributes X, Y, and Z,
>
> If A, B, C, and D have the attributes X and Y
>   and B, C, and D have the attribute Z,
>
> Then A probably has the attribute Z.

Since even some general systems theorists admit they have difficulty understanding analogies (as Boulding confesses) and even more difficulty in discriminating good from poor analogies, students of management systems ought to be at least dimly aware of the criteria involved in analogizing. The following criteria discussed in the literature should shed some light on the subject.[7]

Criterion 1. The Number of Entities Between Which the Analogies Are Said to Hold.   The greater the number of entities involved in the analogy, the higher the probability. An example may help. If I advise you not to take your personal computer to be serviced by the G.Y.P. Computer Service because I had mine there once and it came back noticeably unaltered, you might say that I was being somewhat hasty in judgment. However, if I told you that on four other occasions I had taken it there and each time the work was unsatisfactory and the charges outrageous, the conclusion would enjoy a higher degree of probability than before. Furthermore, if I told you that three other members of our personal computer group also had similar unhappy experiences with the work of the G.Y.P. Computer Service, the conclusion would enjoy a still higher degree of probability. The greater the number of entities involved, the higher the probability. (Note that there is no mathematical ratio between the number of instances and the probability of the conclusion.)

Criterion 2. The Number of Attributes in Which the Entities Involved Are Said to Be Analogous.   The greater the number of attributes or respects in which the entities are analogous, the higher the probability that they will share some other analogous attribute. If porpoises resemble humans in having lungs, warm blood, and hair, they will also *probably* share with the human species a four-chambered heart, a like pattern of nerves and blood vessels, etc. (Again, one does not have available a simple mathematical relationship between the

---

[7] The treatment of this topic leans heavily on Irving M. Copi's discussion of analogy in his *Introduction to Logic,* 7th ed. (New York: Macmillan, 1986), pp. 403–14.

number of points of resemblance in the premises and the probability of the conclusion.)

Criterion 3. Strength of the Conclusion Relative to the Premises. The more assertive the conclusion relative to the premises, the lower its probability. Copi gives this illustration. If Jones has a new car and gets 23 miles to the gallon, Smith may infer *with some degree of probability* that her car of the same make, model, and year will also give *good* mileage. However, Smith may infer more specific conclusions from the same premises, each with its own degree of probability. If she concludes that her car will get *over 20* miles to the gallon, the conclusion may enjoy a high degree of probability.

If she infers that her car will get *over 21* miles to the gallon, the conclusion is less probable. If she concludes that her car will get *exactly 23* miles to the gallon, as does Jones's car, the conclusion is considerably weakened. In other words, the probability of her conclusion being true is very low.

Criterion 4. The Number of Disanalogies. The greater the differences between the instances mentioned in the premises and the instance involved in the conclusion, the weaker the argument by analogy and the less probable the conclusion. An example: If we are told that Jones, for the most part, drives his car to and from his place of work at a steady pace on the tollroad and that Smith nearly every day almost exclusively uses her car in the stop-and-go city traffic, the disanalogy between the instance in the premises and that of the conclusion will reduce the probability of the conclusion.

Criterion 5. Dissimilarities in the Attributes of the Entities Involved. The more dissimilar the instances mentioned in the premises, the greater the probability of the conclusion. If a particular medication has successfully alleviated symptoms of a particular type of disease in various groups of patients, differing among themselves in age, sex, economic and social class, ethnicity, body build, racial stock, sexual mores, etc., the probability is very high that it will be successful in other categories of patients.

Criterion 6. Relevance. One overriding criterion that embraces all of the above is that of relevance. In the car example given above, the fact that Jones's car resembles Smith's in having the same number of cylinders, weight, and horsepower lends credence to the proposition that it will perform similarly, giving similar good mileage. However, if the similarities were confined to the same color, the same number and types of gauges on the dashboard, the same types of safety glass and hubcaps, and the same style of upholstery, the argument by analogy would be notoriously weakened since none of these is relevant to the mileage factor.

Relevance is indeed important. An argument based on a single relevant analogy evidenced in only one instance is worth more than a dozen irrelevant similarities in scores of instances. An analogy is relevant to establishing the presence of a given attribute if the other attributes have some causal relationship to the given one. If my neighbor has his house insulated and his fuel bills go down, I can with a high degree of probability expect my fuel bills to go down too if I have my house insulated. The reason for this is that insulation is indeed relevant to the size of fuel bills, since it is causally related to fuel consumption. Relevance thus is tied to our knowledge of causal connections discoverable through observation or experimentation. Herein lies ample room for honest differences of opinion.

Isomorphism

All isomorphisms are analogies, but not all analogies are isomorphisms. This is true because isomorphisms are a subclass of analogies. Every subclass enjoys the properties of the generalized class plus one or more properties of its own. Isomorphisms (iso-like + morphe-form, structure) imply a *likeness of structure,* which is a very particular type of likeness.

The principle of isomorphism was discovered by the German chemist Eihard Mitscherlich around the year 1820. He formulated the principle that isomorphic structures have similar chemical formulas.

Two systems can be said to be *isomorphic* to each other if and only if (1) a one-to-one correspondence can be established between the elements of one and those of the other, and (2) all the relations defined on the elements of one system also hold among the corresponding elements of the other.[8]

The first condition stipulates that the elements of one system can be mutually and uniquely assigned to the elements of another system. The second condition does the same for relationships. The elements of a characteristic relationship in one system can be mutually and uniquely assigned to the corresponding elements of a characteristic relationship in the other system.

Perhaps the better to grasp the significance of this definition of isomorphism, one ought to derive some corollaries. Klir[9] has already done so and has given us these four:

Corollary 1.   Two finite isomorphic systems necessarily have the same number of elements. Consequently, there is a one-to-one correspondence of the elements in the systems.

Corollary 2.   Isomorphic systems are symmetrical. ($S^1iS^2 = S^2iS^1$). Read: If system S^1 is isomorphic to system S^2, then system S^2 is isomorphic to system S^1.

Corollary 3.   Isomorphic relations are reflexive, i.e., are isomorphic to themselves. (SiS).

Corollary 4.   Isomorphic relations are transitive. If S^1iS^2, and S^2iS^3, then S^1iS^3.

A state map and the state it represents are often said to be isomorphic[10] from the geometrical point of view. To every element on the map (road, railroadcrossing, city monument, airport, county seat, etc.) there is a unique corresponding

---

[8] Strictly speaking, there is a difference between isomorphic and homomorphic systems. The difference is that in homomorphic systems one of the systems has more elements than the other, while in isomorphic systems the number of elements in both systems is the same. Often this distinction is not observed and both types are called isomorphic systems.

[9] George J. Klir and Miroslav Valachi, *Cybernetic Modeling* (Princeton, N.J.: Van Nostrand, 1967), p. 110.

[10] See footnote 8.

element in the state. Also the geometrical relationships between the elements on the map are the same as those between the corresponding elements in the state. A distance of 3 inches between two points on the map corresponds to a distance of 75 miles when the scale is 1 inch = 25 miles.

Of course, there are a lot more elements in the state than are indicated on the usual state map. However, for every element on the map, there is a corresponding element in the state, and for every relationship of elements on the map, there is a corresponding relationship of the elements in the state.

HIERARCHY OF SYSTEMS

While an obvious hierarchy of systems exists (the ultimate system being the universe), still, almost any system can be divided and subdivided into subsystems and subsubsystems depending on the particular resolution level desired. This nesting of systems one within another can be seen in nature as well as in man-made systems. The universe, for example, includes subsystems of galaxies of stars which in turn include the solar system, and so on. The inventory system of a firm is a subsystem of the production system, which in turn is a subsystem of the firm, which in turn is a subsystem of the industry, which in turn is a subsystem of the economic system.

The amount of nesting of systems within systems employed in any analysis will depend on the nature of the problem being investigated, the depth of analysis sought, and the particular framework employed. Perhaps the failure to adequately solve many organizational and institutional problems may be the tendency to concentrate on too restricted a system. What should be regarded as but a subsystem is taken as the system, with the result that the significant interrelationships of the system with other subsystems are either overlooked or completely ignored. What constitutes the environment of the subsystem should, for practical reasons and for a realistic solution, be part and parcel of the system itself.

All of this can be put succinctly in propositional form:

1. A system is always made up of other systems.
2. Given a certain system, another system can always be found that comprises it, except for the Universal System, which comprises all others.
3. Given two systems, the one system comprising the other can be called the high-level system in relation to the system it comprises, which is called the low-level system.
4. A hierarchy of systems exists whereby lower-level systems are comprised into high-level systems.
5. The low-level systems are in turn made up of other systems and can, therefore, be considered the high-level system for the lower-level systems to be found in it.[11]

---

[11] John P. van Gigch, *Applied General Systems Theory,* 2nd ed. (New York: Harper & Row, 1978), p. 376.

Figure 2–3 depicts a system of hierarchies, a "system within a system," for a business organization. One could readily apply the five previously mentioned hierarchical propositions to this and to other contrived social systems. In fact, one might go even further—one might try to identify the larger system in Boulding's somewhat abstract classification to which this system belongs.

In a business organization like that depicted above, one can identify many possible subsystems for purposes of analysis or design. The choices about organizational subsystems are many because systems and subsystems are "contrived" by humans. This range of choices, however, does not hold for the subsystems of a biological organism, for these are givens. Social insects, like the ant and the bee, also have hierarchies, but these are not contrived: they are innate properties of the species. It is because social systems are contrived systems— contrived by social groups—that hierarchies are instituted. The "system within systems" must be carefully planned and implemented (contrived) if all is to function as a system—as a whole.

Several interesting points are made by Simon in his discussion of hierarchy.[12] Simon, like others, defines a hierarchy as a system that is composed of interrelated subsystems, each of which, in turn, is hierarchic in structure until the lowest level of subsystem is reached. In physics atomic particles may be treated as a subsystem; in biology the cell; in society the family; in organizations a division or department.

Decomposable Systems

Simon also distinguishes between interactions *among* subsystems and interactions *within* subsystems. The latter involve interactions among the parts/objects of the subsystem. Typically there is more interaction among the parts of the subsystem than between the subsystems and their system. If this is so, then each part of the subsystem can be treated *as if* it were independent of the others. Such systems are *decomposable* into the subsystems comprising them. An example might be the university and the various colleges and departments of which it is comprised. There is far more interaction among the members of a department than among the departments of the several colleges.

Nearly Decomposable and Nondecomposable Systems

A *nearly decomposable* system is one in which the interactions among the subsystems, though not negligible, are still rather weak. *Nondecomposable* systems directly influence other systems. They are dependent on other systems.

Two propositions are developed from this discussion of nearly decomposable

---

[12]Herbert A. Simon, *The Sciences of the Artificial* (Cambridge, Mass.: MIT Press, 1970), pp. 84–87.

FIGURE 2–3 Contrived Hierarchies in Social Systems

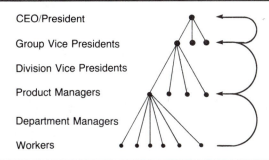

CEO/President

Group Vice Presidents

Division Vice Presidents

Product Managers

Department Managers

Workers

systems. First, in a nearly decomposable system the short-run behavior of each subsystem is approximately independent of the short-run behavior of the other subsystems. Second, in a nearly decomposable system the long-run behavior of any of the components is dependent on the behavior of the other subsystems in only an *aggregate* way.

The implication of the concept of decomposability should now be obvious. It allows one to understand the nature of a system without becoming immersed in the details of the component interactions. In social systems one's information-processing capacity is limited, and therefore we should be more concerned with the interactions of the subsystems with the system than with the interactions of the subsystems with one another.

Little information is lost by using the hierarchical concept. Indeed, hierarchy allows one better to understand the many subsystems through the process of simplification. Other techniques for reducing complexity will be dealt with in the chapter on cybernetics.

ADAPTABILITY OF SYSTEMS

The question is often asked, Why do some organizations thrive and grow (survive) while others fail? To be sure, nothing here is constant; everything changes. Even the ancient Greek philosophers recognized this truism. Organizations that fail to change fail to survive. The organizations that exist today are the survivors of earlier systems. Except for the social historian, seldom do we study organizations that do not survive. While the major organizations of today are exceedingly large in size (measured either by sales, assets, or the number of employees), most of them are relatively new, the oldest being, perhaps, only a few hundred years old. Organizations like the Roman Catholic Church, which is nearly 2,000 years old, are very rare indeed. The survival of an organization is thus not a trivial property of systems behavior. In fact, it is this very property of organizations that allows us to study them in the first place.

Systems have sometimes been classified as being adaptive or nonadaptive. Adaptability of a system is its ability to learn and to alter its internal operations in response to changes within itself or in its environment. Let us consider this more at length.

Systems can adapt either in the short run or in the long run.[13] In the short run the system can respond to changes externally induced — that is, it can respond to stimuli arising in the environment. For example, as the price of gasoline increased, the consumer responded by demanding of the auto manufacturers lighter and more efficient cars. When they did not furnish these, the consumers turned to the products of foreign car manufacturers. When the sale of foreign-made cars seriously eroded the market share of American car manufacturers, these belatedly responded to the increase in consumer demand for lighter and more cost-efficient cars by changing the size and weight of the cars they would turn out. This is a response of an organization to a change in its environment. The system can also respond to changes in an object within the system itself — e.g., a change occurring in packaging a product or in the sales organization or in the method of producing a product. These short-run adaptations should be viewed as functional, since there is no change in the structure or structural attributes of the system. In the short run, time is so short that structural changes cannot be made. This is similar to the short run in the field of economics, where firms can only adjust inputs and outputs and cannot increase in size. It is good to remember that function pertains to "what you do" and structure to "what you are." Changes in products, suppliers, production technologies, and the like are all functional, while changes in centralization or decentralization or organizational charts as well as mergers and acquisitions are all examples of structural changes.

In the long run a system can respond to changes either by structurally modifying itself or by structurally modifying the environment. By altering an object within the system itself, the system can be thought of as *evolving,* much as in the Darwinian theory of evolution. The system modifies its structure the better to cope with the changed or changing environment. In other instances, the system seeks to modify its environment in much the same way that business organizations seek to change their environment — by lobbying their legislators who make the laws, by advertising, by acquisitions and mergers. This type of structural change has been called Singerian, after the author Edgar A. Singer.

To sum up these forms of adaptability:

Functional (short-run):
  Internal — induced internally.
  External — induced externally.

Structural (long-run):
  Darwinian — changing the system's structure.
  Singerian — changing the environment's structure.

---

[13]Francisco Sagasti, "A Conceptual and Taxonomic Framework for the Analysis of Adaptive Behavior," *General Systems* 15 (1970), pp. 153–55.

A prerequisite for adaptability of a system is its familiarity with its environment, whether it be natural or man-made. Just as individuals must first acquire environmental information through their senses before adapting to any changes in the environment, so too must the organization. Adaptive systems therefore are simply those systems fully cognizant of their environment and "able and willing" to adapt themselves to it. In the same way, all living systems must possess this characteristic if they are to survive.

In order to provide greater adaptability, firms often employ the strategy of diversification as well as designing a flexible organizational structure. This latter element implies a decentralization policy by which the response time of a firm to react to a rapidly changing environment is typically less than for a highly centralized structure.

As will be shown later, the ability and willingness of a firm to acquire information concerning the state of its environment is a critical factor in how well it can adapt. It is only through the acquisition of information that firms can learn of threats or of opportunities existing "out there" in the environment. These threats may take the form of new competitors, new regulations, new products, or new processes. Opportunities may take the form of learning of, or creating, new needs of consumers, taking advantage of newly developed technologies of the firm, or a number of other forms. The research of many pharmaceutical firms and the resultant new drugs are illustrative of opportunities for the firm. Such new technology is conducive to increasing the firms' adaptability.

Adaptability can hardly be overemphasized, in that some firms that paid little attention to their environment have failed to survive. While the consequences have not been as severe in other cases, market positions have become less secure. One need only compare the list of *Fortune's* top 100 companies of 20 years ago with the top 100 of today to be convinced of this fact. Some organizations like du Pont and General Electric have responded to massive economic and social changes of past decades through a transformation process of both structure and outputs. These firms possess adaptability. The earnings of many of the so-called progressive firms of today are from products unknown 10 years ago. It can be said of such firms that they are in close touch with their environment.

EVOLUTION OF SOCIAL SYSTEMS

Social organizations, like natural systems, have a history. They are born, they evolve, they mature, and they die. Like many natural systems, they thrive better in some environments than in others. Like organisms, they must adapt to their changing environments, and in some instances they must do so quite rapidly. Their failure to do so can be as catastrophic as with organisms. Witness the extinction of some of the dominant forms of sea life as well as some large mammals. For social organizations the time span is much more compressed. A decade, even a few years, may be all the difference between a viable organization that has adapted and one that has failed to adapt to its environment.

Another difference between social organizations and natural systems is that man's technology allows him to help the adaptation process along, and for this reason organizations are often termed sociotechnical organizations. However, even with man's tool-making[14] ability, there are alarming parallels of demises in the social systems, as with plant and animal systems. The Greek Empire, the Roman Empire, the Mayan Empire, and those of Europe's recent past — all attest to the fragility and uncertainties of organizations. From a more recent vantage point, the names of once-familiar companies like Baldwin Locomotive, U. S. Cordage Company, Auburn Motors, and others have faded from social consciousness.

Like organisms, organizations can also suffer from internal problems. Like them, organizations get sick and suffer from a variety of problems. Often the internal parts need to be revitalized or even replaced.

Interestingly, there are many researchers who have studied the evolution of large industrial organizations. The collective results of their studies have led to the development of what has since been called a *contingency view* of organizations. Such contingencies naturally include many variables. A contingency view of an organization's design, for example, assumes that the type of organizational structure depends on the nature of the organization's environment, technology, and the needs of the members of the organization. A contingency theory of leadership focuses on the situation that the leader currently faces. Types of employees and the size of the organization have also been mentioned as contingency variables. As implied, there is no uniform list of organizational variables that apply in and out of season. What may have been good at one stage of an organization's development may not be good at its next stage of development. This point will be made more fully as we now examine the various stages of growth of social organizations.

GROWTH OF SOCIAL SYSTEMS

Greiner has developed a model that describes how organizations change over time and how these changes affect management practices and organizational structures.[15] He suggests that ongoing organizations move through five developmental phases, each of which is made up of two stages that he labels *evolution* and *revolution.* Here evolution is taken to mean prolonged periods of growth where there are no major upheavals in organizational practices. Greiner regards each phase as the result of the previous phase and the cause of the succeeding phase. Each evolutionary stage causes its own revolution. Revolution, according to Greiner, refers to those periods in the organization's history where there is substantial turmoil. An example of evolution causing revolution would be the occurrence of centralization. Here decision making would be carried out at the top levels of the organization. However, lower-level managers, displeased with this practice, would begin to make demands for decentralization. These demands

---

[14] See Irene Taviss, *Our Tool-Making Society* (Englewood Cliffs, N.J.: Prentice-Hall, 1972).

[15] Larry E. Greiner, "Evolution and Revolution as Organizations Grow," *Harvard Business Review,* July-August 1972, pp. 37–46.

would eventually lead to revolution. Solving this problem would require a change in structure.

As could be expected, structure, the reward and control systems, together with managerial practices would all vary appropriately as the organization aged and grew in size. In nearly all social organizations, time institutionalizes practices and procedures, and often the factors that prevent change become cast and rigid. Organizations unable to break out of the rigid mold of policies and procedures in time become outmoded. This is especially true if the industry itself is experiencing rapid change. Organizations unable or unwilling to adapt to take advantage of opportunities or to react to crises facing them are bound to fail.

Greiner feels that each growth phase possesses a dominant managerial style that in turn affects the rate and direction of growth, and that each revolutionary stage displays a major management problem that must be overcome before growth can continue.

Figure 2–4 describes the five stages of evolution and revolution through which organizations pass, and Table 2–1 outlines the organizational practices characteristic of each phase.

Phase One.   This is the phase of evolutionary *creativity* followed by one of *leadership* crisis. During an organization's early growth, its activities usually center about the development of innovative products and markets. Typically, the organization is small and communications are informal. As the organization grows during its evolutionary stage, developing products and markets, it hires more and more people, thus creating a need for professional management. Top management, however, continues to stress the technical aspects of work as it has always done and fails to direct its attention to the managerial aspects of running the organization. It is reluctant to loosen the reins of control and surrender them to others obviously unfamiliar with the expertise that has worked so well in the past.

Although Greiner does not give any specific examples of companies in this first phase, several come to mind that have experienced these growing pains in their early development, companies such as Head Ski Company, Strum Ruger & Company, Polaroid, Collins Radio, and scores of others founded by an individual high on technical competence. Firms that do not adapt to the ensuing management crisis have a high likelihood of eventually failing or of being bought out by firms that can provide the necessary professional managerial expertise.

Phase Two.   This is the phase of *directed* evolutionary *growth* followed by an *autonomy* crisis. The organization, through its professional leadership, applies many of the current management techniques like specialization, the use of incentives, budgets, and work standards. A hierarchical structure is developed with formal communication channels. The firm, as a result, becomes increasingly bureaucratized, and the various organizational functions – production, marketing, accounting, research, etc. – become distinct and separate functions with narrow lines of authority. The hierarchy becomes more layered, permeated with less and less communication.

FIGURE 2–4 The Five Phases of Growth

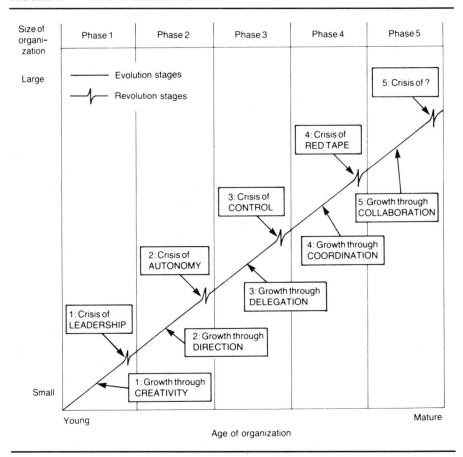

Lower-level managers who know more concerning the internal operations and market factors affecting the firm than top managers are stifled by this heavily centralized hierarchical dominance. Lower-level managers, accustomed to success, want the autonomy to carry out their assignments as they see fit. If this need for autonomy is frustrated, the workers will become alienated, and morale will deteriorate. A new revolutionary period is at hand.

Phase Three. This is the phase of evolutionary growth through *delegation* followed by a *control* crisis. The crisis in the previous phase is solved through delegation, and once again a period of sustained growth can be expected. Managers are put in charge of autonomous units with commensurate authority and responsibility. With looser reins, the divisions are free to develop new products and new markets. Top managers in their long-range planning are looking at new acquisitions and generally manage by exception. Lower-level managers are now free to "do their thing." There is less communication from the top down;

TABLE 2–1   Organization Practices during Evolution in the Five Phases of Growth

Category	Phase 1	Phase 2	Phase 3	Phase 4	Phase 5
Management focus	Make and sell	Efficiency of operations	Expansion of market	Consolidation of organization	Problem solving and innovation
Organization structure	Informal	Centralized and functional	Decentralized and geographical	Line-staff and product groups	Matrix of teams
Top-management style	Individualistic and entrepreneurial	Directive	Delegative	Watchdog	Participative
Control system	Market results	Standards and cost centers	Reports and profit centers	Plans and investment centers	Mutual goal setting
Management reward emphasis	Ownership	Salary and merit increases	Individual bonus	Profit sharing and stock options	Team bonus

SOURCE: Larry E. Greiner, "Evolution and Revolution as Organizations Grow," *Harvard Business Review*, July-August 1972, p. 45.

motivation is improved, and it finally begins to appear that the divisions are running their own companies. This situation top management increasingly interprets as a loss of control. It again attempts to centralize operations, usually unsuccessfully, bringing on another crisis.

Phase Four.   This is the phase of evolutionary growth through *coordination* followed by a crisis in *bureaucracy*. Growth in this phase is achieved by introducing new formal systems allowing for better coordination and control. At headquarters, extra staff is hired to aid in this coordination. The decentralized units are now merged into product groups with appropriate responsibility for *results*. The new staff work with the line personnel, reviewing their programs and giving them advice. These product groups become investment or profit centers that compete with other groups for needed resources. Bonuses and profit sharing follow. Some functions, however, like data processing, are centralized.

With the pressure for results come tensions between line personnel and the new headquarters personnel staff. The rationales required of line personnel demand more and more paperwork. A proliferation of rules and regulations follows. These, in turn, interfere with the work output. Line-staff conflict now emerges in full bloom. The staff members at headquarters are now perceived as "watchdogs," not as facilitators of line operations. With problem solving relegated to second place after rules compliance, innovation is stifled and activities supplant results. The organization has become rigid and inflexible. The stage is now set for another crisis.

Fifth Phase. This is the phase of evolutionary growth through *collaboration* followed by another crisis of *unknown origin.* In response to the rigidity problem in the previous phase, the organization will adopt more behavioral-oriented techniques. The word now is *teamwork,* and techniques like collaboration, conflict resolution, organizational development, participative management, and team building are in vogue. Teams may be interdivisional or may follow the matrix pattern. Problem solving, once again the focus of top management, now takes on a holistic organizational form. Headquarter staff is trimmed and those remaining are now put forward as "team consultants." Training for managers is heavily behavioral, with major emphasis being put on team performance.

One can only speculate on the crisis that follows. Greiner suggests that the revolution could center around the *psychological saturation* of the employees. Intensive teamwork can dissipate employee efforts on the one hand, while on the other some may find the new behavioral concepts and techniques incompatible with their personality structure. They may long for the good old days when organizations did not need to concern themselves so much with people considerations. They may leave the organization, thus creating a turnover crisis, or, as Greiner suggests, "burnout" could be the basis of the next revolution, with possible solutions being sabbaticals or the use of other revitalization techniques. Many of the modern techniques, such as a four-day workweek, job security, flexitime, etc., reflect the current interest on quality of worklife.

Many organizations today are in this fifth phase of growth, and many of them may not be able to anticipate the direction and form of the ensuing crisis. In his formulation, Greiner offers these three points:

1. There is no one organizational structure that will serve a firm in and out of season. Organizations can outgrow a given structure, and just because a particular structure was the correct one at an earlier time period in the organization's development does not make it the right one at the present stage of the organization's life cycle. The common label for this view is the *contingency approach.*

2. As an organization grows, periods of comfortable growth and smooth operations will alternate with periods of turbulence. Managers must be made aware that these turbulent periods will require new structures and that they may even have to take themselves out of leadership positions.

3. Solutions to present problems breed new problems. An awareness of this should help modern managers to recognize the various stages of organizational growth and to anticipate the problems that particular solutions pose. A firm may consciously choose *not to grow* when it is aware of the consequences of further growth. This is precisely the alternative desired by some managers who want by

all means to hold on to the control they now enjoy or who wish to continue in their favorite hobby.

In constructing an organizational growth model, Greiner considers the following five dimensions essential. These were extrapolated from a number of studies.

1. Age of the organization.
2. Size of the organization.
3. Stage of evolution.
4. Stage of revolution.
5. Growth rate of industry.

The five phases outlined here by Greiner are but approximations. Still, a knowledge of these phases should help managers manage better in the future, having attained a better understanding of the organization's past history.

IMPLICATIONS OF SYSTEMS THEORY FOR ORGANIZATIONS

The tenets of general systems theory were simplistically related to an example of the organization earlier in this chapter. Even from such an unpretentious effort one should have become aware of the potential usefulness of this approach. In the process of conceptualizing goals, structure of tasks, regulatory mechanisms, environment, interdependencies of components, boundaries, subsystems, inputs, and their transformation into outputs, all begin to take on more significant meaning.

Indeed, it is only through such conscious recognition of the organization as a system that one can begin to realize the full complexity that must be managed. And, while each of the tenets of general systems theory and organizational relatedness will be examined in detail in the forthcoming chapters, it is useful to briefly mention some of the potential benefits of systems thinking to the managers of organizations:

1. It frees the manager from viewing the task at hand from a narrow functional viewpoint and indeed coerces the manager to identify other subsystems which are either inputs or outputs to the system. Many corporations utilize products or services of other organizations such as auditors, bankers, brokers, consultants, suppliers, and so on which are external to the organization, but which nevertheless affect the performance of the organization. This identification of systems and subsystems is imperative, since cooperation is required from many segments far beyond the boundaries of the internal organization.

2. It permits the manager to view his or her goals as being related to a larger set of goals of the organization. It is the task of managers to understand not only their own goals but how these goals are integrated with broader goals which make the organization a system. The manager must realize that the summation of the

goals at this level of the organization should be equal to or greater than the goals of the next level. Viewing goals in such a manner focuses attention on the interrelatedness of tasks that must be carried out by the different members of the organization.

3. It permits the organization to structure the subsystems in a manner consistent with systems goals. More specifically, it can take advantage of specialization within the system and subsystems. Viewing the organization as a system emphasizes the fact that the goals of the subsystem must be designed to be compatible with overall systems goals.

4. The systems approach with its goal attainment model allows for evaluation of organizational and subsystems effectiveness. Measurement in this type of model is against specific objectives. While certain implicit assumptions are made in the goal-attainment model – e.g., that organizations have goals and that such goals can be identified and progress toward them measured – these assumptions are not formidable in most situations. A detailed discussion on goals and attendant problems will be presented in subsequent chapters.

SUMMARY

Many a contemporary concept has its historical roots in the early history of civilization. So too with the systems concept. The idea that objects should be viewed as wholes, that they are endowed with intrinsic goals, that the whole is more than the sum of its parts – these can all be traced back to Aristotle. However, it was von Bertalanffy, the acknowledged "father" of general systems theory, who disseminated and made the concept acceptable among scientists.

Three decades ago a small group of similarly minded scholars from a variety of disciplines founded what is now known as the Society for General Systems Research dedicated to discovering isomorphisms in their fields. Assumptions were stated, laws sought, propositions advanced, characteristics delineated, classification schemes outlined, models constructed, applications attempted; and inevitably critics were born, intellectual battles waged, and progress made.

The methodology that evolved during this developmental stage truly revolutionized the study of organizations. To treat organizations embedded in an environment (the open systems approach) was indeed a clear and present departure from the well-entrenched and traditionally acceptable closed systems approach.

Interestingly enough, while systems theorists were busy searching for their roots, so too were other scientists busily engaged in digging around for the origins of concepts that they used. The same decade that saw Bertalanffy's contribution also marked the discovery that all feedback control systems were identical in structure, which qualified as one of the more celebrated generalized laws of General Systems Theory. This discovery, or perhaps rediscovery, added substance and impetus to the "holy" quest for a general systems theory. And ever since, there has been no dearth of knights to go forth to rescue the laws of science imprisoned, as it were, in specialized disciplines and to make them available for all.

The analytical approach to problems has dominated the field of science as *the* approved methodology ever since the time of Galileo. Only in the 1930s was the systems approach brought forth for handling problems of open systems.

Boulding formulated what can be called the postulates of General Systems Theory (GST). All of them deal with order. In GST order, regularity and purposiveness (nonrandomness) are to be preferred to their opposites. Orderliness it is that makes the world one and "good." There is order in orderliness. And for establishing order, quantification is useful. The search for order involves the search for its empirical referents. In short, GST involves a search for an order of order, for a law of laws.

Litterer's 10 properties of GST were then considered. While perhaps not a comprehensive listing of all properties, still, the formulation does begin to do justice to the topic. The chief properties of GST are: differentiation, entropy, equifinality, goal seeking, hierarchy, holism, inputs and outputs, interrelationships and interdependencies of objects and of their attributes, regulation (control), and transformation.

In his classic article on GST, Boulding presents a nine-level ordering of systems on the basis of complexity. As with Guttman's scalogram, each higher level incorporates the features of those below it. Boulding intended his schematic primarily as an aid in assessing the gap between our theoretical models and the current level of empirical knowledge.

The classification of systems into open and closed has served us well. Open systems are those that are open to their environment — systems that import resources from the environment. Closed systems are self-contained, with no interaction with the environment. Open systems are sometimes treated as closed systems for purposes of an experiment. In isomorphic systems there is a one-to-one correspondence, not only between the elements of one system with those of another but also between the relationships of one system with those of another. The discovery of the isomorphy of concepts, laws, and models in various fields of endeavor is one of the objectives of the Society for General Systems Research. Isomorphisms are often discovered through analogical thinking. An analogy is the assertion that things that resemble each other in some respects will also resemble one another in some further respect. The one general and five specific criteria for appraising analogical arguments are the number of entities, the number of attributes, the number of disanalogies, the number of dissimilarities and the strength of the conclusion relative to that of the premises, and finally relevance.

When using the systems approach to problems, one can view a problem from various levels of resolution. When doing so, the amount of nesting or piling up of systems in a hierarchy will depend on the nature of the problem, the depth of analysis intended, and the particular framework used. The amount of interaction within or among subsystems will determine whether a system is decomposable, nearly decomposable, or nondecomposable. Decomposable systems have more interaction among the subsystems comprising them than with the system itself.

Consequently, in these the components can be considered independent. In a nearly decomposable system the interaction within the subsystems predominates. The interactions among the subsystems, though weak, are not negligible. Nondecomposable systems directly influence other systems. They are dependent on other systems.

In order to survive, systems must be adaptable. Adaptability is the ability of a system to learn and to alter its behavior in response to changes either within itself or within the environment. Adaptability should be considered in the short run and in the long run. Short-run adaptability concerns functional changes, while long-run adaptability concerns structural changes either in the system itself (Darwinian) or in the environment (Singerian). The capacity of the system to adapt presupposes that the system has information on the state of the environment.

Both natural and social systems evolve much as a child evolves from infancy and babyhood through youth to middle and old age. Greiner distinguishes five developmental phases, each of which is made up of two stages, a stage of evolution and one of revolution. These five phases are:

1. Creativity------------------→leadership.
2. Direction------------------→autonomy.
3. Delegation----------------→control.
4. Coordination------------→bureaucracy.
5. Collaboration-----------→unknown???

The benefits of systems thinking for the future manager are many. Systems thinking, as it were, discards a manager's blinders, enabling the person to view the organization holistically. He or she begins to be aware of the many interrelationships and interdependencies of the system's components. It also helps the manager to line up goals, to integrate goals of the immediate level with higher- and lower-level goals. It also makes it possible to structure the subsystems in such a way that they will be consistent with the goals of the overall system. Finally, systems thinking allows one to evaluate organizational effectiveness in a realistic manner.

KEY WORDS

Interrelationships	Differentiation
Holism	Equifinality
Goal-Seeking	Isomorphic Systems
Inputs and Outputs	Analogies
Transformation Process	Decomposable Systems
Entropy	Adaptability
Regulation	Evolution and Growth
Hierarchy	Nesting of Hierarchies

REVIEW QUESTIONS

1. Apply the appropriate criteria presented as the hallmarks of systems thinking to either an educational system, a correctional institution, or a professional football team.

2. "Systems thinkers violate the very laws they accuse the analytical proponents of violating: they determine the boundaries of the system and then proceed to break the system down in order to see how it works. In fact, if one did not use the analytical approach, the only whole system would be the universe itself. To try to study the universe as a whole would be foolish since one could never learn how it works without relying on the analytical approach." Comment.

3. It has been more than 35 years since Boulding first set out his classification scheme of systems. What effect has recent research in the behavioral sciences together with advances in technology had on our understanding of the various levels?

4. Why is it that some firms which have never heard of the systems approach or the tenets of general systems theory have been successful in spite of their lack of knowledge?

5. A recent farm implement company adopted an inventory-control system from another company and was unconcerned with the factors that comprised the system. Their only comment was "As long as it works, that's the only thing we're interested in." What systems principles might backfire on this company?

6. Boulding's hierarchy of systems was done according to the criterion of complexity, yet one can question whether social organizations are more complex than the human individual. How would one go about measuring complexity?

7. One of the criticisms made of general systems theory is that it fails to be predictive and only explains what has happened "post factum." Explain.

8. Some scholars have stated that general systems theorists delight in finding analogies, e.g., the growth of moss and the growth of populations. Yet there are no meaningful similarities between the two objects being compared. Comment.

9. Why do the systems theorists delight in regularity? What is meant by "law of laws"?

10. Some systems theorists question the actual existence of closed systems. However, in the text the usefulness of this concept is illustrated. From your study of the sciences, what other mental constructs can you name that have proved valuable to scientists even though their referents may not actually exist?

11. Since life began on this planet, over 90 percent of all species that ever lived are now extinct. And of those that have survived, only a few species like those of the cockroach and other insects have been around since the Carboniferous

period, about 350 million years. Human organizations by comparison have a much more minuscule existence. Most businesses fail, and one finds it hard to single out a business that has been around for more than a few hundred years. Organizations like the Roman Catholic Church, which is not quite 2000 years old, are of the greatest rarity. Survival then must be more than a trivial property of any system. How does a system go about the business of surviving?

12. Look up the Fortune 500 list of the largest firms from 30 years ago and compare it with the most recent one. List those firms that no longer exist and try to find out whether these either did not adapt to changes in their environment and so ceased to exist or were forced to adapt since they were the victims of mergers and leveraged buy-outs. How many of these were in the automobile industry? the oil industry? the food industry?

Additional References

Bertalanffy, Ludwig von. "The Theory of Open Systems in Physics and Biology." *Science* 3 (1950), pp. 23–29.

————. "General Systems Theory – a Critical Review." *General Systems* 7 (1962), pp. 1–20.

————. *Perspectives in General System Theory: Scientific-Philosophical Studies.* Edited by Edgar Taschdjian. New York: George Braziller, 1975.

Boulding, Kenneth E. *The Meaning of the 20th Century: The Great Frustration.* New York: Harper & Row, 1964. Chap. 7: "The Entropy Trap."

————. "The Universe as a General System. Fourth Annual Ludwig von Bertalanffy Memorial Lecture." *Behavioral Science* 22 (July 1977), pp. 299–306.

Bowler, T. Downing. *General Systems Thinking: Its Scope and Applicability.* New York: Elsevier, 1980.

Churchman, C. West. "On Whole Systems: The Anatomy of Teleology." Space Science Laboratory, University of California at Berkeley, August 1968.

Dillon, John A., Jr. *Foundations of General Systems Theory.* Seaside, Calif.: Intersystems Publications, 1981.

Grinker, Roy R., Sr. "In Memory of Ludwig von Bertalanffy's Contributions to Psychiatry." *Behavioral Science* 21 (July 1976), pp. 207–18.

Klir, George J. *An Approach to General Systems Theory.* Princeton, N.J.: Van Nostrand, 1967.

————. *Trends in General Systems Theory.* New York: John Wiley & Sons, 1972.

Lange, Oskar. *Whole and Parts: A General Theory of System Behavior.* Oxford: Pergamon Press, 1965.

Laszlo, Ervin. *The Systems View of the World.* New York: George Braziller, 1972.

————. "The Meaning and Significance of General Systems Theory." *Behavioral Science* 20 (January 1975), pp. 9–24.

Miller, James G. "Second Annual Ludwig von Bertalanffy Memorial Lecture." *Behavioral Science* 21 (July 1976), pp. 219–27.

Rapoport, Anatol. "General Systems Theory: A Bridge between Two Cultures. Third Annual Ludwig von Bertalanffy Memorial Lecture." *Behavioral Science* 21 (July 1976), pp. 228–39.

Rifkin, Jeremy and Ted Howard. *Entropy: A New World View.* New York: Bantam Books, 1980.

Rosen, Robert. "Sixth Annual Ludwig von Bertalanffy Memorial Lecture." *Behavioral Science* 24 (July 1979), pp. 238–49.

PART TWO

Various Approaches to Systems Thinking

The central thesis of cybernetics might be expressed thus: that there are natural laws governing the behaviour of large interactive systems — in the flesh, in the metal, in the social and economic fabric. These laws have to do with self-regulation and self-organization. They constitute the "management principle" by which systems grow and are stable, learn and adjust, adapt and evolve. These seemingly diverse systems are one, in cybernetic eyes, because they manifest viable behaviour — which is to say behaviour conducive to survival.

Stafford Beer

PART OUTLINE

- Connective Summary
- Chapter 3 Cybernetics
- Chapter 4 Cybernetic Principles and Applications
- Chapter 5 Particularized Approaches to Systems Thinking

CONNECTIVE SUMMARY

Having considered the origin and nature of systems thinking in Part One, and having developed a vocabulary of the most frequently encountered terms, we now can pass to a closer exposition of the concept of cybernetics.

Chapter Three investigates cybernetics in general. The overriding objective of this chapter is to expose the student to the science of control and communication, its origin, its logic, its historical developments, and the tools needed for studying exceedingly complex probabilistic systems which constitute the domain of cybernetics.

There are three main characteristics of cybernetic systems: (1) extreme complexity, (2) probabilism, and (3) self-regulation. The corresponding tools or techniques for dealing with these properties of cybernetics systems are: (1) the black box, (2) information theory, and (3) feedback control. This chapter deals with two of these — complexity and the black box, and self-regulation and feedback control. Probabilism and information theory are dealt with in Chapter Six.

Chapter Four is a continuation of the discussion of cybernetics and presents several control principles. These two chapters are given primary emphasis in this section.

The remaining chapter in Part Two of this text looks at several other particularized approaches to systems thinking. Of the many that could have been selected, systems engineering, operations research, and systems analysis merit our serious attention, for they apparently qualify as bona fide systems approaches. Some may object that the two particularized approaches, namely, operations research and systems analysis, are not legitimate disciplines, having their roots in electrical engineering and in other "sordid" sciences. One cannot but admit that many of the useful disciplines today are hybrids, like physical chemistry, biomechanics, mathematical biology, etc. The value of any discipline lies, not in its ancestry, but in its utility for problem solving. Systems analysis is a product of space technology and has proven useful for the study of systems. Operations research is discussed here principally as a tool to aid the researcher in the quest to unlock the secrets of systems. Although other approaches could also have been added, still, one need not apologize for the inclusion of the approaches that have been both time-tested and popular with researchers.

CHAPTER 3

Cybernetics

*Benjamin Franklin's reply to a lady who queried the usefulness of his work on electricity:
"Madam, what use is a newborn baby?"*

Arthur Koestler, *The Ghost in the Machine*

CHAPTER OUTLINE

INTRODUCTION

Previously cybernetics was noted as a particularized approach to systems thinking. The cybernetic approach meets the requirements of systems both conceptually and operationally, as the engineering sciences amply testify. However, cybernetics may also be viewed as one of the generalized laws of general systems theory. It merits this appellation since it is concerned with feedback processes of all kinds.

In Chapter Two we presented Boulding's nine-level classification of systems based upon the sole criterion of complexity. Here we consider Beer's classification scheme, which uses the two criteria of *complexity* and *predictability*.[1]

With respect to the first criterion, Beer uses three subclasses: simple, complex, and exceedingly complex. A simple system is one which has few components and few interrelationships; similarly, a system which is richly interconnected and highly elaborate is complex, and an exceedingly complex system is one which cannot be described in a precise and detailed fashion.

The second criterion concerns the system's deterministic or probabilistic nature. In the former, the parts interact in a perfectly predictable way, while in the latter the system is not predetermined in its behavior, although what may likely occur can be described.

The six categories of the two criteria, one threefold and the other twofold, are presented in Figure 3–1. Although Beer is clear in his admonition that these bands are hazy and that they represent merely bands of likelihood, still such a scheme has value since his grouping is done according to the kinds of *control* to which they are susceptible. Not all categories are of equal difficulty and of equal importance.

Deterministic systems are of little interest because behavior is predetermined and because they do not include the organization as does an open system. As shown in Figure 3–1, examples of this type of system include the pulley, billiards, a typewriter, most machines in the organizations, the movement of parts on an assembly line, the automatic processing of checks in a bank, and so on. In each of the above examples, the output of the system is controlled by management of the input to the system.

From simple deterministic systems one moves to complex deterministic ones, the singular difference being the degree of complexity involved. The computer is illustrative of this class of system in that it is much more complex than the previously mentioned systems but still operates in a perfectly predictable manner. The point made earlier that the band separating the categories is hazy is demonstrated by the fact that to a computer specialist the computer may not be complex. In a similar manner, the automobile engine is complex for many, but again for a mechanic it is a simple deterministic system. In all of the above

---

[1] Stafford Beer, *Cybernetics and Management* (New York: John Wiley & Sons, 1964), p. 18. Warren Weaver had previously distinguished three states or types of complexity: organized simplicity, organized complexity, and disorganized complexity in "Science and Complexity," *American Science* 36 (1948), pp. 536–44.

FIGURE 3–1 A Classification of Systems Based on Susceptibility to Control

Complexity / *Predictability*	*Simple*	*Complex*	*Exceedingly Complex*
Deterministic (one state of nature)	Pulley Billiards Typewriter	Computer Planetary system	Empty set
Type of control required	Control of inputs	Control of inputs	Control of inputs
Probabilistic (many states of nature)	Quality control Machine break- downs Games of chance	Inventory levels All conditional behavior Sales	Firm Humans Economy
Type of control required	Statistical	Operations research	Cybernetic

Adapted from Stafford Beer, *Cybernetics and Management,* Science Edition (New York: John Wiley, 1964), p. 18.

examples, there is only a single state of nature for the system, which is determined by the structural arrangement of the elements composing it. If these are in the proper configuration, the system will operate in a predetermined pattern.

If one were to introduce a second state of nature into each of the above systems, they would become probabilistic. As seen from Figure 3–1, probabilistic systems can range from the simplest games of chance, such as the flipping of a coin, in which only two possible states can exist, to the organization, in which multiple states are possible.

In this simple probabilistic system, the additional examples of quality control and machine breakdowns are presented. Because humans are introduced into the production system, and, of course, because humans can exhibit many states of nature, quality becomes a variable factor. It is for this reason that quality-control techniques are applied to ensure that a certain state of nature will prevail. Likewise, the wear of parts in a machine necessitates periodic maintenance. The usage rate to a large extent determines the time interval that the machine will be functional (probability of breakdown). In all the above examples, simple statistical techniques can be employed to control the system.

As the complexity of a probabilistic system and the number of states of nature increase, prediction and control of systems behavior become extremely difficult. Thus, while in deterministic systems control of the inputs will provide prediction of the outputs, in probabilistic systems control of the inputs will provide only a range of possible outputs.

The last category of exceedingly complex, probabilistic systems includes the firm, the individual, and the economy, all of which can exhibit variable states of nature. The firm, being composed of multiple subsystems, interacts with other external systems such as the government, competitors, unions, suppliers, and banks. The interaction of the various internal departments and components of an organization and its external subsystem is so intricate and dynamic that the system is impossible to define in detail.

What, then, is of concern are those systems which exhibit probability and complexity. As noted in Figure 3–1, simple probabilistic systems are controlled through statistical methods, while complex probabilistic systems are dealt with through more sophisticated methods of operations research. These tools serve adequately in dealing with systems exhibiting a measure of complexity, but in treating exceedingly complex systems which lack definability, they are deficient. Highly complex systems will not yield to the traditional analytical approach because of the morass of indefinable detail; yet these too must be controlled. The technique employed when dealing with extreme complexity is that of the black box. A later section will treat this in detail.

There can be but little doubt that only a few of the systems encountered in the workaday world are of the deterministic type. Most are probabilistic in both structure and behavior. Any system operating within a margin of error is probabilistic and therefore must be treated statistically.

In addition to the two characteristics of probabilism and complexity, Beer includes one additional characteristic of cybernetic systems, and that is self-regulation.

The *self-regulatory feature* of cybernetic systems is essential if systems are to maintain their structure. Control must, therefore, operate from within, utilizing the margin of error as the means of control.

For each of the above characteristics, specialized tools are available for defining, operating, and controlling systems. These, together with the tools of analysis, are presented in Figure 3–2.

However, before examining each of these characteristics in detail, it may be useful to trace the development of cybernetics from its inception.

HISTORICAL ANTECEDENTS

For many years now, automatic control systems, which have largely been confined to governors, servomechanisms, and the like, have had their greatest impact in the field of engineering. This is not to be wondered at, for ever since the 1790s when James Watt invented his "governor"—the mechanical regulator for stabilizing the speed of rotation of the steam engine—the field of cybernetics has been almost wholly dominated by the mechanical engineer. Even today many of the guidance and control systems employed in missiles are based on fundamentally the same principles enunciated decades ago. While it is true that automatic control systems are used more and more each year, still relatively little application of such systems outside the realm of mechanical devices takes place. Until the recent contributions to cybernetics by such men as Norbert Wiener, W. Ross Ashby, and Stafford Beer, to name a few, the all-important idea of feedback, so vital to a cybernetic system, has

FIGURE 3–2   Characteristics and Tools for Analysis of Cybernetic Systems

Characteristics of a System	*Tools for Analysis*
Extreme complexity	Black box
Probabilism	Information theory
Self-regulation	Feedback principle

only with difficulty been transferred to the political, economic, social, and managerial fields.

Historically, cybernetics dates from the time of Plato, who, in his *Republic,* used the term *kybernetike* (a Greek term meaning "the art of steersmanship") both in the literal sense of piloting a vessel and in the metaphorical sense of piloting the ship of state—i.e., the art of government. From this Greek root was derived the Latin word *gubernator,* which, too, possessed the dual interpretation, although its predominant meaning was that of a political pilot. From the Latin, the English word *governor* is derived. It was not until Watt termed his mechanical regulator a "governor" that the metaphorical sense gave way to the literal mechanical sense. It was this that in 1947 provided the motivation for Norbert Wiener to coin the term *cybernetics* for designating a field of studies that would have universal application. With this the term has now come full circle.

In more recent times the science of cybernetics has been much abused by writers who equated it with electronic computers, automation, operations research, and a host of other tools. Cybernetics is none of these, nor is it a theory of machines, although it derives from a particular type of mechanism (regulators). In his classic text, Wiener defines cybernetics as the science of control and communication in the animal and the machine.[2] It is quite evident that Wiener intended cybernetics to be concerned with universal principles applicable not only to engineering systems but also to living systems.

> In giving the definition of Cybernetics in the original book, I classed communication and control together. Why did I do this? When I communicate with another person, I impart a message to him, and when he communicates back with me he returns a related message which contains information primarily accessible to him and not to me. When I control the actions of another person, I communicate a message to him, and although this message is in the imperative mood, the technique of communication does not differ from that of a message of fact. Furthermore, if my control is to be effective I must take cognizance of any messages from him which may indicate that the order is understood and has been obeyed. . . .
>
> When I give an order to a machine, the situation is not essentially different from that which arises when I give an order to a person. In other words, as far as my

---

[2] Norbert Wiener, *Cybernetics, or Control and Communication in the Animal and Machine* (New York: John Wiley & Sons, 1948).

consciousness goes, I am aware of the order that has gone out and of the signal of compliance that has come back . . . Thus the theory of control in engineering, whether human or animal or mechanical, is a chapter in the theory of messages.[3]

This view of control can be profitably applied at the theoretical level of any system and to diverse disciplines in both large and small systems. Wiener further states, "It is the purpose of Cybernetics to develop a language and techniques that will enable us indeed to attack the problem of control and communication in general, but also to find the proper repertory of ideas and techniques to classify their particular manifestations under certain concepts."[4]

Many of the concepts of cybernetics as applied to physical systems are relevant for an understanding of social groups as well. Wiener certainly anticipated this, and, while cautioning against abuse of cybernetics in areas lacking mathematical analysis, he pointed out that the application of cybernetic concepts to society does not require that social relations be fully described mathematically, but only that mathematical analogies be used to describe some of their functions. By clarifying formal aspects of social relations, cybernetics can contribute something useful to the science of society.[5]

Apropos to this, Charles Dechert remarks:

> More recent definitions of cybernetics almost invariably include social organizations as one of the categories of systems to which this science is relevant. Indeed Bigelow has generalized to the extent of calling cybernetics the effort to understand the behavior of complex systems. He pointed out that cybernetics is essentially interdisciplinary and that a focus at the systems level, dependent upon mixed teams of professionals in a variety of sciences, brings one rapidly to the frontiers of knowledge in several areas. This is certainly true of the social sciences.[6]

Besides the element of control, the other central concept in cybernetics is that of communication. Communication is concerned with information transfer, both between the system and its environment and also among the parts of the system. The cybernetic concept of information ranges farther afield than it does in other disciplines. It includes not only electrical impulses as in engineering, signals sent to the brain as in human beings, cardinal values as in mathematics; it embraces all carriers of information. While information theory can be considered as a special tool dealing with quantitative aspects of information, the use here of the term *information* will be much less restrictive. At present, information theory is somewhat limited in application, although attempts to extend it to other disciplines have not been wanting.

---

[3] From *The Human Use of Human Beings* by Norbert Weiner. Copyright 1950, 1954 by Norbert Wiener. Reprinted by permission of Houghton Mifflin Company, pp. 16-17.

[4] Ibid. p. 17.

[5] Norbert Wiener, *God and Golem, Inc.* (Cambridge, Mass: MIT Press, 1964), p. 88.

[6] Charles R. Dechert, "The Development of Cybernetics," in P. Schoderbek, ed., *Management Systems,* 2d ed. (New York: John Wiley & Sons, 1971), p. 74. Originally printed in *The American Behavioral Scientist* (June 1965), pp. 15-20.

While some may find fault with a presentation of cybernetics as a tool for analyzing *all* purposeful behavior, nevertheless, the present writers are convinced that cybernetics can provide a better and a fuller understanding of the system at hand. While social scientists, mathematicians, and other researchers may restrict their definitions to certain domains with the rigidities so necessary for scientific investigation, these writers believe that systems thinking ought to be extended to the many problems encountered in the living firm and not only to those fabricated in the laboratory. The domain of systems thinking and cybernetics should be enlarged as much as possible and not harnessed within narrow and restrictive disciplinary limits.

FEEDBACK AS A MEANS OF CONTROL

Feedback control systems are neither new nor rare. While their historical antecedents date back at least 2,000 years,[7] it was only in the 20th century, and indeed within the past decades, that their underlying principles have been exploited. In some respects it is remarkable that it took so long to unravel the central ideas involved in feedback control systems. Yet the recognition that some systems utilized common principles, and that similar systems could be constructed, was indeed a profound and important discovery. Once this was realized, the first approximation to a theory of feedback systems was soon in coming, growing by quantum leaps in the past two decades.

The feedback control system is characterized by its closed-loop structure. Such a system can be defined as one "which tends to maintain a prescribed relationship of one system variable to another by comparing functions of these variables and using the difference as a means of control."[8] The same source also defines feedback as "the transmission of a signal from a later to an earlier stage." For purposes of discussion, a distinction will be made between automatic and manual feedback control systems. In the former, a closed-loop feedback exists which is executed by the system, while in the latter a human operator is needed by the system to close the loop, i.e., to take some course of action on the basis of the feedback information received. Although both types are important, still, because business organizations utilize people in the feedback control system, more attention will be devoted to manual feedback control systems. The automatic closed-loop feedback systems, however, serve as excellent examples for illustrating the feedback concept.

---

[7] Otto Mayr, in his extremely thorough book entitled *The Origins of Feedback Control* (Cambridge, Mass.: MIT Press, 1969), traces the evolution of the concept of feedback through three separate ancestral lines: the water clock, the thermostat, and mechanisms for controlling windmills. The water clock is the earliest description of a feedback device on record and dates from the third century B.C. The thermostat has a more recent history, having been invented in the early 17th century by Cornelius Drebbel. Devices for the automatic control of windmills were invented in the 18th century.

[8] A.I.I.E. Committee Report, "Proposed Symbols and Terms for Feedback Control Systems," *Electrical Engineering* 70 (1951), p. 909.

An easy-to-understand example of feedback is the mechanical governor invented by James Watt for his steam engine. Should the engine speed up past the preset level, the governor reduces the supply of steam, thus decreasing the speed. Should the engine slow down, the governor admits more steam into the engine, thus increasing its speed. In this way the governor keeps the engine running at a uniform speed.

Perhaps of greater interest for men and women are the feedback mechanisms in the human body. Some feedback processes are involved in maintaining respiration, circulation, and digestion; others in maintaining body temperature at a uniform level. The insulin-producing pancreas keeps the glucose concentration in the blood at a normal level by secreting insulin when the glucose level rises above the norm. Even such an apparently simple movement as picking something up from the floor involves feedback. When the hand, for instance, reaches for a distant object, data on its spatial position is continuously fed back to the brain, both by the eyes and by position-sensing nerves in the arm. The brain then processes (transforms) the data into information that is used to guide the hand to the object.

The Block Diagram and Closed-Loop Systems

Because researchers in different disciplines lack a common language for communicating with one another, attempts have recently been made in the field of feedback control systems for the adoption of standardized symbols and terminology. The block diagram is a basic linguistic tool for illustrating functionally the components of a control system. It is precisely this type of representation that has revealed the underlying similarity of seemingly unrelated systems. The blocks themselves represent things which must be done and not the physical entities, equipment, and the like. In the block diagram, four basic symbols are employed: the arrow, the block, the transfer function, and the circle.

1. The arrow (Figure 3–3) denotes the signal or command, which is some physical quantity acting in the direction of the arrow. It is not the flow of energy which is being noted but rather the flow of information which shows the causal relationship that exists. This input is variously termed the command, the signal, the desired value, or the independent variable and is often represented as a mathematical variable.

2. In a block diagram, all system variables are linked to each other through the functional *block* (Figure 3–4). This block is a symbol for the mathematical operation on the input signal to the block which produces the output. Signals coming into the block are independent, but signals leaving the block are dependent, since these are the outputs or effects.

3. Included within the block is the transfer function, which is the mathematical operation to be performed on the block. The output is then the transfer function multiplied by some input (Figure 3–5). When a signal has

FIGURE 3–3   The Command Signal

x

two separate effects which go to two different points, this is termed a branch point, as in Figure 3–6.

4. The circle with a cross represents a summation point where a comparison is made between two quantities, the command signal or desired state, and the feedback signal or actual state. It is here that the signals are added or subtracted. In Figure 3–7, θ_i is an input signal being fed into the system and θ_o is the output of the system. This summation point is also referred to in the literature as the error detector, the comparator, or the measurement point.

In the simplest closed-loop system, shown in Figure 3–7, the output is fed back to the summation point, where it is compared with the input signal θ_i. This comparison function is one of the requirements of an automatic control system where the command signal is compared with the variable being controlled. The difference is used as the means of control; it is this difference which is termed *negative feedback*. In negative feedback, subtraction takes place at the comparator. As a signal travels around the loop, its sign must be reversed, since to have a closed loop without a reversal of signs would make the system unstable. Thus, the reversal of signs is associated with negative feedback.

Figure 3–8 depicts a closed-loop feedback system. The output in this is

FIGURE 3–4   The Block

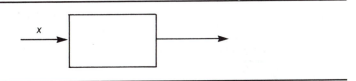

x

FIGURE 3–5   Block with Transfer Function

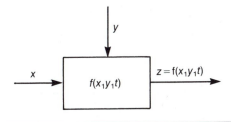

y

x

$f(x_1y_1t)$

$z = f(x_1y_1t)$

FIGURE 3–6   Branch Points

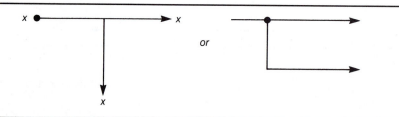

obtained by the multiplication of the transfer function (in this case κ) by the input to the block (ϵ).

The system in the above figure may be described mathematically by the following set of equations:

$$\epsilon = \theta_i - \theta_o$$
$$\theta_o = \kappa\epsilon$$

which can be reduced to the single equation

$$\theta_o = \frac{\kappa\theta_i}{1 + \kappa}$$

FIGURE 3-7   Summation Point

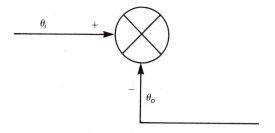

FIGURE 3-8   Closed-Loop Feedback System

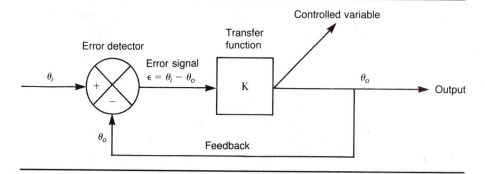

When κ is large, θ_o approximates θ_i.

While no one would claim that the following example of an economic system corresponds to reality, still it can provide some insights into the control mechanisms utilized by the government for correcting unstable conditions. Since many econometric simulation models of the economy do in fact deal with real-life problems, from this premise the economy is a fitting subject for cybernetic control.

Although several general treatments of the economy from the feedback control approach exist, the following one by Porter[9] serves well for illustrative purposes.

Let

S_d = Desired level of spending

S_a = Actual level of spending

$\epsilon = S_d - S_a$

Assume R = interest level = $(R_O + r)$, in which R_O = standard level of interest and r = $-\kappa\epsilon$, is the change in interest, dependent upon ϵ, which provides the control.

The simplified equations of the economic system can now be written

$$\epsilon = S_d - S_a$$
$$R = R_O - \kappa\epsilon$$

The corresponding block diagram (Figure 3–9) is the following:

FIGURE 3–9 A Simple Economic Control System

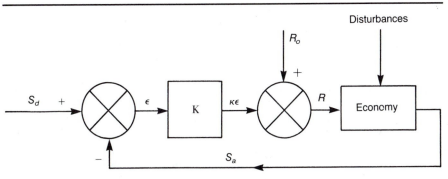

Open-Loop Systems

An open-loop system is one in which the output of the system is not coupled to the input for measurement. Mechanical examples of open-loop systems are the water softener in a home, the washing machine or dryer, the automatic sprinkler

---

[9] Arthur Porter, *Cybernetics Simplified* (New York: Barnes and Noble, 1970), pp. 14-15. [Cf. + and − in Figure 3–7.]

system, the traffic light, automatic light switches, and the toaster. In all of these cases the output is not compared with the reference input, but, instead, for each of the reference inputs there exists a corresponding fixed operating condition. Most of the above operate on a time basis. Open-loop systems can be depicted as in Figure 3–10.

Perhaps all of the above could be made into closed-loop systems, if they met the previously mentioned criteria for such systems. The individual examining the dryness or cleanliness of the clothes and comparing it with some standard, the person testing the amount of moisture in the ground and comparing it with a standard, a man checking the amount of daylight and relating it to a standard – all provide for measurement against a goal.

Organizations are basically open-loop systems when viewed without people. However, with the appearance of the human operator as a controller who compares the input and output and makes corrections based on the differences between actual and planned performance, they become closed systems. Using individuals to close systems presents some difficulties, but it is both necessary and desirable. The major difficulty is that it is next to impossible to describe the human behavior of an individual in a mathematical equation and, even if one could, it would still be difficult to adjust behavior for learning. Although scientific rigor is lessened when the system cannot be mathematicized, still there is value in studying the system. Inventory control systems, for example, are obviously closed-loop control systems, for the actual inventory level (the output of the system) is compared with the desired inventory level. If an error is detected, then the production rate is adjusted by some individual so that the inventory level is again at the desired level.

Positive Feedback

Although the preponderance of attention in the literature has been given to negative feedback systems, some mention should be made of positive feedback. Positive feedback systems utilize part of their output as inputs to the same system in such a way that they are, in fact, *deviation-amplifying* rather than deviation-counteracting systems. All growth processes involve positive feedback systems, since a part of the output is amplified. This is true both for mechanical systems and for human organizational systems. In mechanical systems, power steering and power brakes are common examples of positive feedback systems.

In social systems such as organizations, the term *positive feedback* is usually interpreted as "good news" as contrasted with negative feedback, which is associated with "bad news." These colloquial expressions are very misleading for students of systems. Positive feedback mechanisms are growth-promoting devices, while negative feedbacks are control-maintaining processes. The activities of the marketing or promotion subsystem of an organization perform growth processes by attempting to enlarge any positive difference between accomplished usage of the organization's products or services and the goal. On the other hand, the activities of the accounting, quality-control, and industrial or public-relations subsystems perform control functions in that their purpose is to minimize

FIGURE 3-10   Open-Loop System

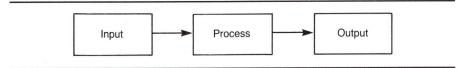

deviations between set standards (budgets, morale, corporate image, and so on) and actual performance.

The concepts of positive and negative feedbacks are extremely useful in understanding organizational or system behavior.

Examples of positive feedback loops can be found not only in the physical arena but in the social arena as well, one of the most famous examples being the great stock market crash of 1929.[10] What happened was this. In the 1920s interest rates were low and were purposely kept that way by the Federal Reserve Board. With low interest rates, it was comparatively easy to borrow money. With the borrowed money, it was easy to buy stock on the stock exchange, since one did not need to have the full price of the stock to purchase it. One could buy stock on margin. At the time, one need pay only a very small percent of the cost of the stock — the stock itself could be held as collateral. Speculators in stocks had a field day. To slow down the uncontrolled speculation, the Federal Reserve Board decided to raise the interest rates, thus making it harder to borrow money for buying stocks. With lessened demand for stocks, stock prices fell; so too did the value of the collateral. Seeing the drop in prices, those holding stock as collateral decided to cash in the stock lest its value fall even further. Now with more sellers than buyers in the stock market, it was to be expected that the stock prices would fall even more, depreciating the collateral even more. As a result, more and more stock was dumped on the market, leading to a further drop in prices. And so it went until stock prices eventually dropped to less than 10 percent of the 1929 highs. It took government intervention to finally halt this downward spiraling in the mid 1930s.

In this finite world, no positive feedback mechanism can go on indefinitely. Eventually either something outside itself must bring it to a halt, or the system itself will be destroyed. Hence positive feedback by its very nature is a temporary phenomenon. Positive feedback loops are therefore subject to negative feedback mechanisms that will cause the self-acceleration (deviation-amplifying phenomenon) to grind to a halt.

The Block Diagram and General Systems Theory

As problems become more complex and the use of computers for simulation and problem solving becomes more widespread, the need for a conceptual

---

[10] See Robert U. Ayres, *Uncertain Futures: Challenges for Decision Makers* (New York: John Wiley & Sons, 1979), pp. 31-33.

framework, both for the definition of the problem and for its solution, becomes more acute. A critical step in this process is the design of the system, which in turn requires a decision regarding the structure of the system. The structural aspects of the system must therefore be specified. The block diagram serves admirably for the identification of the structual relationships of the system. It is at this point that, as Mesarovic states, the block diagram and general systems theory come together, for general systems theory can assist in the structural considerations by "preserving the simplicity of the block diagram while introducing the precision of mathematics."[11] For complex systems, he holds, a general systems model is a necessary step between the block diagram and the detailed mathematical model, as shown in Figure 3–11.

First-Order Feedback Systems (Automatic Goal Attainment)

The closed-loop control feedback systems discussed in the previous section are all feedback systems of the first order because the system is monitored against an external goal (Figure 3–12). It is given one particular command, which it is to

FIGURE 3–11   Relationship of the Block Diagram and General Systems Theory

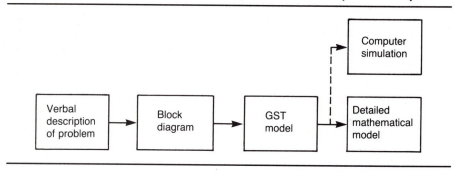

FIGURE 3–12   First-Order Feedback System

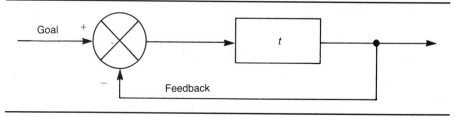

[11] Mihajlo D. Mesarovic, "General Systems Theory and its Mathematical Formulation," paper presented at the 1967 Systems Science and Cybernetics Conference, IEEE, Boston, October 11-13, 1967, p. 13.

carry out irrespective of changes in the environment. In goal-directed systems which operate on the principle of negative feedback, the system is maintained by correcting deviations from the goal. There is no other choice available to the system but to correct the deviation. Thus, the purpose of a first-order feedback system is to maintain the system at a desired state of equilibrium. The system cannot make any conditional response. It has no memory, nor does it have available any alternative action. In this type of control system the operation is clearly circular, since, after the comparison against the standard, a recycling must take place. Every first-order feedback system always operates in this manner irrespective of changes in the environment.

An everyday example of a feedback system for automatic goal attainment is the familiar house thermostat coupled to a furnace or an air conditioner. The temperature of the room is set, say, at 68 degrees Fahrenheit. This is the goal that the system is to attain. If the room temperature falls below this ideal setting, the thermostat actuates the furnace to send up more heat. When the temperature begins to go beyond this ideal setting, the thermostat cuts off the heat.

The thermostat's basic principle is that one of its components expands or contracts in a significant way during a temperature change. Often a thermostat uses a bimetallic strip made up of two metal pieces with differing coefficients of thermal expansion. As the room temperature changes, the lengths of the two pieces also change at differing rates, causing the strip to bend. The bending causes the strip to either make or break an electrical circuit.

In general, any system which has a *single goal* — such as a temperature control system in a room, a cruise control system in a car, or a light control system in a home — can assume only one of two states, the ON or OFF state. There are no additional choices possible. It's either on or off, period.

Second-Order Feedback Systems (Automatic Goal Changer)

When a system contains a memory unit and can initiate alternative courses of action in response to changed external conditions and can choose the best alternative for the particular set of conditions, it is said to be a feedback system of the second order. A memory includes all the facilities in the feedback loops available to the system for storing or recalling data from the past. In an organization, such a memory would include the personnel, policies, filing systems, and so on. This feedback of information from the past is done for decision making in the present.[12]

The second-order system has the ability to change its goals by changing the behavior of the system. In other words, goal changing is part of the feedback process itself. In Figure 3–13 a feedback control circuit with a memory device is

---

[12] See Karl W. Deutsch, *The Nerves of Government* (New York: Free Press, 1966), for an excellent exposition of types of feedback. See especially chapters 11 and 12.

FIGURE 3–13   Feedback Circuit with Memory Device

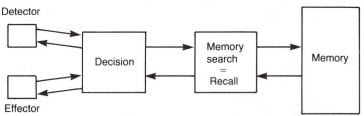

By adding a memory and more complicated feedback loops, an organization can have more control over its own activities. In this case a series of alternatives for action is built into the system if external conditions (detected by the detector) change. An example is the automatic switching of a telephone exchange.

SOURCE: Adapted from C. West Churchman et al., *Introduction to Operations Research* (New York: John Wiley & Sons, 1957). Reprinted by permission of John Wiley & Sons.

shown. The detector depicts the state of the variable being controlled, while the effector produces the required change. Churchman explains that a telephone exchange is one automatic goal-changing unit.[13]

There are many examples of this type in which, if goal A is attained, priority is shifted to goal B, and so on. When goal B is attained, there is a shift back to goal A or on to goal C. Any system that can change goals is said to be autonomous. Goal changing is dependent upon memory. When there is no more memory—i.e., when either the system is cut off from all past information or the information has ceased to be effective—the system can no longer change goals. Such a system loses control over its behavior and acts simply as an automaton. The better the memory and the greater the ability to recall past information, the more autonomous the system. The ability to store and recall information, allowing the system to choose alternative courses of action in response to environmental changes, is termed learning. Learning can be defined as the internal arrangement of resources that are still relevant to goal seeking. This can mean the addition of another channel of communication, a change of information put in the memory, a change in the control process, or any number of similar actions.

The chapter in this text entitled "Artificial Intelligence" deals mainly with expert systems. These systems provide some excellent examples of second-order feedback systems. As will be seen, these systems have had until now a limited degree of learning capacity and some ability to switch goals

---

[13] C. West Churchman et al., *Introduction to Operations Research* (New York: John Wiley & Sons, 1957), p. 81.

depending on the environmental conditions encountered, but as progress continues this capacity for learning and for goal changing will be notably improved.

Third-Order Feedback Systems (Reflective Goal Changer)

A third-order feedback system is one that can reflect upon its past decision making. It not only collects and stores information in its memory but also examines its memory and formulates new courses of action. Obviously, such a system refers to both the individual and the organization. The organization as a system can direct its growth by changing its goals, terminating certain activities, initiating new activities, engaging in research, continually searching its memory for vital information, modifying the value system of its personnel, or changing the firm's operating patterns. The third-order feedback system not only is autonomous, it also possesses a consciousness. Figure 3–14 shows a possible configuration of this type.[14]

A third-order feedback system is also termed a *self-organizing* system. It has a learning capacity, which means that it can learn from its past errors. The fact that it is self-organizing implies that it is constantly adapting itself through altering its products or services, its structure, its policies and processes. Information is the vehicle that provides the impetus for the self-organizing and self-adapting system. Required are not only *feedback* systems that depict the present state of the organization but also a *feedforward* system which provides information on the projected future state of the environment. Included in such an assessment is the obvious need for information concerning the interaction of the organization with the environment.

Lest one be inclined to limit the usefulness of cybernetic systems to control purposes only, something should be said here on the viability of the self-organizing system. While cybernetic systems are most often employed and best understood from a mechanical viewpoint or when viewed as a basic control tool, still the concept can be applied at all resolution levels. What may be needed, however, is further development in the sensing capabilities of the detector for introducing *anticipatory* controls. While the concept is not new, the present emphasis is. Anticipatory controls have long been employed in maintaining the quality of machine-produced parts. The operator does not wait until the quality falls outside the control limits before making machine adjustments. Rather, the operator anticipates the loss of control and makes adjustments *before* the machine begins turning out defective parts.

Conceptually this is no different from anticipating changes in the environment. For example, there is little doubt that the mean population of the United States is increasing. This is verifiable. Also, the number of new births per capita

---

[14]Churchman et al., *Operations Research,* p. 84.

FIGURE 3–14   Additional Memory Refinements

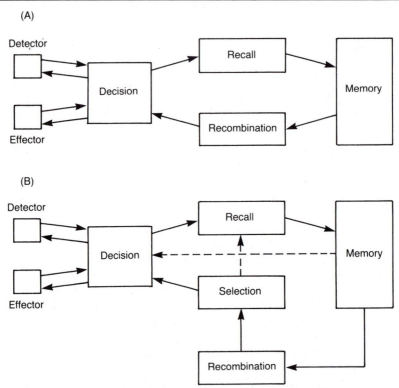

(A) If information in the memory can be recombined and new alternatives produced for action (by the machine or organization itself), the unit becomes more versatile and autonomous. This device makes simple predictions. (B) Development of a consciousness. If many memories can be combined, and if from the many combinations a few can be selected for further consideration, further recombination, etc., the unit will have reached a still higher level of versatility or autonomy. The dashed lines indicate comparisons of what is going on with what has happened in the past and what might occur in the future (second- and third-order predictions). In many organizations, these comparisons are poorly made.

SOURCE: C. West Churchman, et al., *Introduction to Operations Research* (New York: John Wiley & Sons, 1957). Reprinted by permission of John Wiley & Sons.

is decreasing. Given these statistics, companies marketing products designed for children face a decreased market share, all else being equal. Intelligence gathering (scanning) of changes in the environment should uncover this and aid in anticipatory decisioning. The number of companies are legion that have failed to anticipate change and to adapt. In all of them, managers reacted to crises rather than anticipating them.

Control and the Law of Requisite Variety

The more complex a system, the more difficult it is to understand and control it. The more complex a system, the more difficult it is to define its structure (its interrelationships) and consequently, the more difficult to predict its behavior. As the components of a system increase in number, the interrelationships typically increase, and the system is said to possess more variety than it did initially. In moving from a simple organization with few employees having few interdependencies to a more complex one with many employees, the variety of uncertainty increases. When one asks, "How can uncertainty be reduced?" the answer is, "Through information." Information extinguishes variety, and the reduction of variety is one of the techniques of control, not because it simplifies the system to be controlled, but because it makes the system more predictable.[15] Therefore, what is required is the same amount of variety in the control mechanism as there is in the system being controlled. This important principle Ashby has called the *Law of Requisite Variety.*[16] If there is enough permutation variety to provide for a one-to-one transfer from the control mechanism to the system, then there is "requisite" variety. As Ashby states, "Only variety can destroy variety." This fundamental concept has very general applicability in all control systems. Beer, in discussing the above law, has this to say:

> Often one hears the optimistic demand: "give me a simple control system; one that cannot go wrong." The trouble with such "simple" controls is that they have insufficient variety to cope with the variety in the environment. Thus, so far from not going wrong, they cannot go right. Only variety in the control mechanism can deal successfully with variety in the system being controlled.[17]

Essentially this means that, if one is to control a system, there must be as many actions available to the systems controller as there are states in the system. This concept as presented here may be new to the business executive, but it is a familiar one to the practitioner. Decision rules operate on this premise. Managers at all levels of the organization attempt to determine courses of actions based on certain outcomes of previous actions or on their competitors' actions, and so on.

The statement was made above, that the controller should have at least as many alternatives as the system can exhibit. In the case of a machine, if one knows all the possible causes of stoppage and can take corrective action, the control mechanism (the controller) possesses requisite variety. If, however, one does not, then one does not have control of the system. The general public, for example, does not have control of the automobile in the sense that it does not know the malfunction permutations – the many things that can go wrong.

---

[15] Stafford Beer, *Cybernetics and Management,* Science Edition (New York: John Wiley & Sons, 1964), p. 44.

[16] W. R. Ashby, *Introduction to Cybernetics* (New York: John Wiley & Sons, 1963).

[17] Beer, *Cybernetics and Management,* p. 50.

Thus, it should be clear that we can better control a system when we have variety in the control mechanism. However, we may also better control a system when we can simplify, partition, or otherwise reduce the variety in the system itself. This is precisely what policies attempt to do in an organization. Thus, rather than having sales personnel deal individually with every instance as it occurs, a credit policy serves as a guideline for all sales inquiries. In this instance the credit policy serves as a regulator whose function is to block the flow of variety into the system. In the following diagram (Figure 3–15), I represents some inputs to a system pursuant to a goal. D represents disturbances that may occur within or without the system, while R is the regulator whose function is to block the transmission of variety to some outcome O.

Or – to take another example – in preparing for an important football game, a team will study the game films of its upcoming opponents to determine the pattern of tactics which the team is likely to employ. The proliferation of variety which the opposing team is capable of employing is obviously great, but, nevertheless, the coaches will deduce some patterns with a high probability of occurrence. This would be especially true if the opposing team's "power" is centered around a few individuals. The team, in an effort to control the situation (game), will try to counteract this variety by adjusting its own resources. As is well known, the cybernetician's strategy of requisite variety often works. In this example we are obviously dealing with incomplete information respecting the variety that is possible; however, this is essentially the same for the manager attempting to control a business organization. A firm will typically acquire (scout) environmental information (competition, political factors, the economy, the state of technology, labor activities, and so on) in order to reduce the uncertainty of its operation. It is the task of the manager to scan the environment in order to better control the system.

Ashby's Law of Requisite Variety is an important element for organizations since it provides valuable guidelines about the best fit of the organization for its environment. Essentially the variety in the environment must be met by equal variety in the organization; variety in problems must be met by variety in solutions. This law calls attention to the specific need for methods of ascertaining variety in the environment. The ultimate solution would have a TV tuned to a "variety channel" that periodically, if not continuously, informs the organization of significant environmental happenings. It would also be nice to have a TV program guide to serve as a feedforward mechanism.

These are indeed available to the adapting organization. Both scanning and forecasting provide us with the necessary program notes so that organizations can develop the requisite variety they need to cope with the variety found in the environment.

That the law of requisite variety is of universal application in control systems must not be too obvious; otherwise there would be fewer systems that go out of control each year. One is led to believe that in many cases this principle is not even

FIGURE 3–15   Input-Output Model

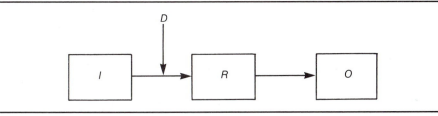

adverted to. It would be difficult to find systems analysts who claim that federal agencies generally have requisite variety in their control mechanism. Or take the controversy over manned and unmanned space flights. One side states that manned flights provide greater variety for counteracting all possible disturbances to the system. On the other side, the greatly increased costs that the taxpayers would have to pay for unmanned systems with comparable high variety are generally underplayed. The greater the systems variety, the greater the control variety, and the greater the costs.

Ashby's Law of Requisite Variety is at work in many daily situations. One controls a business meeting by limiting the variety of topics to be discussed by the use of an agenda. On a camping trip, one keeps in mind the motto: Be prepared. Because of the variety of system contingencies in the camping situation, one tries to employ requisite variety in the control system (first-aid kits, lotions, pills, food, and so forth).

Seldom in the real world do we possess requisite variety; yet we operate as if we had total control. Basically what occurs is this: we attempt to develop variety only for those factors that have a high probability of occurrence. Factors with a low probability of occurrence are given but scant attention. Thus, before we go on a long trip by car, we have the car checked for those things most likely to go awry (with the highest probability of malfunctioning). Parts with but a low probability of malfunctioning are not even considered, unless evidently faulty.

COMPLEXITY AND THE "BLACK BOX"

Earlier in this chapter the cybernetic system was defined in terms of extreme complexity, probabilism, and self-regulation, and the analytical tools corresponding to each of these systems characteristics were outlined. Thus, self-regulation in a cybernetic system is best understood by employing the analytic tool of the feedback principle. Probabilism is best handled through the vocabulary and conceptual tools of probability theory or its modern equivalent, information theory. Because of its importance in understanding organizational behavior, an entire chapter will be devoted to the logic, principles, and foundations of the subject. This section will be devoted to the first characteristic of a cybernetic system and its corresponding analytic tool; namely, extreme complexity and the black box.[18]

Complexity

The explanation of the term *complexity* can be approached from many different viewpoints. From the mathematical viewpoint, complexity can best be understood as a statistical concept. More precisely, complexity can best be explained in terms of the probability of a system's being in a specific state at a given time.[19] From a nonquantitative viewpoint, complexity can be defined as the quality or property of a system which is the outcome of the combined interaction of four main determinants. These four determinants are: (1) the *number of elements* comprising the system; (2) the *attributes* of the specified elements of the system; (3) the *number of interactions* among the specified elements of the system; and (4) the *degree of organization* inherent in the system; i.e., the existence or lack of predetermined rules and regulations which guide the interactions of the elements and/or which specify the attributes of the system's elements.

Most attempts at measuring the complexity of a given system usually concentrate on two criteria: the number of elements and the number of interactions among the elements. This is especially true in classical statistics situations. This kind of measure of complexity is very superficial and, to some extent, misleading. Confining oneself to these two dimensions of complexity will lead one to classify a car engine as a very complex system. There are indeed a large number of elements and an equally large number of interactions among all the parts of a car engine. By the same token, one would be inclined to classify a two-person interaction as a very simple system, for there are only two elements and only two possible interactions involved.

If one were to incorporate the other two determinants of complexity into one's attempt to measure it—namely, the attributes of the elements and the degree of organization—then one would arrive at a different conclusion. Concerning the example of a car engine, one would observe that the interactions must obey certain rules and follow a certain sequence. One would also observe that the attributes of the system's elements are predetermined. By using all these four criteria of complexity one must conclude that the car engine is, in fact, a very simple system.

---

[18] The investigation of the black box approach to complexity is based upon the following works: Ashby, *Introduction to Cybernetics;* W. R. Ashby, *Design for a Brain* (London: Chapman and Hall, Ltd., and Science Paperbacks, Butler and Tanner, Ltd., 1960); Beer, *Cybernetics and Management;* S. Beer, *Management Science, The Business Use of Operations Research* (New York: Doubleday, 1968); and H. A. Simon, *The Sciences of the Artificial* (Cambridge, Mass.: MIT Press, 1970).

[19] Beer, *Cybernetics and Management,* p. 36.

The seemingly simple system of the two-person interaction is indeed a complex system, since the attributes of each element are not predetermined and since the degree of organization, despite the existence of some rules of human interaction, is very low. The elements, in other words, have a free will in obeying or disregarding the rules of human conduct. In this case the ultimate outcome of the interaction — that is to say, the degree of predictability of the final state of the interaction — is uncertain. One must therefore conclude that this two-person system is indeed complex.

The relationship between these four determinants of complexity (the number of elements, the attributes of the elements, the number of interactions, and the degree of organization in the system) and the degree of complexity can be illustrated by using the so-called span-of-control principle. This principle states that "no supervisor can supervise directly the work of more than five or, at the most, six subordinates whose work interlocks."[20] The rationale of this principle is the increased complexity which accompanies the increase in the number of the subordinates for each supervisor. This complexity is equated with the number of direct and cross relationships between the different members of the group which increase by a geometrical progression. Thus, a superior interacting with seven subordinates who also interact with each other will generate 490 potential relationships — an enormous complexity indeed.[21]

This is a misleading measure of complexity, however. In order to gain a meaningful measure of the complexity involved in the span of control, one must consider in addition to the above two criteria — namely the number of elements (members of the group) and their interactions — the attributes of each member, as well as the organization of the task involved. By considering these two additional criteria, one can arrive at a different set of possible states of the group/system. If the task is highly routinized and at the same time the members of the group are well trained, then, assuming no intentional attempts to overburden the superior, the system will be fairly simple to the extent that most of the possible interactions will not be exercised by the subordinates. In addition, there will be a set of rules and procedures which will tend to reduce considerably the possible number of interactions.

---

[20] F. Urwick, *Scientific Prinicples and Organization* (New York: American Management Association, 1938), p. 8.

[21] The number of relationships is derived by using the well-known Graicunas formula:

$$C = n \left[\frac{2_n}{2} + n - 1 \right]$$

where

n = Number of employees reporting to a superior

C = Number of potential relationships

A supervisor attempting to supervise two energy experts, one advocating coal as the most promising future energy source, the other explaining the benefits of fusion as a source of energy, would be confronted with a much more complex system than a colleague who supervises 20 oil engineers. Complexity is indeed a relative concept which is determined by the interaction of all four determinants and not just by the mere number of elements and their possible interactions.

The Black Box

In explaining the complexity of a system in the above example, no attempt was made to describe or define in detail the elements, processes, interactions, and states of these systems. Only the number of inputs and the number of potential relationships were specified. No speculations were made regarding the nature of the process responsible for producing one state or another. In other words, as far as the process of the system is concerned, the system was presented as undefinable in detail. This is the same thing as saying that the system has been treated as a black box.

The problem of the black box first arose in electrical engineering. The engineer is given a sealed black box with terminals for input to which he may apply any voltages, shocks, or other disturbances. The box also has output terminals from which he may observe whatever he can. He is to deduce the contents of the black box.

Figure 3–16 depicts a black box. When the experimenter presses the various buttons at the front of the box, the bulbs on the top can be made to come on in different combinations. The engineer can thus deduce the contents of the box even without seeing what is going on inside.

Although the black-box technique originated with electrical engineers, the present range of application is far wider. The physician studying a patient with brain damage may be trying—by giving tests and observing the responses—to deduce something of the neurological problem. In 1976 a dozen or so people attending an American Legion convention in Philadelphia were stricken with an unknown disease. Many of them died, and because of the circumstances, the disease was labeled Legionnaire's disease. Many medical researchers toiled long and hard before the cause of the disease was discovered. The process of finding out the cause resembled the black-box technique. In effect, the researchers tried, by manipulating the various inputs and coding the corresponding outputs, to ascertain some regularity in the observed relationships. In this way they could eventually pinpoint the cause of the disease. The same could be said of AIDS (acquired immune deficiency syndrome), whose existence was recognized in the early 80s (1982) but for which a cure has not yet been found.

The psychologist, psychiatrist, or business consultant employs the black-box technique whenever he or she attempts to study anomalies in the behavior of the individual or firm by testing certain input functions of the system and by recording the changes in the composition of the outputs. The researcher manipulates the input and classifies the output.

FIGURE 3–16   The Black Box

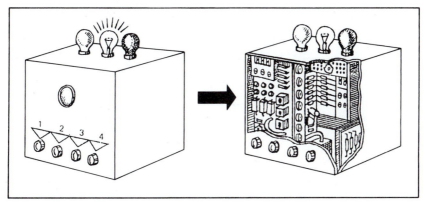

This is a black box. By repeatedly pressing the buttons, the lamps on top can be made to flash on in different combinations, making it possible to establish the laws of input-output relationships without knowing what goes on inside the box.

SOURCE: V. Pekelis, *Cybernetics A to Z* (Moscow: MIR Publishers, 1974), p. 50.

If the account of the black-box technique were to stop here, one might get the impression that what is involved. is but the classical conditioning or stimulus-and-response technique of the early psychologist or the cause-and-effect approach of the analytic thinker. Nothing could be further from the truth than to equate the black-box technique of input manipulation and output classification with either stimulus and response or cause and effect. Both the latter techniques assume fairly simple situations consisting of two-term causal relationships that are more often than not fabricated by the observer/experimenter. While here the observer attributes certain responses or effects to certain stimuli or causes, the theory of the black box is simply the study of the relations between the experimenter and the object, as well as the study of what information comes from the object, and how it is obtained.

The black-box technique is illustrated in Figure 3–17. This figure shows that an experimenter who thus acts on the box and is also affected by the box and the recording apparatus (i.e., the protocol), has become coupled with the box, so that the two together form a system with feedback. When a generous length of record has been obtained, the experimenter will examine it for regularities in the behavior of the system represented by the box.

Let us explain the above line of reasoning by a simple example taken from Ashby.[22] Suppose that a system has two possible input states, *a* or *b,* and four possible output states, *f, g, h,* or *j.* A typical representation of the various input and output states might be that which appears in Table 3–1a. One might view the

---

[22] Ashby, *Introduction to Cybernetics,* pp. 86-92.

FIGURE 3–17 The Black-Box Technique

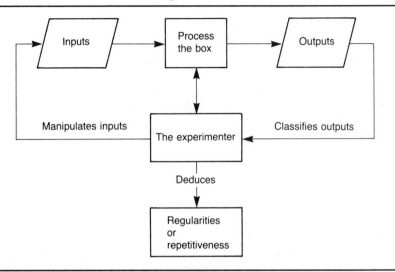

bottom row in the table as a paper tape that has a sequential record of the input and output data of the black box under investigation. This sequential record is known as a *protocol.*

As the protocol tape reveals, in trial 1 with input *a,* the output *g* was obtained; in the second trial with the same input, the output *j* was gotten; in the third trial with the same input, the output was now *f.* And so on for the rest of the data. When examining the protocol of the 17 recorded trials, one can look for the regularities, the repetitiveness of the black box's behavior. Of course, the more regular patterns we can detect in the protocol record, the more we will know of the workings of the black box.

Let us now examine the above protocol items in detail. To begin with, we note that for all but the last protocol item, we have an item that immediately follows. For convenience sake, the item that immediately follows a previous protocol item is referred to as a protocol transition value and its second element is termed the output protocol transition value. We have thus 17 protocol items and 16 protocol transition values.

Now in the body of Table 3–1b we have listed the *output protocol transition*

TABLE 3–1a

Time Trials	1	2	3	4	5	6	7	8	9	10	11	12	13	14	15	16	17
Inputs a or b	a	a	a	a	a	b	b	b	a	a	b	a	b	b	a	b	a
Outputs f, g, h, j	g	j	f	f	f	f	h	h	h	j	f	h	j	f	h	j	f
Protocol	ag	aj	af	af	af	bf	bh	bh	ah	aj	bf	ah	bj	bf	ah	bj	af

SOURCE: Adapted from Ashby, *Introduction to Cybernetics* (New York: J. Wiley & Sons, 1963), p. 89.

TABLE 3–1b

	f	*g*	*h*	*j*
a	fff	j	jjj	ff
b	hhh	—	hh	ff

TABLE 3–1c

	f	*g*	*h*	*j*
a	f	j	j	f
b	h	—	h	f

SOURCE: Adapted from Ashby, *Introduction to Cybernetics* (New York: J. Wiley & Sons, 1963), p. 89.

values for every possible protocol item (*af, ag, ah, aj,* and *bf, bg, bh, bj*). One can verify this by making a table with the heading *f g h j* on the top line and *a* and *b* at the left margin of two successive lines as in Table 3–1b. Now by going back to the protocol tape and for each item writing down the letter of its output protocol transition value under the appropriate column and row, one can derive a table identical with Table 3–1b. For example, the first item *ag* is followed by *aj.* Write the *j* under the *g* column in the *a* row. The second item *aj* is followed by *af.* Write the *f* under the *j* column in the first row. Do this for all of the *a* items. For the *b* items one enters the transition values on the *b* row. Thus for trial 6 item *bf,* whose protocol transition value is *bh,* write the *h* under the *f* column in the second row. And so on.

Since all of the entries in any one cell in Table 3–1b are the same (for *af* we have three *f*'s, for *bf* we have three *h*'s and so on), we can further generalize our findings and so simplify our presentation by noting the entry just once. Thus Table 3–1c shows us that after every *ah* protocol item, we always get an output protocol transition value of *j,* after every *bh* we always get one with *h,* and so on.

In this way we have discovered that in the entire string of protocol items, all of their *output* transition values are *single-valued.* Table 3–1c clearly demonstrates this, since no cell has more than one type of output transition value. On the other hand, the *a* and *b* inputs here are *multivalued.* Table 3–1c shows that in row *a* there is more than one output protocol transition value—here *f* and *j;* for input *b* (row *b*) there are also two output transition values—here *f* and *h.* Table 3–1c also shows that not only is there no *bg* protocol item on the entire protocol tape but also that there are no output protocol transition values of *g* at all.

After ascertaining the multivalued nature of the inputs and the single-valued nature of the outputs, and the nonoccurrence of any *g* output protocol transition value, we can now predict any output transition value, given a protocol item. Furthermore, we have learned all this by simply observing the black box's behavior and analyzing the sequence of the protocol items. No previous knowledge of the black box was needed. We deduced the nature of the black box from its actual behavior—"only this and nothing more."

Another protocol can be displayed as follows. Suppose you were studying a "box" that has fallen from an unidentified flying object. You might write down something like this:

Time	State
11:18	Did nothing—box emitted steady buzzing sound at 240 hertz frequency.
11:19	Pushed button marked "K"—buzzing sound rose to 480 hertz and continued at that level.
11:20	Accidentally pushed button labeled "!"—temperature of the box rose by 20° C.
11:21	. . . .

One of the merits of the black-box technique is that it provides the best antidote against the tendency of the investigator to oversimplify a complex phenomenon by breaking it into smaller parts. The black-box technique for dealing with complexity represents a selection procedure based on a series of dichotomies. In other words, the investigator of a complex situation manipulates the inputs to the black box and classifies the outputs into certain distinct classes based upon the degree of similarity of the output state. The investigator then converts each class into a "many-to-one" transformation. In this way, the observer obtains a black box with a binary output and a large number of input variables which permute themselves and their interconnections to represent *one* output state.[23]

To sum up, the black-box technique involves the following sequential steps: (1) input manipulation, (2) output classification, and, finally (3) many-to-one transformations.

The input manipulations over an extended number of trials reveal (in the output classification as recorded in the protocol) certain similarities or repetitiveness. These similarities are in turn converted into legitimate many-to-one transformations that act as implicit control devices; these many-to-one transformations account for the reduction in the system's variety without unnecessary simplifications.

Nature is full of examples manifesting the black-box technique for dealing with complexity. There are mechanisms common to all life, such as the hereditary apparatus—the genetic structure with its hundreds of genes; occasional variations in the nature of genes through mutation; the distributing and combining of gene variations by sexual recombination. There is the principle of natural selection, whereby favorable (i.e., survival-promoting) mutations gradually become incorporated as normal elements in the gene complex, and so on.

In the industrial world the same process is at work. Were the manager of one of today's complex industrial enterprises to attempt to comprehend all possible combinations of the elementary units of this system, he or she would simply be overwhelmed by the detail. By assuming the system to be undefinable in detail, and by applying the black-box approach, the manager does succeed in formulating

---

[23] Beer, *Cybernetics and Management;* also, *Management Science.*

enough many-to-one transformations (policies) — to use Beer's expression — kill the variety of the system and suppress its concomitant dangers. The modern manager's most indispensable tool, the computer, operates in accordance with the same mechanism of input manipulation, output classification, and many-to-one transformations. Indeed, for the majority of managers the computer is the almost perfect example of the black box both in its figurative and in its literal sense.

Reducing Complexity

Simon in discussing complex systems uses the aspect of redundancy in order to simplify the system.[24] Given the following array of letters:

$$
\begin{array}{cccccccc}
A & B & M & N & R & S & H & I \\
C & D & O & P & T & U & J & K \\
M & N & A & B & H & I & R & S \\
O & P & C & D & J & K & T & U \\
R & S & H & I & A & B & M & N \\
T & U & J & K & C & D & O & P \\
H & I & R & S & M & N & A & B \\
J & K & T & U & O & P & C & D
\end{array}
$$

To reduce the redundancy in the above system, let us designate the various constituent arrays by letters. Thus,

$$
\left|\begin{array}{c} AB \\ CD \end{array}\right| = a
\qquad
\left|\begin{array}{c} MN \\ OP \end{array}\right| = m
\qquad
\left|\begin{array}{c} RS \\ TU \end{array}\right| = r
\qquad
\left|\begin{array}{c} HI \\ JK \end{array}\right| = h
$$

Also
$$
\left|\begin{array}{c} ABMN \\ CDOP \end{array}\right| = am
\quad \text{and} \quad
\left|\begin{array}{c} RSHI \\ TUJK \end{array}\right| = rh
$$

Then
$$
\left|\begin{array}{c} ABMN \\ CDOP \\ MNAB \\ OPCD \end{array}\right| = \frac{am}{ma} = w
\qquad
\left|\begin{array}{c} RSHI \\ TUJK \\ HIRS \\ JKTU \end{array}\right| = \frac{rh}{hr} = x
$$

The entire system can then be represented thus:

$$
S = \left|\begin{array}{c} wx \\ \overline{} \\ xw \end{array}\right|
$$

---

[24] Simon, *Sciences of the Artificial,* pp. 109-10.

This analogy can be usefully applied to the business situation. Say, for example, that a firm in attempting to assess the impact of a possible price reduction in its products can from past experience predict that competitors will do the same. This in effect reduces complexity in the system. Simon also makes the important point that hierarchic systems are composed of only a few different kinds of subsystems but arranged in different ways. If this is known, then again complexity can be reduced. It is precisely this factor that makes the field of cybernetics applicable to all types of systems. The control aspect of a thermostat is similar to those employed in complex space explorations, albeit an obvious difference in degree of complexity is involved.

SUMMARY

Cybernetics, which deals with feedback processes of all kinds, is characterized by complexity, probabilism, and self-regulation. These elements are analyzable by the black-box technique, information theory, and the feedback principle respectively. The concept of cybernetics as the science of communication and control is applicable to highly complex problems, especially those which do not readily lend themselves to traditional analytical approaches.

A feedback control system with its closed-loop structure can employ positive as well as negative feedback, though negative feedback is the more familiar of the two. The use of block diagramming as a common language for cybernetic applications was illustrated in this chapter, and its suitability for identifying structural relationships in a system was pointed out. Various kinds of feedback systems were noted: the first-order feedback system (the system is monitored against an external goal); the second-order system (which can store alternatives in memory and choose the best among them); and the third-order system (which can formulate new courses of action).

Ashby's Law of Requisite Variety indicates that, with increases in the complexity of a system, the variety of uncertainties also increases and, in order to control a complex system, the amount of variety in the control mechanism must equal that in the system itself. A simple control system is possible only where the variety in the system has itself been simplified.

The final section of this chapter concluded with a brief treatment linking complexity with the black-box technique. The latter, which involves input manipulation, output classification, and many-to-one transformations, finds its almost perfect example in the computer.

Although cybernetics is the science of communication *and* control, only the control aspect was treated here. It is left to a later chapter to take up the element of communication.

KEY WORDS

Feedback	Positive Feedback
Closed Loop System	Negative Feedback
Open Loop System	Cybernetics

Control

First-Order Feedback

Second-Order Feedback

Third-Order Feedback

Law of Requisite Variety

Black Box

Complexity

Probabilistic Systems

Deterministic Systems

Block Diagram

Span of Control

Protocol

REVIEW QUESTIONS

1. What is cybernetics? Briefly present the historical developments or circumstances which led to the formulation of the science of cybernetics.

2. According to Beer, there are three characteristics of a cybernetic system. What are these, and what are the tools of analysis to deal with each of these characteristics?

3. Explain the classification framework presented in Figure 1 and illustrate the six categories of systems with examples other than the ones provided in the text.

4. How does feedback operate, and what role does it perform in a control system?

5. Is control possible without negative feedback? Cite some control systems which do not possess negative feedback as a means for control.

6. What is positive feedback and what is its function in a system?

7. Show how the concept of requisite variety is important in the marketing subsystem of an organization.

8. Present several examples of first-, second-, and third-order feedback subsystems of an organization.

9. How would you go about assessing the complexity of a given system?

10. What is the black-box technique, and how is it used in understanding the behavior of complex systems?

Additional References

Beer, Stafford. *Brain of the Firm,* 2d ed. New York: John Wiley & Sons, 1981.

Brix, V. H. "Systems and Cybernetics: A Methodology for Human Systems Management." *Human Systems Management* 1, no. 1 (February 1980), pp. 53–61.

————. "Control, Bureaucracy, and Power." *Human Systems Management* 2, no. 4 (December 1981), pp. 316–21.

Clemson, Barry. *Cybernetics: A New Management Tool.* Kent: Abacus Press, 1984.

"The Concept of Feedback and Cybernetics." *American Business Computing Association Bulletin* 36, no. 3, (September 1973), pp. 16–19.

Cook, Norman D. "An Isomorphism of Control in Natural, Social, and Cybernetic Systems." *Journal of Cybernetics* 10, no. 1–3 (January-March 1980), pp. 29–39.

Crosson, Fredrick J. and Kenneth M. Sayre (eds.). *Philosophy and Cybernetics.* New York: Simon & Schuster, 1967.

Dechert, Charles R. (ed.). *The Social Impact of Cybernetics.* New York: Simon & Schuster, 1966.

Dewan, Edmond M. *Cybernetics and the Management of Large Systems.* New York: Spartan Books, 1969.

Dord, Richard C. *Modern Control Systems.* 4th ed. Reading, Mass: Addison-Wesley, 1986.

Foerster, Heinz von (ed.). *Cybernetics.* New York: Josiah Macy, 1953.

Greniewsky, Henry. *Cybernetics without Mathematics.* New York: Pergamon Press, 1960.

Guidbaud, Georges T. *What is Cybernetics?* New York: Grove Press, 1960.

Klir, Jiri. *Architecture of Systems Problem Solving.* New York: Plenus Press, 1985.

————. "Complexity: Some General Observations." *Systems Research* 2, no. 2 (1985), pp. 131–40.

Klir, Jiri, and Miroslav Valachi. *Cybernetic Modeling.* Princeton, N.J.: Van Nostrand, 1967.

Livas, Javier. "Law: A Challenge for Cybernetics." *Human Systems Management* 3, no. 3 (September 1982), pp. 215–18.

Lusk, Edward J. "Control Systems Analysis: An Evaluation Systems Perspective." *Human Systems Management* 2, no. 4 (December 1981), pp. 205–93.

Morgan, Gareth. "Cybernetics and Organization Theory: Epistemology or Technique." *Human Relations* 35, no. 7 (July 1982), pp. 521–37.

Osborn, Richard N., James G. Hunt, and Robert S. Bussom. "On Getting Your Own Way in Organizational Design – An Empirical Illustration of Requisite Variety." *Organization and Administration Sciences* 8, nos. 2 and 3 (Summer–Fall 1977), pp. 294–310.

Phillips, Charles, and Royce D. Harbor. *Feedback Control Systems.* Englewood Cliffs, N. J.: Prentice Hall, 1988.

Pickett, Robert T. *Feedback Control for Technicians.* Englewood Cliffs, N.J.: Prentice-Hall, 1988.

Richards, Laurence. "Cybernetics and the Management Science Process." *OMEGA* 8, no. 1 (1980), pp. 71–80.

Sethi, Suresh P., and Gerald L. Thompson. *Optimal Control Theory.* Boston: M. Mijhoff, 1981.

Strank, R.H.D. *Management Principles and Practice: A Cybernetic Approach.* New York: Gordon and Breach, 1983.

Valentinuzzi, Maximo. *The Organs of Equilibrium and Orientation as a Control System.* New York: Harwood Academic Publishers, 1980.

Vande Vegte, John. *Feedback Control Systems.* Englewood Cliffs, N.J.: Prentice-Hall, 1986.

Weir, Michael. *Goal-Directed Behavior.* New York: Gordon Breach Science Publishers, 1984.

CHAPTER 4

Cybernetic Principles and Applications

For if cybernetics is the science of control, and if management might be described as the profession of control, there ought to be a topic called management cybernetics — and indeed there is. It is the activity that applies the findings of fundamental cybernetics to the domain of management control.

Stafford Beer

CHAPTER OUTLINE

MANAGEMENT CYBERNETICS AND CITY GOVERNMENT
MANAGEMENT CYBERNETICS IN POLITICAL LIFE
SUMMARY AND CONCLUSIONS
KEY WORDS
REVIEW QUESTIONS
ADDITIONAL REFERENCES

INTRODUCTION

There is little doubt that the nerve system of any organization must employ the warp and woof of cybernetics. We need only reflect on the routines of our daily lives to appreciate the impact of cybernetics on what we do.

We are automatically awakened by an alarm, a buzzer, or strains of music. Even on the coldest winter morning, we awaken in a room that has been preheated by an automatic furnace. After turning off our automatic electric blanket, we wash or shower with hot or warm water maintained by an automatic water heater. We put coffee into an automatic coffee maker, our bacon in a microwave oven. Perhaps our eggs are cooked with the aid of an automatic timer. After breakfasting, we put the dishes in an automatic dishwasher. We may even preprogram the oven for preparing the evening meal and may turn on the automatic telephone-answering device in our den.

We get into our car that may have an automatic transmission, automatic speed control, and automatic temperature control. Before we drive out of the garage, we activate the automatic garage opener and again press the activator to automatically close the garage door after us.

We drive to work through a maze of automatic traffic lights. After parking the car in the company garage, we step into an elevator that automatically stops at our office floor. Once inside the office, we are surrounded by further cybernetic devices purchased and maintained for our comfort, convenience, and efficiency.

The purpose of this chapter is twofold: (1) to highlight the basic principles of cybernetics and to relate them to the management of enterprises, and (2) to provide a brief description of several real-life situations in which the science of cybernetics is being applied as a managerial conceptual and practical framework.

To accomplish these two aims, this chapter begins with a recapitulation of the subject of cybernetics as it relates to the art and science of management. Two main points will emerge from this recapitulation: (1) a cybernetic view of an enterprise, and (2) a cybernetic view of management. These two points taken together provide considerable insight into the governing principles of cybernetics per se and of management cybernetics in particular. Both subjects have often been misinterpreted, and to some extent misunderstood, as a result of the mysticism unnecessarily attached to them, as well as the highly mathematical treatment previously accorded the subject of cybernetics[1] and the erroneous association of cybernetics with science fiction.

The chapter then moves on to applications of management cybernetics to the management of modern enterprises. The description of applications of

---

[1] See for example, N. Wiener, *Cybernetics* (Cambridge, Mass.: MIT Press, 1961); J. Klir and M. Valachi, *Cybernetic Modeling* (Princeton, N.J.: Van Nostrand, 1967); O. Lange, *Wholes and Parts: A General Theory of System Behavior* (Oxford, England: Pergamon Press, 1965).

management cybernetics begins with micro and unfolds into macro considerations. Thus, the workings of management cybernetics are first examined in
connection withthe management of the operations of a micro enterprise such
as a manufacturing firm. This description is then extended to the management
of the firm as a whole.

Having thus described the application of management cybernetics for the
micro enterprise (firm), the discussion continues with the management of a larger
enterprise, such as a city. Under the heading "Management Cybernetics and City
Government," cybernetic principles employed to assist the management of a city
government are highlighted. Management cybernetics is applied to political life in
the final section.

CYBERNETICS AND MANAGEMENT (REVISITED)

In this book, attempts were made to relate the subject of cybernetics to the
process of managing modern enterprises. These were primarily illustrative
examples of possibilities for diffusing cybernetic thinking into the management
process. As such, these examples were primarily intended to give certain
cybernetic principles a more or less commonsensical explanation by associating
them with conventional administrative principles and practices. Now, however,
the time has come to discuss more at length the fundamentals of this hybrid
discipline, management cybernetics.

All too infrequently the literature draws a dichotomy between general
systems theory (GST) and cybernetics rather than viewing them as integrated.[2]
The integration of GST and cybernetics was, of course, implicit in many writings,
including Bertalanffy's, as he clearly indicated:

> It appears that in development and evolution dynamic interaction (open system)
> precedes mechanization (structured arrangements particularly of a feedback
> nature). In a similar way, G.S.T. can logically be considered the more general theory;
> it includes systems with feedback [cybernetics] constraints as a special case, but this
> assertion would not be true vice versa. It need not be emphasized that this statement
> is a program for future systematization and integration of G.S.T. rather than a theory
> presently achieved.[3]

In view of this goal of "future systematization and integration," management cybernetics is proposed as the unifying framework or theory which
integrates GST and cybernetics into a coherent scheme for dealing with control
and communication, as well as with the evolution of complex dynamic open
systems.

---

[2] L. von Bertalanffy, "General Systems Theory — A Critical Review," in W. Buckley, *Modern Systems Research for the Behavioral Scientist* (Chicago: Aldine Publishing, 1967), pp. 16–17.

[3] L. von Bertalanffy, "General Systems Theory," p. 19.

THE BASIC ELEMENTS OF A CONTROL SYSTEM

Implicit controllers are subsystems whose main function is to keep some behavioral variables of the focal or operating system within predetermined limits. They consist of four basic elements which are themselves subsystems. These basic elements are:[4]

1. A control object, or the variable to be controlled.
2. A detector, or scanning subsystem.
3. A comparator.
4. An effector, or action-taking subsystem.

These four basic subsystems of the control system, along with their functional interrelationships and their relationship with the operating system, are depicted in Figure 4–1.

The student will notice similarities in this figure with those in Figure 1–5, which depicts an organization with its resources and its environment. The broken lines enclose the area of the system under consideration, here the control system of the organization. It too has its inputs, processes, and outputs with the outputs of one system or subsystem serving as the serial inputs of another. In Figure 4–1 a time element has been associated with the outputs. The various points in time are indicated by t sub 0 through t sub 8. If the system depicted were a production system, it would be evident from the figure that some of these timed outputs have exceeded the upper control limit. These control objects are fed into the detector, then to the comparator and finally to the effector before being fed back into the overall system. What one sees at the top right of the figure would really be a blown-up view of the results of the comparator — that element of the control system that compares the magnitude of the control objects against a predetermined standard.

In previous figures the control function was shown merely as a feedback line going from the output of the system back to its input. In the present diagram the control function is depicted as a system within a system with its own subsystems. Besides, one can readily see that, as far as the control system is concerned, the source of disturbance is the operating system itself, whose behavior the control system is supposed to be regulating.

Control Object

A control object is the variable of the system's behavior chosen for monitoring and control. The choice of the control object is the most important consideration in studying and designing a control system. Variations in the states of the control object — i.e., its behavior — become the stimuli which trigger the functioning of the

---

[4] A much more detailed discussion of the basic elements of a control system can be found in R. A. Johnson, F. E. Kast and J. E. Rosenzweig, *The Theory and Management of Systems,* 2d ed. (New York: McGraw-Hill, 1967).

FIGURE 4–1   The Major Elements of a Control System

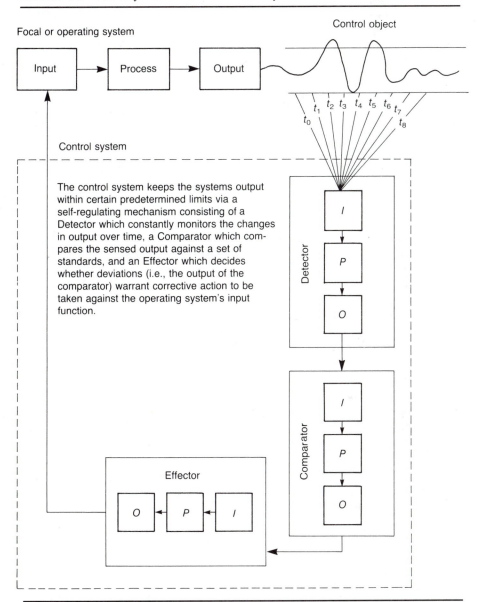

control system. Without these variations, the system has no reason for existence. Since in reality there is never a perfect match between desired and actual outcome, variations will always exist; ergo, the need for control.

From the foregoing it should be clear that great care and much thought ought to be given to "what must be controlled." Let it suffice to point out the rather obvious observation that the control object must be chosen from the system's output variables. Well-balanced quantitative and qualitative attributes of the system's output should provide the best choice of control variables. Focusing on controlling system output does not necessarily imply an ex post facto account of system behavior. Feedback control systems can just as easily function as anticipatory mechanisms rather than ex post facto corrective devices.

Detector

The structure and function of a detector or scanning system has been the subject of an entire previous chapter. There the entire organization was conceptualized as a scanning system. The only point that must be repeated is that scanning systems feed on information. Again, as in the case of the control object, the detector operates on the principle of selective acquisition, evaluation, and transmission of information. As such, a detector system is another name for a management information system (MIS).

Frequency, capacity, efficiency, accuracy, and cost of detector devices are some of the important aspects that an administrator must reckon with.

Comparator

The output of the scanning system constitutes the energizing input of the comparator. Its function is to compare the magnitude of the control object against the predetermined standard or norm. The results of this comparison are then tabulated in a chronological and ascending or descending order of magnitude of the difference between actual performance and the standard. This protocol of deviations becomes the input to the activating system.

Note that if there are significant differences between the output and the goal, the system is said to be "out of control." This could mean that the goal formulated is unrealistic, unsuitable for the system's capabilities. Either the goal itself must be charged, or the design characteristics of the system must be altered. For example, if the production goal cannot be met, the goal must be altered or, if kept, either more people must be put on the production line or more equipment employed.

Effector

The effector is a true decision maker. It evaluates alternative courses of corrective action in light of the significance of the deviations transmitted by the comparator. On the basis of this comparison, the system's output is classified as being in control or out of control. Once the status of the system's output is determined to be out of control, then the benefits of bringing it under control are compared with the estimated cost of implementing the proposed corrective action(s).

These corrective measures might take the form of examining the accuracy of the detector and of the comparator, the feasibility of the goal being pursued, or the optimal combination of the inputs of the focal system—that is, the efficiency of the "process" of the operating system. In other words, the output of the activating system can be a corrective action which is aimed at investigating the controllability of the operating system and/or the controllability of the controller itself.

STABILITY AND INSTABILITY OF THE CONTROL OBJECT

The control object can take either of two states: (1) it can be stable, or (2) it can be unstable. Both states are necessary for system survival. While stability is the ultimate long-run goal of the system, short-run instability is necessary for system adaptation and learning. The system, in other words, pursues a long-run stability via short-run changes in its behavior manifested in its output's deviations from a standard.

Let us briefly explore the nature of stability and instability as well as some of the reasons for instability. In general terms, stability is defined as the tendency of a system to return to its original position after a disturbance is removed. In our systems nomenclature, stability is the state of the system's control object which exhibits at time t_1 a return to the inital state t_0 after an input disturbance has been removed. Were the system's control object not able to return to or recover the initial state, then the system's behavior would exhibit instability. The input disturbance may be initiated by the feedback loop, or it may be a direct input from the system's environment. The particular behavior pattern that the system will exhibit is dependent on the quality of the feedback control system (detector, comparator, and effector) in terms of sensitivity and accuracy of the detector and comparator as well as the time required to transmit the error message from the detector to the effector. Oversensitive and very swift feedback control systems may contribute as much to instability as do inert and sluggish ones.

Time delay is the most important factor for instability of social systems such as business enterprises and governments. Although the application of information technology management information systems (MIS) and electronic data processing (EDP) has made considerable progress toward accelerating the transmission of information from the detector to the effector, as well as expediting the comparison and evaluation of information inside the comparator, still, the impact of the corrective action upon the control object's behavior is felt after a considerable time lag.

Continuous oscillations of the kind exhibited in Figure 4–2 are the results of two characteristics of feedback systems: (1) the time delays in response at some frequency add up to half a period of oscillation, and (2) the feedback effect is sufficiently large at this frequency.[5]

---

[5] Arnold Tustin, "Feedback," in *Automatic Control* (New York: Simon & Schuster, 1955).

FIGURE 4–2   Control Object's Behavior

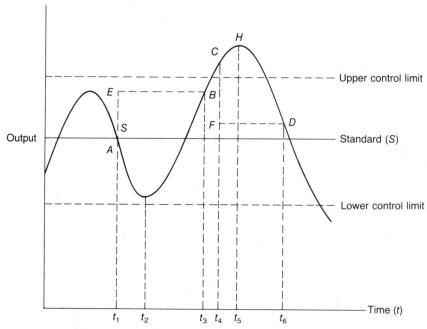

A: Point where direction of output is recognized.
B: Corrective input is added.
C: Error is noted — directive to remove resources.
D: Resources are removed.
EB and FD: Information time lag.

SOURCE: Adapted from R. A. Johnson, F. E. Kast, and J. E. Rosenzweig, *The Theory and Management of Systems,* 2nd ed. (New York: McGraw Hill, 1967), p. 89.

Figure 4–2 demonstrates the behavior of the system's output, which is controlled by a feedback system characterized by a one-half cycle time delay.[6] When a time delay of that magnitude exists, the impact of the corrective action designed to counteract the deviation comes at a time when this deviation is of a considerably different magnitude, although it has the same direction. This causes the system to overcorrect. In Figure 4–2 a deviation of the magnitude equal to SA is detected at time t_1. At time t_2, new inputs are added to bring the output back to the standard (S). The impact of this corrective action upon the system's output is not felt until t_3. By that time the actual system's output is at point C—i.e., after a time lag equal to $t_4 - t_3$. The detector senses this new deviation and initiates

[6] Johnson, Kast, and Rosenzweig, *Theory and Management of Systems,* p. 89.

new corrective action. Because of the one-half-cycle time lag, the actual system's output has oscillated above the upper limits at the point H. New corrective action initiated, aimed at bringing the output back to the standard, will be felt at time t_6

This basic principle of time delay and its impact upon the system's control behavior is illustrated very clearly in Tustin's diagrams, which appear in Figure 4–3.

It might seem from the above brief description of the function of the basic elements of the control system that the task is a formidable one when measured in terms of cost and/or time. In conventional control systems, this might indeed be the case. In cybernetic control systems, however, this is definitely not true. The reason, of course, is that this task is performed as part of the normal operation of the system and requires no extra effort.

CONTROL PRINCIPLES

The principles governing control functions in a cybernetic system are universal and simple. These principles as formulated by Beer read as follows:

Control Principle I

Implicit controllers depend for their success on two vital tricks. The first is the *continuous and automatic comparison* of some behavioral characteristic of the system against a standard. The second is the *continuous and automatic feedback* of corrective action.[7]

Thus, according to Control Principle I, implicit controllers are engaged in both detector and comparison activities, as well as in corrective action. This is, of course, common to all control systems. However, what is unique in the case of implicit controllers is the prerequisite that these functions are *continuous and automatic.* That is to say, detecting, comparing, and correcting activities are not initiated periodically, nor are they imposed upon the control system from outside, but they are rather executed from within in a perpetual manner.

With implicit controllers it is the very act of going out of control that brings the variables back into control. Together, the continuous and automatic comparison of the behavioral characteristic against a standard, and the continuous and automatic feedback of corrective action makes this possible. Implicit controllers can be found both in machines and in humans. It is possible to program machines to easily self-correct. Many functions of the human body are self-correcting. The maintenance of body temperature, blood count, the pH of

---

[7] S. Beer, *Management Science: The Business Use of Operations Research* (New York: Doubleday, 1968), p. 147.

FIGURE 4–3 Oscillations in Feedback Systems

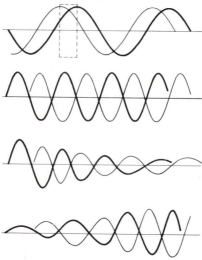

Oscillation is inherent in all feedback systems. The drawing at top shows that when a regular oscillation is introduced into the input of a system (lighter line), it is followed somewhat later by a corresponding variation in the output of the system. The dotted rectangle indicates the lag that will prevail between equivalent phases of the input and the output curves. In the three drawings below, the input is assumed to be a feedback from the output. The first of the three shows a state of stable oscillation, which results when the feedback signal (thinner line) is opposite in phase to the disturbance of a system and calls for corrective action equal in amplitude. The oscillation is damped and may be made to disappear when, as in the next drawing, the feedback is less than the output. Unstable oscillation is caused by a feedback signal that induces corrective action greater than the error and thus amplifies the original disturbance.

SOURCE: Adapted from Arnold Tustin, "Feedback," in a Scientific American book, *Automatic Control* (New York: Simon & Schuster, 1955), pp. 20–21.

electrolytes, body-fluid retention, etc. are all controlled continuously and automatically. Social organizations, on the other hand, are typically neither able nor willing because of costs to have continuous and automatic control. The few systems that do manage to control on a continuous and automatic basis are, not surprisingly, financed by the federal government—like the North American Air Defense Command (NORAD) headquartered in Colorado.

Control Principle II

In implicit governors, *control is synonymous with communication.* Control is achieved as a result of transmission of information. Thus, to be in control is to communicate. Or, in Norbert Wiener's original words, "Control . . . is nothing but the sending of messages which effectively change the behavior of the recipient. . . ."[8]

This is indeed the most basic and universal principle of cybernetics. The realization that control and communication are two sides of the same coin motivated Wiener to use them as the subtitle of his classic pioneering work *Cybernetics*. This is where one reads: *to control is to communicate and vice versa*!

It is evident from the above principles of control (I and II) that the system whose behavior is subject to this type of control becomes literally a slave to its own purpose. Since every deviation from standard behavior is autonomously and automatically communicated (through the sequential activities of the detector, comparator, and effector), the more frequently out-of-control situations occur, the more frequently communication takes place and consequently the more corrective action is taken. It is for this reason that implicit controllers are also referred to in the literature as "servomechanisms" (*servo* means "slave").

This observation allows us to formulate another basic principle of cybernetic control, originally conceived by S. Beer: Control Principle III.

Control Principle III

In implicit controllers, variables are brought back into control *in the act of* and *by the act of* going out of control.[9]

This principle follows directly from our explanation of the basic structure of the control system. It will be recalled that what triggers the detector subsystem is the existence and magnitude of the deviation (d) between the goal and actual performance, i.e., the output of the operating or focal system. It follows that the more frequently deviations occur, the more frequent will be the communication between the detector and the comparator. In addition, the more frequent and more substantial the magnitude of the deviation, the more likely it is that corrective action will be initiated and executed.

From the foregoing discussion of the basic principles of cybernetic control, the following question is inescapable: Given the unique nature of a cybernetic system, what kinds of demands do these control principles impose upon goal-directed systems?

The most important demand facing the system is that it be an adaptive learning system. In other words, the function of the implicit controller demands that the operating system eventually learn that being in control is as necessary a condition for its survival as its growth capabilities.

---

[8] From *The Human Use of Human Beings* by Norbert Wiener. Copyright 1950, 1954 by Norbert Wiener. Reprinted by permission of Houghton Mifflin Company.

[9] Beer, *Management Science,* p. 147.

THE ROLE OF THE ADMINISTRATOR AS AN IMPLICIT GOVERNOR

The basic conviction underlying the ideas developed in the previous chapters of this work is that an administrator or manager is essentially a decision maker. Most modern literature in organization or management theory would agree with this contention.[10]

The foregoing development of the basic principles of cybernetic control, along with their accompanying discussions, should have provided an obvious hint that in cybernetics, decision making and control are two very closely related, if not identical, activities. This allows us to state that *decision making and control are similar if not identical managerial activities.* Both activities are initiated and maintained through communication.

The relationship between information acquisition, evaluation, and dissemination (communication) will be explained in Chapter 6 of this book. The discussion of the basic principles of cybernetics has also emphasized the dependence of control on communication. It thus appears that the common denominator between the two basic managerial activities of decision and control is communication.

The cybernetic framework or way of looking at decision, control, and communication is of tremendous importance for administrators of modern enterprises, for the precise reason that enterprises consist of human beings who are by definition communicative. In communicating, humans decide; in deciding, they communicate; in communicating, they control; in controlling, they communicate; . . . and the cycle goes on as long as the enterprises remain living entities.

Management Principle

The role of the manager as an implicit controller and as the focal point in the self-adapting organization should not be overlooked. It is the manager who sets goals, who communicates; it is the manager who provides leadership and motivation for subordinates. It should be kept in mind that the manager is not merely an added component of the system, an appendage, but an integral part of the system, affecting its structure and organizational climate.

We shall now attempt to report briefly on some actually operating control systems which function in accordance with the basic principles of cybernetics explained previously. To preserve a modicum of authenticity, and to avoid as much as possible the imposition on the reader of our own biased interpretation, we have chosen to present the crux of these systems freely in the language of the original

---

[10]R. M. Cyert and J. G. March, *A Behavioral Theory of the Firm* (Englewood Cliffs, N.J.: Prentice-Hall, 1963); J. G. March and H. A. Simon, *Organizations* (New York: John Wiley & Sons, 1958); J. G. March, *Handbook of Organizations* (Chicago: Rand McNally, 1965); H. A. Simon, *Administrative Behavior* (New York: Macmillan, 1959); F. E. Kast and J. E. Rosenzweig, *Contingency Views of Organization and Management* (Chicago: Science Research Associates, 1973).

designer. We hope that in doing so we can present the material in a comprehensible and logical framework.

We will therefore begin with the illustration of the application of management cybernetics to specific operations of the firm. However, it must be kept in mind that the difficulty of designing implicit controllers increases as one goes from a production and inventory-control system to a system for the whole enterprise. The reason for this difficulty is that our understanding of the detailed structure and function of the enterprise itself and of the environment, as well as our understanding of the interactions between the two, are still very meager.[11]

Even though the task of designing implicit controllers for larger systems seems exceedingly difficult, it is by no means impossible. Moreover, the benefits to be derived from the dependable operation of such control devices far outweigh their costs. In addition, one has little choice in deciding whether or not to design such systems. One must in varied degrees abide by the laws governing natural behavior. This is especially true today, when the physical or natural environment seems to more or less dictate the systems' designs.

MANAGEMENT CYBERNETICS IN CONTROLLING A MANUFACTURING FIRM

It must be pointed out at the outset that the greatest successes are to be found in the management of the operations rather than in the management of the so-called human side of the enterprise. The reason for this is that operations deal primarily with material flows and processes and, to a large extent, are quantifiable and deterministic. The human element is less important to the outcome, and decision making is more routinized. In cybernetic terms, there is less variety in such systems and therefore less variety is needed for controlling such systems. Because of this, the field of cybernetics has been dominated by the engineer. Only recently has cybernetic control of organizational activities been applied to the human element. Systems of this type will be discussed in the section of this chapter entitled People Control Systems.

Production- and Inventory-Control Systems

In his application of cybernetics to the manufacturing process, Beer has utilized the familiar block diagram shown in Figure 4–4 to depict a production-control system.[12] The input (θ_i) into the production system is the raw material, and its output (θ_o) is the product. Work flows through the system at the rate of (μ). θ_L represents the arrival of new orders. There are two control loops, as can readily be seen. The major loop of $\theta_i - \theta_o$ is the difference between the desired state and the actual state—in other words, the error (ϵ). This error is fed back into the

---

[11] S. Beer, *Decision and Control* (New York: John Wiley & Sons, 1967), pp. 301–2.
[12] S. Beer, *Cybernetics and Management* (New York: John Wiley & Sons, 1959), pp. 171–72.

FIGURE 4-4   Production-Control System as a Cybernetic System

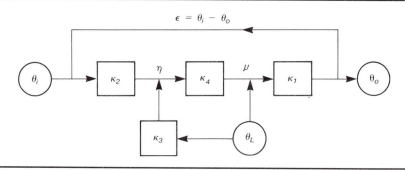

system through the operator (κ_2), who makes the appropriate rate of adjustment with respect to the goal. The planned level (η) becomes the actual rate (μ) by operator (κ_4). However, the actual rate of production is also modifed by previous rates and new orders (θ_L). The new orders constitute an input affecting the planned rate via operator (κ_3). This load directly affects the output, and its action is represented by operator (κ_1). This order level represents the second control loop.

The output of this system may be described mathematically:

$$\theta_o = \kappa_1 \,(\mu - \theta_L) \tag{1}$$

The output of the system (θ_o) is the rate of production as affected by the order load (θ_L), the influence being represented by the primary operator κ_1. The error feedback (ϵ) is, of course, the difference between output and input. The actual production rate is being determined by the planned rate (η) through its operator κ_4. The planned rate (η) itself is influenced by the error (ϵ) with its operator (κ_2) and the order load (θ_L) with its operator κ_3.

The relationships are presented in equations 2, 3, and 4.

$$\epsilon = \theta_i - \theta_o \tag{2}$$

$$\mu = \kappa_4 \eta \tag{3}$$

$$\eta = \kappa_2 \epsilon + \kappa_3 \theta_L \tag{4}$$

The four equations effectively define the system. With the treatment of the variables as functions of time, it is possible to complete the analysis. For the interested reader, the complete mathematical analysis may be followed in Beer[13] and especially in Simon's original work,[14] where the mathematical relationships are worked out in greater detail.

---

[13] Beer, *Cybernetics and Management;* p. 172.
[14] H. A. Simon, *Models of Man* (New York: John Wiley & Sons), 1957.

In the previous chapter in Figure 3–11 the relationship between the block diagram and general systems theory was depicted. One can readily see how in the above example we move from a verbal description of the production system to a block diagram depicting the interrelationships of the various elements in the system and finally to the mathematical analysis.

People-Control Systems

Were one to ask how managers control their subordinates, the answer might well be *through implicit controllers.* The technique that most closely approximates the actions of implicit controllers in people-control systems is MBO, management by objectives. We say "approximates" since in human systems — social organizations — continuous and automatic monitoring, comparison, and feedback of human subjects are rare and difficult. Perhaps a cardiac intensive care unit in our larger hospitals offers the closest approximation to implicit controllers, but then the success of these CIC units derives from their being cybernetic in nature.

Management by objectives, first enunciated in 1954 by Peter Drucker, was designed to measure the contribution of both a department and an individual to the system (organization) by a careful and explicit statement of the particular goals to be accomplished. Management by objectives calls for an identification of the results in light of the originally planned goals and expected results. When objectives are defined in terms of the results to be achieved, then one generally has a fairly good notion of what must occur in the system.

In the decades since it has been introduced, MBO has been found to be a powerful management control tool and a highly successful one, too. Perhaps its very success is the source of one of its chief problems: difficulty in explaining how it works. Practitioners have written extensively on the procedural aspects of the technique, nor have they been loath to point out the do's and don'ts to be observed if success is to be achieved. The literature is replete with guidelines — guidelines for setting goals, for measuring performance, for adjusting goals, etc. However, if MBO is viewed as a cybernetic system, then how it works becomes immeasurably easier to understand, and many of the seemingly contradictory research findings can more readily be reconciled. MBO is basically a cybernetic system, and its success must be due to its cybernetic nature.

In its most basic form, management by objectives includes the following procedures:

1. An individual writes down the objectives that he or she is to accomplish in the next time frame and specifies how the results will be measured.

2. The objectives are submitted to the individual's immediate superior for review. Out of this review comes a set of objectives to which the subordinate makes a commitment.

3. Evaluation of performance is carried out in light of the previously agreed-upon objectives. Modifications in the individual's behavior may occur because of variances between the results achieved and the results expected.

The model of the MBO process shown in Figure 4–5 illustrates the similarities of goal setting in MBO to goal striving or goal achievement in cybernetics. Goals depicted by θ_i are the inputs to the MBO process and represent the standard to be maintained or achieved. The transfer function which transforms inputs into outputs (in this instance performed by individuals) is affected by a number of internal and external factors (disturbances). While these disturbances (identified in the research literature as intervening variables) are important, since they have been shown to affect the level of goal attainment, their discussion would be beyond the scope of this work.

Critical to goal maintenance or goal attainment is feedback on system performance, represented by θ_o. In a recent factor-analytic study of MBO, the amount, type and frequency of feedback had extremely high loadings on the first factor.[15] This is not at all surprising, since feedback is the essential ingredient needed for control. As in every cybernetic system, control requires the comparison of actual performance against planned performance. Deviations from planned goals suggest the need for corrective action. The elements of this MBO model are, therefore, seen to be identical with those in the cybernetic model presented earlier. Although one can easily add complexity to both models, the purpose here is merely to show that the underlying tenets of the cybernetic model are also found in the MBO process.

Any system which can reflect upon its goals, search its memory for past behavior, and change its goals is a third-order feedback system. Such is the management-by-objectives system. In the MBO system, deviation from a goal forces one to reexamine the reasons why the goal was not attained. This reexamination could lead to the formulation of a new goal, which could then result in a change in the environment or a change in the system.

The foregoing should make clear that management by objectives is applicable to situations other than those of business organizations. When one considers that all living systems are teleological (goal striving), it isn't too surprising that management by objectives has such wide applicability. Management by objectives, instead of being a revolutionary tool, can be regarded rather as the application of well-tested principles of cybernetics to the management of human resources of an enterprise. Indeed, some may even consider MBO as a systems approach, since it is vested with many of the characteristics of a system.

Although the benefits of management by objectives as applied in industry today reach far beyond the expectations of its early practitioners, the system has been and will continue to be a basic control tool. The iterative cycle is goal formulation, feedback, measurement, corrective action, goal formulation, and so on. The point being made here is simply that the vastly popular management-by-objectives system, which is utilized throughout the world today, is really a simple cybernetic system. Recognition is made of the fact that goals in this system are not self-maintaining ones but rather third-order feedbacks.

---

[15] John C. Aplin and Peter Schoderbek, "A Cybernetic Model of the MBO Process," *Journal of Cybernetics* 10 (1980): pp. 19–28.

FIGURE 4–5   Simplified Cybernetic Model of the Management-by-Objectives Process

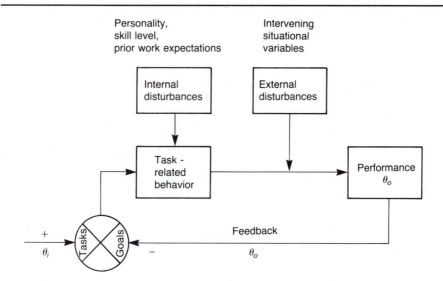

Multiple Goals

Cybernetic systems perhaps have their greatest value when applied to control operations. In the discussion, the impression may have been given that only one factor was being controlled. In real life, there is often a multiplicity of goals. The question then arises: How do we control when there are multiple goals? Two means are available. The first of these is simply to have multiple detectors and feedback loops. The system may even be expanded to include multiple detectors and actions. This can help account for the fact that certain actions may result, not from a deviation from a single goal, but from deviations from more than one goal. Figure 4–6 illustrates this point. While the actual condition of a single goal, say x_3, may never be maximized, the overall system objectives (x_1, x_2, x_3, x_4, etc.) may indeed be. Management of course, has many objectives, and these must be handled by means other than maximization. One cannot maximize profit and let customer services deteriorate, nor can one disregard union-management relations in the pursuit of profits. There are approximately five key areas with organizational objectives: market standing, profitability, research and development, productivity, and financial resources. In each of these areas there are distinct objectives, and any attempt to optimize one alone will ensure the suboptimization of the others. In practice, a firm accepts less than optimization and seeks an "acceptable" level of performance in each major area. In reality, the number of independent organizational goals is rather small. These are usually spelled out in terms of acceptable and/or desired levels of performance rather than the optimal.

Before discussing the second way that multiple goals can be handled, we wish to return to a concept mentioned earlier in the text, namely, that of systems

FIGURE 4–6 Multiple-Goal, Multiple-Detector, and Multiple-Action System

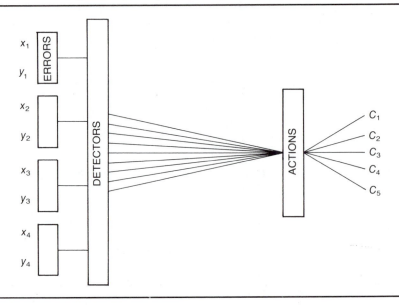

hierarchy. We saw that systems can be divided and subdivided into subsystems and into subsubsystems depending on the resolution level desired. Obviously, one can set up control systems in the same manner. Practically speaking, this means that at the plant level there will be systems and subsystems. The inventory system, while a system in and of itself, is also a subsystem of production. The same with quality control; so too the scheduling system, the maintenance system, the safety materials–handling system, etc. Each of these subsystems is controllable as a total system with its own goals and type of feedback.

At the plant-manager level, the system to be controlled takes on a different form of control. It may take the form of productivity, which is basically costs over the number of products produced. This productivity ratio is the result of the outputs of many of the previously mentioned subsystems.

At the top level of management, ratio analysis is a popular control tool. Ratio analysis is an appropriate control tool, not only for one or the other subsystem, but for the entire organization. Some of the ratios traditionally employed for this purpose are those of profit to sales, profit to total assets, total-assets turnover, inventory turnover, fixed-assets turnover, debt to equity, and current and quick liquidity.

Note that the concept of cybernetics can be applied at each of the various organizational levels. At each level there is a monitoring of results, their comparison with predetermined goals, and sometimes even daily feedback reports that make the comparisons possible.

The chapter "Organizational Effectiveness" has a section entitled Suboptimization of Goals in which this subject will be touched upon again.

The Whole Enterprise Control System

The cybernetic model for a control system for the whole enterprise is a rather complex one. This complexity, however, is dictated by the very nature of the enterprise as an organic system that functions within a dynamic, complex environment. To the extent that real-life enterprises are complex (in the sense of organized complexities), and to the extent that the environment is also complex, the law of requisite variety dictates that a control system designed to guarantee survival of an enterprise must also be complex.

In addition to complexity, a second characteristic of a cybernetic whole enterprise control system is that it should be an open system. This means that the system must encompass both the enterprise per se and its environment. It follows that this kind of control system focuses on the interaction between these two subentities and not on the entities themselves, as most conventional financial controls do.

The third characteristic of a cybernetic whole enterprise control system is that it must be a self-regulating or closed-loop system. This is, of course, a basic prerequisite for all implicit controllers. Self-regulation does not necessarily mean that all control activities will be completely autonomous from other managerial activities; rather it implies that the control system constitutes the suprasystem of a multiplicity of quasi-independent control domains, and is, therefore, itself quasi-independent. Self-regulation implies autonomy within a structure.[16]

The necessity of structure constitutes the fourth characteristic of a cybernetic whole enterprise control system. The nature of this structure has been characterized throughout this book as being that of a hierarchical order, that is to say, a hierarchy of feedback control systems of different orders constituting a hierarchy of homeostats. As was explained earlier, this hierarchy is not imposed upon the control system but is implicitly informed. In other words, the functions of the local control subsystems (production, inventory, employment, accounting, and other control systems) are mediated centrally to avoid suboptimizations. Suboptimizations occur, it will be recalled, when certain improvements in the operating system's performance do not contribute toward increases in the performance of the whole system.

Although some authors have attempted to depict a total integrated model for controlling organizations, we will not do so here. Such models, to do justice to the matter, require explanations far beyond our space limitations. It should suffice to say that any total enterprise model would have to include complexity and the means to deal with it. This obviously means use of the black box for unexplained processes. Policies and information would be two major areas through which to reduce variety and complexity. Control systems would have to be generously employed for the purpose of self-regulation at all levels of the organization. As noted above, a hierarchical structure would be used both in the layering of

---

[16]This is the same autonomy that profit centers enjoy. Within certain broad market and financial policies set by the headquarters, the centers can operate as they see fit.

information and in the structuring of the subsystems. Information about the environment would be sought both for the feedforward policies as well as for blocking disturbances. In short, the total control system would consist of many smaller integrated systems. Just as no totally integrated information system exists today for any organization, neither can such a model of the whole enterprise be validated. Certainly attempts have been made to provide for holistic integrated systems, and notable successes have been achieved, but the "holy grail" of one omnipotent system will probably remain just as elusive tomorrow as it is today and has been in the past. Perhaps the answer to this is not the form and substance of cybernetic control systems but rather the cybernetic manager who can comprehend the major elements for controlling organizations.

Stafford Beer in his recent text entitled *Diagnosing the System for Organizations* provides his most complete cybernetic treatment of the organization. He claims that the Viable System Model (VSM) has had wide and varied applications in organizations. It is a hands-on application of his model and therefore difficult to summarize. Beer asks each reader to choose an organization and then proceeds to walk the reader through his model. This technique requires the reader to identify his or her organization's environment, its boundaries, system levels, and so on. For the seasoned cybernetician, terms like purposefulness, complexity, requisite variety, channel capacity, homeostasis, closed loop systems, comparator, and feedback seem like old friends. One experiences a good feeling because here at last one sees the whole puzzle instead of just the pieces.

Beer's disciples, who come from many lands, have made extensive use of VSM in everything from neurophysiology to book publishing. For those academicians and practitioners of systems thinking who have experienced frustrations in operationalizing systems concepts, Beer illumines the Diogenesean path.[17]

MANAGEMENT CYBERNETICS AND CITY GOVERNMENT

By way of analogy, Figure 4–7 illustrates feedback control systems of governmental operations.[18] Here the government of New York City is depicted in terms of a block diagram. While one may say that such treatment is too vague to be beneficial, an examination of the feedback process throws some light on the usefulness of the cybernetic treatment. Each of the city offices in effect has objectives, and both the desirability and attainment of these objectives are modified by the feedback. The following constitute sources of feedback for a city government:

1. Direct observation by the mayor.
2. Information provided by subordinates.

---

[17]Stafford S. Beer, *Diagnosing the System for Organizations* (Chichester: John Wiley & Sons, 1985).
[18]E. S. Savas, "City Halls and Cybernetics," in E. M. Dewan, ed., *Cybernetics and the Management of Large Systems,* Second Annual Symposium of American Society of Cybernetics (New York: Spartan Books, 1969), pp. 134–35.

FIGURE 4–7   City Government as a Cybernetic System

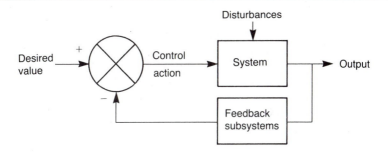

Conventional feedback control system diagram

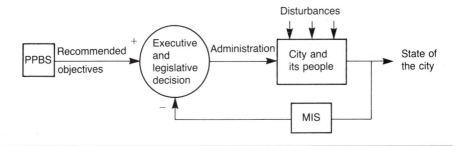

SOURCE: Adapted from E. S. Savas, "City Hall and Cybernetics," in *Cybernetics and the Management of Large Systems,* ed. Edmond M. Dewan, American Society for Cybernetics (New York: Spartan Books, 1969), pp. 134–35.

3. The press.
4. Public officials.
5. Public at large.
   a. Vocal individuals.
   b. Special-interest groups.
   c. Elections.
   d. Civil disorders.[19]

   All of the above are signals, and as such constitute information, even though it may be biased and distorted. When the voters turn down a bond issue, this is feedback; when riots occur in the streets, this is feedback; when officials are defeated in elections, this is feedback. It may be true that a system such as that of a city government lacks quantification because of its extreme complexity. Still, this is the situation as it exists, and it is precisely this type of problem that can benefit from the cybernetic treatment. City government, with its numerous vested interests and at times conflicting objectives, subjected to predictable and unpredictable constraints, reveals its full complexity when studied in such a manner. Regardless of the fact that such an operation defies complete

[19]Savas, *City Halls,* p. 137–38. See also Bulletin, TIMS, 2, no. 4 (August 1972).

identification of its numerous interrelationships, it is nevertheless quite amenable to study through the cybernetic approach.

A more ambitious and sophisticated approach to the study of city management represents the application of industrial dynamics to the investigation of urban social systems, such as the city of Boston. This application is described in Professor Forrester's *Urban Dynamics.*[20] Forrester's book is about the growth processes of an urban area which is conceptualized (and subsequently simulated) as a system of interacting industries, housing, and people.

Interpretations of the simulated results of certain policies and programs designed to secure vitality of an urban area system led Forrester to the following conclusions:

> The city emerges as a social system that creates its own problems. If the internal system remains structured to generate blight, external help will probably fail. If the internal system is changed in the proper way, little outside help will be needed. Recovery through changed internal incentive seems more promising than recovery by direct-action government programs. In complex systems, long-term improvement often inherently conflicts with short-term advantage. The greatest uncertainty for the city is whether or not education and urban leadership can succeed in shifting stress to the long-term actions necessary for internal revitalization and away from efforts for quick results that eventually make conditions worse. Political pressure from outside to help the city emphasize long-term, self-regulating recovery may be far more important than financial assistance.[21]

The relevance of urban dynamics cannot really be overemphasized. It is precisely even more pertinent today, when even the most conservative television commentators advocate limiting the growth of cities as a solution to every urban problem (transportation, crime, poverty, and so forth). It therefore behooves the student and manager of organizations to take a closer look at this significant contribution to the contemporary struggle for the survival of our decaying cities.

MANAGEMENT CYBERNETICS IN POLITICAL LIFE

One cannot but marvel at the fascinating progress which has been made toward the development of a theory of politics, both national and international. Most experts in the field concede that the intellectual models set forth by Karl Deutsch and David Easton represent key milestones in these developments.

Karl Deutsch's *The Nerves of Government: Models of Political Communication and Control* is indeed a classic example of the application of cybernetics to the study and management of government. He states in the preface:

> In the main, these pages offer notions, propositions, and models from the philosophy of science, and specifically from the theory of communication and control — often called by Norbert Wiener's term "cybernetics" — in the hope that these may prove

---

[20] Jay Forrester, *Urban Dynamics* (Cambridge, Mass.: MIT Press, 1969), p. 1.

[21] Forrester, *Urban Dynamics,* p. 9.

relevant to the study of politics, and suggestive and useful in the eventual development of a body of political theory that will be more adequate – or less inadequate – to the problems of the later decades of the twentieth century.

Here again the basic principle of cybernetics, that communication and control are the two sides of the same coin, is evident in the following statement by Deutsch.

> This book concerns itself less with the bones and muscles of the body of politic than with its nerves – its channels of communication . . . [it] suggests that it might be profitable to look upon government somewhat less as a problem of power and somewhat more as a problem of steering; and it tries to show that steering is decisively a matter of communication.[22]

After presenting an historical account of some conventional models for society and politics, the basics of cybernetics, and the role of communication models and political decision systems, Deutsch offers a crude model of control and communication in foreign-policy decisions. The model represents a complete account of the main information flows which are necessary in making foreign policy decisions. Suffice it to say that information-scanning systems, both foreign and domestic, constitute inputs to the system and that internal and external policies constitute the major outputs of the system. Of particular interest is the transformation process. Deutsch here shows the integration of all three types of feedback systems, which were discussed earlier – first-, second-, and third-order systems. The interested student is referred to the original source for a fuller explanation of this example.

Another application of cybernetics to the political arena is depicted in Figure 4–8. Here the demands of the system provide the basic ingredients for the inputs.[23] Responsible authorities (officials) are constantly converting the demands of the populace (raw materials) into some form of suitable output.

In this third-order feedback system, with the feedback loop running from the outputs to the total environment, communications are represented by solid lines connecting the environment with the political system; the arrows indicate the direction of the flow of information in the system; the broken lines indicate that the environments are changing as a result of the outputs.

In the box labeled "political system," one notes that the authorities acquire information about the consequences of previous actions. This information is then taken into account in the formulation of new goals. In reality, there is a constant interchange of information between the officials and the total environment. Indeed, without it no system could survive. Every system must be able to adapt to the threats and opportunities present in the environment, and, in order to do this, it must acquire this information through some feedback process.

---

[22] Karl Deutsch, *The Nerves of Government* (New York: Free Press, 1966).

[23] David Easton, *A Systems Analysis of Political Life* (Chicago: University of Chicago Press, 1965), pp. 17–35.

FIGURE 4–8   A Dynamic Response Model of a Political System

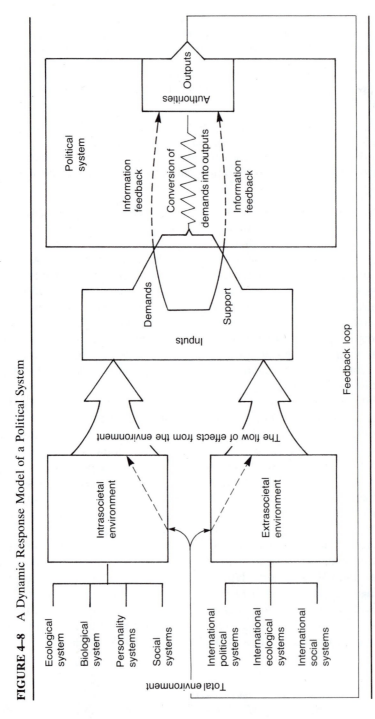

SOURCE: Adapted from David Easton, *A Systems Analysis of Political Life* (New York: John Wiley & Sons, Inc., 1965), by permission of the publisher.

A complete analysis of this political system would have to include a determination of the interactions of the system elements, the range of sensitivity to changes in these elements, and a determination of the processes needed to take advantage of the opportunities and of the processes needed to deal with the threats to the system.

Furthermore, one can readily see even from this rudimentary illustration that the number of systems which affect the political system are indeed many. In fact, the political system is but a subsystem of yet a larger system. Every politician knows that the political system is linked inextricably to the economic system. Every four years the intricate ramifications of the political system with economics, with labor, with minority and religious groups are examined by the pollsters and their constant changes noted for the electorate.

SUMMARY AND CONCLUSIONS

Enterprises, like all organic wholes, have one basic goal: survival. To survive, they must grow and evolve. Evolutionary growth requires that increases in the system's complexity contribute to the system's ability to survive. That is to say, growth processes must be true improvements of the system's performance to cope with ever-increasing and continuously changing environmental complexity. Progress toward that goal is facilitated by implicit governors which control exponential growth and decay processes.

General systems theory studies the general evolutionary growth pattern and principles of biological systems. Cybernetics is concerned with the function of implicit controllers in both natural and man-made systems. Finally, management cybernetics, by combining both general systems theory and cybernetics, determines the growth and evolutionary requirements of enterprises and designs the requisite implicit controllers.

These, then, are the basic principles governing the application of management cybernetics to the management of modern complex enterprises.

The rest of the chapter has been concerned with the review of numerous attempts at applying systems thinking to real-life problems. The authors have attempted to report on several ingenious frameworks devised by serious and inquisitive thinkers whose main objective has been to obtain a better grasp of the world. This exposition obviously reflects our own biases. Furthermore, it is highly selective and condensed. Certainly Wiener, Beer, Deutsch, Forrester, and others would argue that the present authors did not tell everything there is to be known about cybernetics and systems dynamics.

The basic ideas can best be summarized in the sentence: *The behavior of organic systems must, by necessity, be homeostatic.* Homeostatic control is a continuous, automatic corrective action resulting from a continuous and automatic sensing and comparing of the system's output. As such, homeostatic control is an integral part of system behavior. That is to say, it is an implicit control.

In a natural system, this implicit control is designed by nature into the system's basic structure and function. In artificial or man-made mechanical and social systems, these controls must be rationally designed by either the creators or the managers of them, or by both.

The complexity and sophistication of the design of implicit controllers for man-created systems vary in relation to the complexity and sophistication of the operating system whose behavior is to be controlled. This is, of course, true for natural systems as well. The law of requisite variety dictates that variety can only be dealt with through variety.

In natural systems, no matter what their complexity, all three functions of implicit control (i.e., sensing, comparing, and correcting) are performed in a semiautonomous fashion which parallels the goal-directed function of the operating system. To that extent, no extra devices are required. For example, human physiology, the most complex natural system of all, requires no extra extensions of its sensory or motor control devices. All these devices are sufficiently complex to handle the variety inherent in the main functions of the human body.

Unfortunately, the same statement cannot be made regarding human social nature. Here the magnitude and rate of change of the variety generated by person-to-person interaction is considerably more than anyone's sensory and motor devices can handle. This is, indeed, the basic thesis of such popular clichés as, for example, the "overloaded, overstimulated individual," the hero or victim of Alvin Toffler's *Future Shock*. Here natural sensory and motor skills must be supplemented by artificial devices which act as an extension of the insufficient human mechanisms for handling social variety.

A person's role in organizations or enterprises falls under the latter category. This role is essentially sociological in nature. The organizational design of implicit controllers must supplement inborn control devices.

The last decade of the 20th century is destined to form the Age of Cybernation, characterized by the conscious effort toward designing complex social enterprises equipped with implicit governors which guarantee long-range survival. Management cybernetics with its emphasis on systems thinking and information technology is charged with this ultimate task of global human existence.

KEY WORDS

Cybernetics	Comparator
Control Object	Effector
Detector	Multiple Goals
Implicit Controllers	Homeostasis
MBO	Control Principles

REVIEW QUESTIONS

1. Most businesspeople have never heard of cybernetics, yet in spite of this operate their businesses in an efficient manner. Taking a retail store in your town, show how they do employ cybernetic principles, although they are not aware of it. This calls for an identification of the components of the system and a discussion of the functions of the components.

2. The federal government funds many programs dealing with health care, alcoholism, drugs, urban development, and so on. In spite of these many programs, critics argue that they are not doing the job. Cybernetically speaking, what components are missing or ill defined in such systems so that it may be difficult to determine whether the job is being done?

3. The text states the need for continuous and automatic feedback in a cybernetic system. What is meant by continuous? Is this a time dimension, or can it be something else? Show how it can be both, and show how the time dimension can vary with the process of controlling the system.

4. The text discusses the need for a firm to be an adaptive learning system. Is it possible for a firm to reject this principle without negative consequences being incurred?

5. The text states that a system's growth is checked and facilitated by control. Is it possible to grow too fast? Give a particular instance of this situation, and show how control could have been helpful.

6. The authors state that decision making and control are similar if not identical managerial activities. Make the case that these are not similar through the use of examples. After you have done so then refute the case that you have made.

7. Show how the accounting department acts as an implicit controller for the organization. Show the major elements of the control system as their functions.

8. In a university setting, give examples of situations in which it is comparatively easy to establish control systems, or very difficult to do so. Which parts of a college or university might understand cybernetics more easily than others? How does measurement enter into such a system?

9. Much of the material in this chapter dealt with people-control systems. Give some examples of mechanical control systems and identify the principles involved.

10. For the city in which your college or university is located, identify the major control systems employed.

Additional References

Beer, Stafford. *Brain of the Firm.* Hardmondsworth: Allen Lane, Penguin, 1981.

_____. "A Reply to Ulrich's 'Critique of Pure Cybernetic Reason: The Chilean Experience with Cybernetics.' " *Journal of Applied Systems Analysis* 10 (1983).

Berczi, A. "Information as a Factor of Production." *Business Economics* 16 (January 1981), pp. 14–20.

Capra, F. *The Turning Point.* New York: Bantam Books, 1982.

Clutterbuck, David. "Cybernetics Recasts Fisheries Service." *International Management* 33, no. 1 (January 1978), pp. 48–50.

Cummings, Thomas G., and Cary L. Cooper. "A Cybernetic Framework for Studying Occupational Stress." *Human Relations* 32, no. 5 (May 1979), pp. 395–418.

Hardin, Garrett. "What Cybernetics Tells Us about the Marketplace." *Business and Society Review,* Winter 1973, pp. 65–72A.

Heintz, Carl. "Cybernetics Big-5 Accounting Package." *Interface Age* 8, no. 1 (January 1983), pp. 64–71.

Hofstede, Geert. "The Poverty of Management Control Philosophy." *Academy of Management Review* 3, no. 3 (July 1978), pp. 450–61.

Hudson, Leonard C. "A Cybernetic Approach to OD." *Training and Development Journal* 35, no. 4 (April 1981), pp. 86–88.

Kuhn, Arthur J. *Organizational Cybernetics and Business Policy.* Pennsylvania State University, University Park, 1986.

"Marketing Will Vanish if U.S. Enters Cybernetic Transformational Society." *Marketing News* 14, no. 22 (May 1, 1981), pp. 1, 8.

Maruyama, Magorak. "The Second Cybernetics: Deviation-Amplifying Mutual Causal Processes." *American Scientist* 51: pp. 164–79.

Mattiace, John M. "Applied Cybernetics within R and D." *Journal of Systems Management* 28, no. 12 (December 1972), pp. 32–36.

Mantell, Leroy H. "On the Use of Cybernetics in Management." *Management International,* vol. 13, no. 1 (1973), pp. 33–41.

Morgan, Gareth. "Rethinking Corporate Strategy: A Cybernetic Perspective." *Human Relations* 36, no. 4 (April 1983), pp. 345–60.

Morin, Edgar. "Can We Conceive of a Science of Autonomy?" *Human Systems Management* 3, no. 3 (September 1982), pp. 201–6.

Osborn, R. N., J. G. Hunt, and R. Bussom. "On Getting Your Own Way in Organizational Design — an Empirical Illustration of Requisite Variety." *Organization and Administration Sciences* 8, nos. 2 and 3 (Summer–Fall 1977), pp. 294–310.

Rastogi, P. N. *An Introduction to Social and Management Cybernetics.* New Delhi: Affiliated East-West Press, 1979.

Reisig, Gerhard H. R. "Information — System Structure by Communication — Technology Concepts: A Cybernetic Model Approach." *Information Processing and Management* 14, no. 6 (1978), pp. 405–17.

"The Shorter Catechism of Stafford Beer." *Datamation* 28, no. 2 (February 1982), pp. 146–55.

CHAPTER 5

Particularized Approaches to Systems Thinking

It is sheer nonsense to expect that any human being has yet been able to attain such insight into the problems of society that he can really identify the central problems and determine how they should be solved. The systems in which we live are far too complicated as yet for our intellectual powers and technology to understand. Given the limited scope of our capabilities to solve the social problems we face, we have every right to question whether any approach—systems approach, humanist approach, psychoanalytical approach—is the correct approach to the understanding of our society. But a great deal can be learned by allowing a clear statement of an approach to be made in order that its opponents may therefore state their opposition in as cogent a fashion as possible.

C. West Churchman

CHAPTER OUTLINE

INTRODUCTION

In the previous two chapters the field of cybernetics was treated as a particularized approach to systems thinking. By now the reader should realize that all control systems, irrespective of the system designation, employ the same essential elements and that cybernetics can lay no exclusive claim to systems thinking any more than the other techniques and approaches that purport to be systems thinking. To be sure, these approaches include the feedback mechanisms of cybernetics and may therefore be viewed as simple extensions of that discipline, albeit on a grander scale. Opponents of this view would argue that cybernetics is forever committed to follow the design of the system and that therefore the conceptualization of the system is of greater importance. No discussion that we can offer here would convince proponents of either approach, nor would any useful purpose be served by such discussions. Therefore, it is the authors' intention to present in this chapter a brief overview of several other approaches to systems thinking. Many books have been written on their use as systems techniques. The interested reader can comb the literature to his or her heart's content for indepth treatment and application of these particularized systems approaches. The three singled out here for consideration are the engineering approach, the operations research approach, and the systems analysis approach.

SYSTEMS ENGINEERING AS A SYSTEMS APPROACH

The term *systems engineering* was probably first used in the 1950s and is credited to scientists from Bell Laboratories of AT&T. Their 1956 publication entitled *Systems Engineering*, introduced the notion of systems, subsystems, objects, relationships, environment, models, and other traditional systems concepts. The formalization of these concepts was the result of their work experiences on various projects over many years.

Some authors treat systems engineering as a particular systems approach. This seems to be especially true in the field of aerospace engineering, where systems engineering implies not only performance characteristics of the item being designed, but also the configuration specifications. The end item is viewed from the total view of the system. Ramo defines systems engineering as "the invention, design, and integration of the entire assembly of equipment, as distinct from the invention and design of the parts, and geared to optimum accomplishment of a broad product mission."[1]

In Black's view systems engineering begins when a particular system has been broadly defined, its essential tasks segmented into subtasks, and individual systems-engineering groups assigned to carry out the subtasks.[2] According to

---

[1] Simon Ramo, "The Role of the System-Engineering/Technical-Direction Contractor in the Management of Air Forces Systems Acquisition Programs," in Fremont Kast and James Rosenzweig, eds., *Science Technology and Management* (New York: McGraw-Hill, 1962), pp. 187, 334.

[2] G. Black, *The Application of Systems Analysis to Government Operations,* Washington: National Institute of Public Affairs pamphlet, 1966, p. 3.

Harr, "Systems engineering consists of designing plans for the conduct of a systems development program, including its schedule and costs.[3]

Gross and Smith describe systems engineering as "that set of activities dealing with the design of hardware (equipment, computer communication) and design of software, together with some documentation of this design activity.[4]

The engineering approach paralleled the popular project management technique of the 1970s. Systems engineering often employed a variety of approaches among which was cybernetics. Cybernetics found favor especially with scientists involved in the design of electrical and mechanical systems. Also frequently employed in design problems were optimization models. Systems engineering always started with the definition of the problem, and it included such things as systems objective, systems effectiveness, systems alternatives, cost/benefit analysis, systems development, etc. Apparently systems engineering bridged the gap between conceptual systems and operational ones. It was the systems engineer who came down from the mountain to occupy the land below.

OPERATIONS RESEARCH AS A SYSTEMS APPROACH

If one were to single out the one technique that has aroused the most interest in the area of management during the last several decades, it would undoubtedly be that of operations research. Operations research (OR) has enjoyed a remarkable press with reputable and imaginative practitioners reporting from their richly diverse disciplines. Today a professional management meeting would hardly be considered complete unless it included at least one paper on the subject of operations research. A cursory examination of the talks given at recent seminars and symposia would reveal this fascination with that subject. Yet despite its almost total acceptance by management, a most perplexing feature is the disagreement as to the nature of the discipline. More often than not it is defined by its exponents in terms of its activities or with reference to the fields of application. An inkling of this somewhat confusing situation can be gotten by probing some of the pronouncements on the definitional aspect of the problem. OR has been defined many ways. Philip M. Morse of MIT, for instance, defined OR as

> the application of the quantitative, theoretical, and experimental techniques of physical science to a new subject, operations. An operation is the pattern of activity of a group of men and machines doing an assigned, repetitive task.
>
> Horace C. Levinson has defined it as . . . an application of the method and spirit of scientific research to problems that arise in the general area of administration and

---

[3] K. G. Harr Jr., in U.S. Senate, Special Committee on the Utilization of Scientific Manpower of the Committee on Labor and Public Welfare, *Hearing on S. 430 and S. 467,* 90th Congress, First Session, 1967, p. 35.

[4] P. Gross and R. Smith, *Systems Analysis and Design for Management* (New York: Dun-Donnelley Publishing, 1976), p. 36.

organized activities. Its general purpose is to discover the most rational bases for action decisions.[0]

Other writers have been even more ambitious in their formulations. Stafford Beer defines operations research as

> the attack of modern science on complex problems arising in the direction and management of large systems of men, machines, materials, and money in industry, business, government and defense. Its distinctive approach is to develop a scientific model of the system, incorporating measurements of features such as chance and risk, with which to predict and compare the outcomes of alternative decisions, strategies or controls. Its purpose is to help management determine its policy and actions scientifically.[0]

Beer further elucidates the nature of the discipline:

> Operational research, as has been seen, means doing science in the management sphere: the subject is not itself a science; it is a scientific profession. In turning now to the relevance of cybernetics, we encounter a science in its own right.[0]

Other equally notable scholars delineate the limits of OR in equally diffuse terms. The mathematician George Dantzig has this to say:

> Operations research refers to the science of decision and its application. In its broad sense, the word *cybernetics,* the science of control, may be used in its stead.[0]

It comes as no great surprise to note that other writers have equated OR with the systems approach. And as disciplines mature, they often take on labels different from those with which they were born. So it has been with operations research, which has more recently been rechristened Management Science (MS). It would not be easy to say just what differentiates the two titles. Many departments in universities still retain both titles. One suspects that the term *management science* gained in stature at the same time and in much the same way as computers did. In any event, OR and MS seem wedded, for better or for worse, in fashion and out of fashion, till practitioners do them part.

When one reexamines the assumptions underlying systems thinking, one has to admit that, despite its apparent affinity with the systems approach, operations research in the main provides basically a body of computational techniques. That the armamentarium of operations research methods, with its linear and dynamic programming, decision trees, queuing theory, transportation method, network

---

[5] Annesta R. Gardner, "What Is Operations Research?" *Dun's Review and Modern Industry,* December 1955, p. 46.

[6] Stafford Beer, *Decision and Control* (John Wiley & Sons, 1966), p. 92.

[7] Beer, *Decision and Control,* p. 239.

[8] George B. Dantzig, "Operations Research in the World of Today and Tomorrow," Office of Naval Research, January 1965; reprinted in *Operations Research Appreciation,* U.S. Army Management Engineering Training Agency, no date.

analysis, and simulation models, has been dramatically exploited in recent years there can be little doubt. These remain, however, but tools typically utilizing the computer.

The seemingly close identification of operations-research personnel with systems personnel probably stems from the fact that the former are frequently called on to deploy their skills in the design stage of a complex problem. They may be asked to construct models or modify existing ones, or they may be involved in testing the effectiveness or boundaries of a system. In any event, operations-research personnel are generally called on first as trained observers to state the problem in explicit terms.

At the present stage of development, too little is known of operations-research principles that would provide the researcher with a blueprint for solving complex problems. One must admit, nevertheless, that the schematic representations of problems or the ingenious models employed could lead to better solutions. The decisions arrived at by the use of OR techniques can and do give optimal solutions according to formal theory, but they do not necessarily represent the way the human operator behaves.[9] In the final analysis, the utility of a particular course of action must be determined subjectively by the human operator. It is precisely for this reason that the human element is retained in the system, since the operator's behavior cannot be incorporated into the system design. Even the contributions of formal organization theory purporting to describe the performance of the human operator are of little value to the systems designer, since the results tend to center about average performance and not performance of a given operator in a given situation. Since our understanding of the human component in systems is at present inadequate, it is necessary to resort to techniques that lie closer to the empirical world.

The techniques employed by the OR personnel are principally directed to operations management at the lower organizational levels. Of these techniques, a favorite is *simulation,* concerned as it is with the construction of models of real-life situations. Although it has potential for assisting top management, it is in this very area that simulation has achieved but meager and mediocre results. For the most part, business problems are exceedingly complex, and any worthwhile representation of the reality would require exceedingly complex models. In general, models can incorporate only a small segment of the important variables that need to be considered for problem solution. Consequently, the more complex the problem, the more difficult it is to simulate realistically the business situation.

One must not forget that the utility of any model is rather stringently tied to the values assigned to the variables employed. The identification and valuation of

---

[9] Beer lists three major limitations to classical decision theory: mathematical, methodological, and pragmatic. Regarding the mathematical limitation, he says that "it is mathematically impossible to optimize more than one variable of a situation at a time. That is to say, when a mathematical model has been set to maximize profit, or to minimize cost, that is *all* that it can do. If the management has other objectives than this, they have to be handled by other means." See Beer, *Decision and Control,* p. 219.

these variables imply that these can be quantified. This may not always be the case. When dealing with the human equation, operational researchers may believe that it is better to quantify than not to quantify behavior. However, the business executive often believes that personal intuition based on previous experience is as safe a ground for decision making as the constructs of the model builder. For unless the problem under investigation is well structured (and most of them are not), simulation is of little value. The value of simulation seems to diminish in proportion as it deals with behavioral variables. In this no-man's-land the manager, in the absence of scientific rigor, will base judgments on personal observations, experience, and intuition.

What assumptions underlie the applications of operations research? Here, fortunately, there seems to be more general agreement. Probably the most fundamental element in OR is the need to quantify the business problem under study. Without quantification, operations research is unthinkable. Quantification itself implies that the problem is susceptible to rational treatment. There can be no question that the operations researcher is on unassailable ground when the problem under consideration is essentially quantitative and of a repetitive nature. Indeed, this is the one area most susceptible to OR applications. However, decisions that are more of a judgmental nature, less prone to recur with regularity, and more affected by environmental factors are not readily subject to quantification. Likewise, at the upper levels of the organizational hierarchy, where decisions are typically unprogrammed and subjected to undetermined influences of competition, political overtones, changes in income tax structures, regulation by the SEC, FTC, and other government agencies, cold war consequences, labor union maneuvering, irregular economic fluctuations, and so forth, the relevance of operational research for ensuing decisions still needs to be demonstrated.

It is not uncommon for some students of systems to view OR/MS, not as a vehicle for systems analysis, but merely as a set of operational tools. Such critics claim that the field of OR/MS is methodology oriented and as such does not and cannot view problems from an overall systems perspective.

Furthermore, they assert that OR/MS's major value lies in solving simple, trivial problems. This stems from the practice of optimizing a single objective function. Evidence of this, as these opponents of OR/MS point out, is that organizations are abandoning OR/MS teams in favor of management-information specialists. Whether this is true or not is not for us to decide, but there is some evidence that OR teams are less in vogue these days than they were in the late 1960s and early 1970s.

Another argument often put forward for the demise of the OR/MS area is that many of the critical items needed for decision making lack quantification and that the field is not suited to handle nonquantified items. The fact that many decisions are made in response to environmental factors as a short-run adaptation strategy relegates OR problems to assembly-line balancing, product mixes, and a host of other low-level, routine tasks.

Just what is the status of OR as a systems approach in the 1990s? Has it risen phoenix-like with a new array of tools and with a wider range of applications? A perusal of the recent applications of OR shows that OR is still alive and well and tackling problems of massive proportions with an assist from the computer. As to the direction that OR is currently tending, the President of the Operations Research Society recently remarked:

> I do not see OR marching forward with new tools, with new weapons to fight another day. But I do see it becoming more mellow. It will recognize that the process of solution is far more important and essential than the technique and it will develop in the direction of seeking better ways to explore solutions, not merely to find the one solution itself.[5]

It should be remembered that OR was never a science but rather a scientific approach to problem solving. No more, no less. More recently, the criticism has been voiced that OR needs to go beyond its mechanistic rituals and deal with the political and social aspects of problem solving. Such inclusions would indeed supplement the highly technical tools already found in the discipline. However, one must remember that OR applications are most successful when directed to particular types of problems. It was never intended as a systems approach to any and all kinds of problems. The recent criticism of OR is no more relevant than the criticism that could be leveled against, say, computer programming for not tackling the vital social, economic, and health problems of our society. OR is characterized by its concern with operational-level, short-term, managerial-control problems — problems better understood stucturally, problems that are quite readily quanti- fiable. Hardly any social, political or health problems share these same distinctive characteristics.

SYSTEMS ANALYSIS AS A SYSTEMS APPROACH

About the same time that systems engineering became popular, engineers and scientists from the Rand Corporation developed in the 50s and 60s what was to be known as *systems analysis,* a combination of techniques drawn from engineering and other disciplines.[6] These ranged from simple cost analysis to sophisticated computer modeling. Systems analysis was then understood as an attempt to look at the total problem embedded in its environment, to investigate the systems objectives and the criteria for systems effectiveness, and finally to evaluate the alternatives regarding costs and benefits.

The term *systems analysis* was used in the 1960s and 1970s in another way. The designers of information systems adopted this term to designate a set of

---

[5] Roy A. Stainton, "Whither OR?" in Michael C. Jackson and Paul Keys, eds., *New Directions in Management Science* (Aldershot, Hants, England: Gower Publishing Co., 1987), pp. 52–53.

[6] The following are some of the Rand Corporation publications: Newell, A.; Shaw, J.C.; and Simon H. A., *Elements of a Theory of Human Problem Solving* (1957), p. 971; Wohlstetter, A. J., *Systems Analysis Versus Systems Design* (1958), p. 1530; Feigenbaum, E. A., *An Information Processing Theory of Verbal Learning* (1959), p. 1817; Helmer, O., and Rescher, N., *On the Epistemology of the Inexact Sciences* (1960), p. 1955; Hitch, C. J., *On the Choice of Objectives Is Systems Studies* (1960), p. 1955.

procedures by which manual functions performed in an organization could be computerized. This indeed was the assignment of the men and women whose job title was that of a systems analyst. They analyzed inventory systems, payroll systems, employment record systems and so on for the purpose of computerizing them. Systems analysis as conceived by the information specialists of that time was:

> the organized step-by-step study of the detailed procedures for the collection, manipulation and evaluation of data about an organization for the purpose not only of determining what must be done, but also of ascertaining the best ways to improve the functioning of the system.[7]

There are several points worth noting in this definition.

First, systems analysis is primarily concerned with investigating the information system of an organization to determine whether the system is amenable to computerization as well as to prepare the computerization process itself.

Second, systems analysis is primarily an analytic technique: a given system is broken down into its logical components; subsequently, the inputs, processes, outputs, and feedbacks of each subsystem are then defined in terms of the information required for decision making in each of the subsystems. These broad information requirements thus served as guidelines for determining the sources and destinations of data transactions for each subsystem. Data had to be acquired, documented, classified, processed, and evaluated before meaningful information could be obtained.

Third, a systems analysis study such as this resulted in recommendations by the systems analysts for improving or replacing the existing system. These recommendations would necessarily include several systems charts which they designed and which they turned over to programmers. The programmers in turn, after analyzing in exhaustive detail the data transactions, would then develop flow charts and would code them in a programming language that the organization's computer could execute.

Thus, the term *systems analysis* as used here almost always either included the goal of improving a present system or was considered part of the larger domain of systems design.

From that time on the term *systems analysis* has been used in many similar and dissimilar ways. As we shall see presently, the term was used as a synonym for the *systems approach* and for the *general systems theory approach*. Sometimes it was associated with *management science, systems management* and even *systems engineering*. Although authors have the right to define terms as they see fit, still the vast amount of confusion caused by using identical terms for nonidentical approaches could have been obviated if a clear and consistent definition were adopted right from the start. But that was not to be. So let us now examine the other meanings of *systems analysis* found in the literature.

---

[7]Peter Schoderbek et al., *Management Systems: Conceptual Considerations,* 1st ed. (New York: Business Publications, 1975), p. 171.

Malcolm Hoag defines systems analysis as "a systematic examination of a problem of choice in which each step of the analysis is made explicit wherever possible."[8] This view, of course, has much in common with that of information specialists. Others look upon systems analysis as part and parcel of *systems engineering.* Thus Jenkins holds that "the first step in Systems Engineering is Systems Analysis,"[9] and Gross and Smith state in the same vein, "In essence, we could view the systems analysis-design phase as being the 'front end' of 'systems engineering' and 'systems management.' "[10] Others described systems analysis euphorically as the application of the *scientific method* to problems of economic choice.

Those using the method for military purposes outlined the four basic steps of systems analysis:

1. Formulation: clarifying, defining, and limiting the problem.
2. Search: determining the relevant data.
3. Explanation: building a model and exploring its consequences.
4. Interpretation: deriving conclusions.[11]

Perhaps Gross and Smith best present the wide diversity of views when they subsume seven different types under this term, *systems analysis.* (Their intellectual net apparently catches all varieties.)

Type 1: The approach practiced by the Department of Defense which embraces the major steps of determining the objectives of the systems and designing alternative cost-effective ways of achieving these.

Type 2: The approach that would encompass all of the general systems theorists' tools like cybernetics, information theory, set theory, graph theory, game theory, decision theory, and the theory of automata.

Type 3: The approach that would include simulation models of urban growth, such as Forrester's industrial dynamics model and the Club of Rome's world simulation model.

Type 4: The approach that focuses on the public sector and includes techniques such as Planning-Program-Budgeting systems (PPB).

Type 5: The approach that focuses on society and on the organization. Causal models are constructed of the organization as a social system, linking diverse variables of the system and of its environment.

---

[8]Malcolm W. Hoag, "An Introduction to Systems Analysis," in *Systems Analysis,* ed. Stanford L. Optner (Harmondsworth, Middlesex, England: Penguin Books, 1973), p. 37.

[9]Gwilym M. Jenkins, "The Systems Approach," in *Systems Behavior,* ed. John Beishon and Geoff Peters (London: Harper & Row, 1972), p. 65.

[10]Paul Gross and Robert D. Smith, *Systems Analysis and Design for Management* (New York: Dun-Donnelley Publishing, 1976), p. 36.

[11]E. S. Quade, "Military Systems Analysis," in *Systems Analysis,* ed. Stanford L. Optner, pp. 125f.

Type 6: The approach that includes tools such as Program Evaluation Review Technique (PERT) and Critical Path Scheduling (CPM), both of which are commonly considered to be operations research techniques.

Type 7: The approach that would include operations research techniques other than simulation and network analysis. These differ less in substance and more in emphasis from the other types.

A handy distinction Gross and Smith make between operations research and systems analysis is this:

Operations Research	*Systems Analysis*
Concerned with:	Concerned with:
Operational-level	Policy-level
Short-term	Longer-term
Managerial-control problems	Strategic-planning problems
Better understood structurally	Poorly understood structurally
Readily quantifiable problems	Not readily quantifiable problems[17]

Perhaps an 8th type could be added, namely, Management Information Systems (MIS). If one understands MIS as a collection of the organization's information processes that provides decision criteria for the direction and control of organizations, then MIS would undoubtedly qualify as a systems approach. Whether or not MIS is a subclass of systems analysis is debatable. Usually every discipline will defend its position from its own organizational platform. But one should not be beguiled into thinking that the vast technological advances in *computer technology* were what made MIS what it is today. Rather the input of information into the organization's processes is what validates its credentials. Indeed, cost/benefit analysis will show that for many control systems a manual MIS is most efficient.

From all this the reader can readily see that there are many schemes for classifying systems analysis. Some authors view all of the operations research techniques as tools of general systems theory and would include them under that heading. Still others are convinced that systems analysis is but part and parcel of systems engineering. Others hasten to point out that the first applications of systems engineering were in feedback control systems and so could be classed with cybernetics.

Which classification system is correct? Since concepts are by their nature neither right nor wrong but useful or not useful, the question has not been posed correctly. A better formulation would be: Which classification system is most useful? The proper answer to that would be another query: Useful for what?

---

[17]Gross and Smith, *Systems Analysis*, pp. 87–88.

Understanding systems approaches? Clarifying the various techniques? and so on. According to the authors, of all the particularlized approaches to systems thinking, *systems analysis* is by far the most diversified and least unitary.

It is not the purpose of the authors to explore each and every concept or technique that purports to qualify as a systems approach. Indeed there are many that do, and they all have some merit and validity. The ones treated in this text are those that researchers and practitioners alike have favored. Perhaps a preference for this or that approach may well be the result of an individual's disciplinary training rather than any subjective or objective bias. "One wears the clothes one is comfortable with."

SUMMARY

This chapter was concerned with presenting an overview of several of the particularized approaches to systems thinking. It was shown that systems engineering can qualify as a systems approach, even though it uses other approaches like cybernetics and other techniques like optimization models. Systems engineering has been around for several decades and today is prominent in aerospace engineering. One might consider systems engineering as the link between systems theory as such and other operational systems.

Operations research was presented as another particularized approach to systems thinking since it also employs models and calls for the quantification of the interrelationships of the system. While the authors presented a case for the inclusion of OR as a systems approach, in the final analysis they tend to concur with other writers who hold that OR is more of a technique than an approach and as such does not merit the appellation of systems thinking. This point, however, is immaterial, irrelevant to the practitioner who willingly employs any technique or any tool that can help solve a given systems problem.

Systems analysis, according to some, can also be included as a way of systems thinking. It appears that many of the definitions and expositions of systems analysis are so all-encompassing as to include almost everything: one finds it difficult to see what should be excluded. Yet in spite of these mild criticisms, if the various approaches noted in this chapter allow researchers to view their work from a holistic landscape, thus leading to a better understanding of system, then they merit the laudable badge of systems thinking.

KEY WORDS

Systems Engineering   Systems Analysis
Operations Research

REVIEW QUESTIONS

1. What does systems engineering have in common with operations research and with systems analysis?

2. Criticism of operations research revolves around its nonuse for problems of society. What factors inhibit its use in this area?

3. Systems analysis, some say, is as much an art as a science. Would you agree? Why or why not?

4. "The sort of simple explicit model which operations researchers are so proficient in using can certainly reflect most of the significant factors influencing traffic control on the George Washington Bridge, but the proportion of relevant reality which we can represent by any such model or models in studying, say a major foreign-policy decision, appears to be almost trivial" (Charles Hitch). Comment on the above statement.

5. Do the particular techniques used by operations researchers have anything in common? In other words, are there any criteria to assist us in determining whether or not a particular tool should or should not be included in the OR's tool kit?

6. "What then is operational research? There are roughly as many definitions of the subject as there are OR scientists. For these are thoughtful people, and if anyone of them lacks the temerity to formulate a definition, his place is taken by bolder colleagues who have a range of definitions to spare" (Stafford Beer). Comment.

7. How would you perform a systems analysis on the development of a new jet airliner? Would you include factors such as baggage handling at the terminals or parking facilities for automobiles in your analysis? Why or why not?

8. Some students of systems say that systems analysis is really a contradiction in terms since analysis means to break the whole down into its parts and systems means to look at an object as a whole, holistically. Is this contradiction merely a semantic one, or is there something substantively at odds here? How can one reconcile systems analysis with the systems approach?

Additional References

Clark, Raymond T., and Charles A. Prins. *Contemporary Systems Analysis and Design.* Belmont, Ca: Wadsworth Publishing Co., 1986.

Fitzgerald, Jerry, and Ardra F. Fitzgerald. *Fundamentals of Systems Analysis: Using Structural Analysis and Design Techniques.* New York: John Wiley & Sons, 1987.

Kendall, Kenneth, and Julie E. Kendall. *Systems Analysis and Design.* Englewood Cliffs, N.J.: Prentice-Hall, 1988.

Lilien, Gary L. "MS/OR: A Mid-Life Crisis." *Interfaces* 17, no. 2 (March–April 1988), pp. 35–38.

McGuire, C. B., and Roy Radner, eds. *Decision and Organization: A Volume in Honor of Jacob Marschak.* Minneapolis: University of Minnesota Press, 1986.

Optner, Stanford. *Systems Analysis for Business Management.* 3d. ed. Englewood Cliffs, N.J.: Prentice-Hall, 1975.

Peters, Thomas J. "Management Systems: The Language of Organizational Character and Competence." *Organizational Dynamics* 9, no. 1 (Summer 1980), pp. 2–26.

Sacolick, Jacob. "The Role of Operations Research in Systems Analysis." *Interfaces* 10, no. 5, (October 1980), pp. 49–54.

Satty, Thomas L., and Kevin P. Kearns. *Analytical Planning: The Organization of Systems.* Oxford, England: Pergamon Press, 1985.

Simon, Herbert, et al. "Decision Making and Problem Solving." *Interfaces* 17, no. 5 (September–October 1987), pp. 11–31.

Taha, Hamdy A. *Operations Research.* 3d. ed. New York: Macmillan, 1982.

Turner, T.P. "The Application of Operational Research in Government." *Management Services in Government* 365, no. 1 (February 1981), pp. 29–37.

Whitten, Jeffrey L.; Lonnie D. Bently; and Thomas I. M. Ho, *Systems Analysis and Design Methods.* St. Louis: Mosby, 1986.

PART THREE

Environment and Effectiveness

The world is not made up of empirical facts with the addition of the laws of nature: what we call the laws of nature are conceptual devices by which we organize our empirical knowledge and predict the future.

R. B. Braithwaite

PART OUTLINE

CONNECTIVE SUMMARY

In Part Two cybernetics was discussed at some length. There the idea was proposed that it is possible to use cybernetics to help solve complex management problems. Other techniques, including systems engineering, operations research, and systems analysis were presented as particularized approaches to systems thinking.

The thread that holds the organizational fabric together is information. While information can be treated as part and parcel of cybernetics, it merits separate attention because it serves as the language for all control systems. But information is also the linkage between the organization and the environment since the organization is also an ecosystem. To be able to handle it all, managers must be able to know what information is needed for running the system, what in the environment needs to be known for the organism to survive and thrive, how to go about acquiring environmental information, and how to score the game — that is, what criteria to use for assessing effectiveness.

Thus, we move from the methods of conceptualizing a system to some of the structural properties of systems — information, environment, environmental scanning, and organizational effectiveness. Since goals form one of the main approaches to studying organizational effectiveness, a separate chapter is devoted to the notion of goals and to problem identification and solution.

CHAPTER 6

Information

Information is the name for the content of what is exchanged with the outer world as we adjust to it, and make our adjustment felt upon it.

Norbert Wiener

CHAPTER OUTLINE

INTRODUCTION

Modern complex society presents one of the most exciting challenges of our age – the challenge to manage those richly interacting elements of government and industry. The problems of poverty, of pollution, of growth, of unemployment, and of overpopulation all pose forms of crises not adequately dealt with as yet. Like a lanky but awkward adolescent, society has grown enormously; the task at hand is to provide the proper direction, the proper regulation.

In a similar way, the task of the firm in this ever-changing society is also to provide regulation, and the essential factor required for regulation of any system is *information.* Throughout history both the government and the firm have been concerned with the acquisition of information for the purpose of generating change as well as understanding the rudimentary structure of the appropriate bodies. To be sure, both entities possess vast bureaucracies for the collection of information, if for no other purpose than to perpetuate the established order of things. The concept of information underscores the notion that something of value is being communicated to some individual or organization. Since individuals resort to multiple information sources, some type of system that filters, condenses, stores, and transmits all this information must be evolved. Without information there can be no decision making or control. Information ties together all the components of an organization – personnel, machines, money, material. Because information is the lifeblood of any system, a discussion of this subject is in order. But first some definitions.

DATA VERSUS INFORMATION

In the literature, various distinctions have been proposed, but in the final analysis they all hearken back to the original etymology of the terms used. *Data,* which is derived from the Latin verb *do, dare,* meaning "to give," is most fittingly applied to the *unstructured, uninformed* facts so copiously *given out* by the computer. Information, however, is data that have form, structure, or organization. Derived from the Latin verb *informo, informare,* meaning to "give form to," the word *information* etymologically connotes an imposition of organization upon some indeterminate mass or substratum, the imparting of form that gives life and meaning to otherwise lifeless or irrelevant matter. It is most fittingly applied to all data that have been oriented to the user through some form of organization.

Data can be generated indefinitely; they can be stored, retrieved, updated, and again filed. Assuredly, they are a marketable commodity purchased at great costs by both the public and private sectors; however, data of themselves have no intrinsic value. Yet each year the cost for data acquisition grows on the erroneous assumption that data are information. The task of acquiring data presents no obstacles whatsoever, since data are generated as a by-product of every transaction or event. The real problem is data overload, for the government, the firm, as well as for the individual. Even within departments of the government there is no paucity of societal information; rather the problem is one of data overload and data organization. It has been estimated that the American economy produces 1 million pages of new documents every minute, of which some 250 billion pages a year must be stored. Business firms alone store a trillion pieces of paper in 200 million file drawers, and each year they add 175 billion pieces of paper to this enormous amount. This paper level is further raised by the outputs of educational institutions hardly able to digest their own output.

It is not so much a problem of data acquisition as of data organization; not so much of organization as of retrieval; not so much of retrieval as of proper choice; not so much of proper choice as of identification of wants; not so much of identification of wants as of identification of needs. Obviously the problem in information management is not one of gathering, organizing, storing or retrieving data but rather one of determining the necessary information requirements for decision making.

A popular distinction among current writers restricts the label of information to *evaluated* data. Here the orientation is not so much the *function* of informed data as the explicit and specific *circumstances* surrounding the user. Accordingly, the term *data* is used to refer to materials that have not been evaluated for their worth to a specified individual in a particular situation. *Information* refers to inferentially intended material evaluated for a *particular problem,* for a specified individual, at a specific time, and for achieving a definite goal. Thus what constitutes information for one individual in a specific instance may not do so for another or even for the same individual at a different time or for a different problem. Information useful for one manager may well turn out to be totally devoid of value for another. Not only is the particular organizational level important but also the intended functional area. A production manager, for example, is typically unconcerned with sales analysis by product, territory, customer, and so on, while the person in charge of inventory control is little concerned with the conventional accounting reports that affect inventory work only indirectly. Thus, the definition here being considered is that information concerns structured data — data selected and structured with respect to problem, user, time, and place. J. P. van Gigch distinguishes between data, information, and intelligence. Data he understands in the usual sense as the input stimuli in the form of signals and messages which enter the cognitive processes — in other words, "raw data." Information is a subset of data present in the form of knowledge; and intelligence is a subset of information used for decision making or for action. In this distinction the term *information* connotes *potential* information for decision making while the term *intelligence* is reserved exclusively for the information *actually* used in the decision making process.[1]

The process whereby data become information is shown in Figure 6–1. Data in this case may be marketing data, production data, or any other type, even including external data. As mentioned above, data are unstructured, unevaluated facts having little or no meaning. It is only when data are applied to a specific

[1] John P. van Gigch, "Diagnosis and Metamodeling of Systems Failures," *Systems Practice,* vol. 1, no. 1 (1988), p. 35.

FIGURE 6–1   Data Transformation

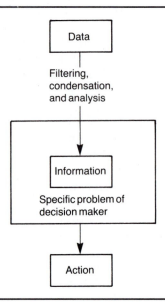

problem (evaluated) that they become information. This distinction is not without merit, for it focuses attention on a most critical problem area of management, which is *data explosion.* While the subject of data overload has been adequately treated in the literature, the magnitude of the problem is one worth noting here. Ackoff, in discussing the misinformation explosion in the 60s, succinctly stated:

> My experience indicates that most managers receive much more data (if not information) than they can possibly absorb even if they spend all of their time trying to do so. Hence, they already suffer an information overload. They must spend a great deal of their time separating the relevant from the irrelevant and searching for the kernels in the relevant documents. For example, I have found that I receive an average of 43 hours of unsolicited reading material each week. The solicited material is half again this amount. I have seen a daily stock status report that consists of approximately 600 pages of computer printout. The report is circulated daily across managers' desks. I've also seen requests for major capital expenditures that come in book size, several of which are distributed to managers each week.[2]

Thus, all information must be viewed as being imbued with relative value. Much of the so-called information utilized in management systems today enjoys a "sacred definiteness" which in reality is subject to wide ranges of both human

---

[2] Russell L. Ackoff, "Management Misinformation Systems," *Management Science,* (December 1967), pp. 147–56.

and institutional errors. Valueless data have in many instances been accepted as information simply because of an emotional investment on the part of the practitioners who have traditionally treated such data in their routine operations. For example, data that constitute information for officers preparing financial statements are not necessarily information for the production manager trying to decide on run length.

THE ECONOMICS OF INFORMATION

Historically, economists have been concerned with the allocation of resources, notably land, raw material, and labor. Optimum allocation of these is expected to lead to an efficient economy. Similarly, organizations too are very much concerned with efficiency through the proper allocation of resources. While it is true that decision making relies heavily on information and its availability, still, with the ever-increasing capability of the computer to generate endless data, the economics of doing so must be realistically considered.

Information is a resource and must be treated as such — a resource having costs and benefits associated with it. Decisions regarding the acquisition of additional information should be treated in the same way as a decision to purchase an additional machine. In other words, a comparison ought to be made of the benefits to be gained from the additional information and of the costs of the purchase. Since information is a commodity that is bought and sold, it ought to be viewed as a good. Not unlike any other good, it can age and become obsolete. Also there can be too much information or too little information. Nor should one forget that information processing — which may include acquisition, storage, transmission, and delivery to the decision maker — requires expenditures of time, resources, and facilities. Truly, information is not a free commodity.

Economists typically use marginal analysis for ascertaining whether or not it is feasible to produce additional goods. According to marginal analysis, a good or service will be consumed in a time period until the marginal cost of the last economic good or service is equal to the marginal utility of that good or service. In the case of information, the firm will continue to acquire information as long as the benefits exceed the cost. As more and more information becomes available, its usefulness decreases — that is, the utility of additional information decreases.

Also, while the utility of additional information decreases, the cost of making the information available — the cost of acquiring it — increases with each additional unit. The optimum amount of information for a manager or organization will be that amount of which the costs of acquiring one additional unit will be equal to the benefit or utility of that unit. This type of marginal analysis applied to information is also known as trade-off analysis. Figure 6–2 graphically portrays marginal analysis of information.

FIGURE 6–2   Marginal Analysis as Applied to Information

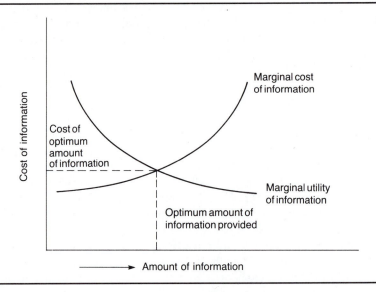

McDonough interestingly treats the supply/demand model of information in his text *Information Economics and Management Systems.*[3] He separates the cost of information and its availability into two distinct phases. In Phase I, only the acquisition of information that is required for problem definition is considered. As shown in Figure 6–3, greater amounts of information are required in the initial study period of the problem than in the latter phase. This assertion agrees with the accepted adage that a clear identification of the problem is half the solution. The leveling off of the curve is an indication of diminishing returns, and the gap between the leveling-off points of both the problem-definition curve and the information-availability curve is the point where the organization or individual decides not to acquire any more information because of either its redundancy or its cost.

Although McDonough does not discuss differential ways of acquiring information, they are implicit in the model as the present authors see it. In other words, before problems are recognized as such by the organization, there must have been some preliminary gathering of information indicating that a potential problem exists. To be sure, some of the preliminary information gathered could just as well have resulted in the decision that there was no problem in the first place.

In a later chapter the various ways of obtaining information on potential problems will be discussed. Different types of information-acquisition modes

---

[3] Adrian McDonough, *Information Economics and Management Systems* (New York: McGraw-Hill, 1963), pp. 80–82.

FIGURE 6–3   Phase I: Problem Definition and Information Availability – Their Assembly during Period of Study

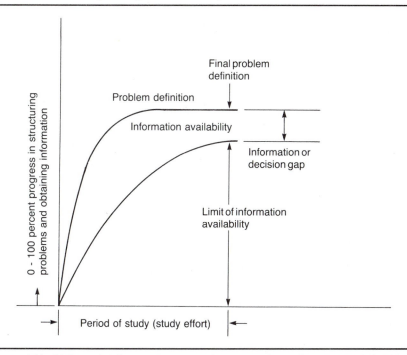

SOURCE: Adrian McDonough, *Information Economics and Management Systems* (New York: McGraw-Hill, 1963), p. 81.

(scanning) will be associated with different situational objectives. McDonough's model brings out an important aspect of information, namely, the timing element, whereas the classic marginal-analysis model assumes that all information is available at one point in time.

Phase II of the model is concerned with the interrelationship between the supply of information and the cost of the study effort. Figure 6–4 depicts this relationship. Obviously the slope of the curves and the rates of change will determine the point where one decides not to acquire further information. The reader will find it interesting to construct curves of differing slopes to see how the value of early information and its costs are associated in different problems.

As with all conceptual models, operational problems crop up whenever attempts are made to apply the models to the firm. This is true of both the marginal analysis model and the models presented above. (It is assumed, of course, that one can quantify the benefits or value of the information.) While the gathering of cost data is not too difficult a task (although seldom done), ascertaining the benefits or utility of information is less than objective. In some instances, it is fairly clear that additional information will allow the manager to make better decisions.

FIGURE 6–4   Phase II: Cost and Value Determination – Their Assembly during Period of Study

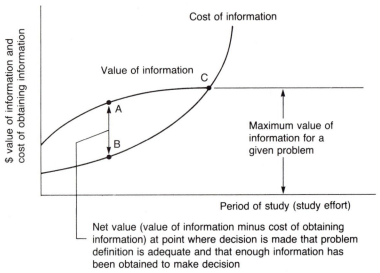

Line AB *represents the maximum net value of information; i.e., diminishing returns set in at this point of the study as costs of obtaining information increase faster than the value of information increases. At Point* C *the value of information is offset completely by the cost of obtaining the information, and there is no net value of the study.*

SOURCE: Adrian McDonough, *Information Economics and Management Systems* (New York: McGraw-Hill, 1963), p. 82.

Although many of the assumptions made regarding the value of information are not often found in the business world, this limitation does not at all vitiate the models, which are indispensable for a proper understanding of the information concept.

EXPECTED VALUE OF INFORMATION

Seldom does an organization have but a single goal; rather it has many goals, some of which concern growth, service, profit, market share, etc. Despite the intricacies of the business world and the many pressures that arise on a daily basis, the manager is expected to make sound and rational decisions. Estimating the expected value of information is a managerial technique that permits a rational approach to decision making.

Central to this theory (often called decision theory) is the determination of the value of perfect information. The purpose of information, in the decision making context, is to reduce the uncertainty surrounding the outcome resulting from a particular decision.

When formulating a decision problem, the decision maker is presented with a number of strategies for achieving a given objective, given various states of nature. The particular strategy selected will determine how successfully the manager will achieve the objective. Simply put, a decision maker chooses a strategy, and a specific state of nature will occur. This state of nature will determine the degree to which the decision maker will achieve the selected objective(s).

In decision theory, the best information possible is that which would eliminate all uncertainty and thus allow the decision maker to predict with certainty the state of nature that will occur. A simple example can be used to illustrate this point.

Assume that a decision maker has available three alternative strategies, A, B, and C, and on the basis of prior (imperfect) information he or she estimates that the outcome (payoff) for each of these alternatives is $100, $50, and $125 respectively. In this situation the decision maker would obviously choose the one alternative (C) with the highest payoff ($125). Now assume that perfect information is then provided that tells the decision maker that the payoff for B is $150 and for C, $50. This information would lead the decision maker to choose strategy B instead of strategy C, thereby increasing the payoff from $50 to $150. The value of perfect information is therefore $100. The expected value of perfect information is the difference between the expected payoff assuming that the decision maker could obtain this perfect information and would select the best alternative given that information, and the expected payoff from selecting the best alternative given the prior information. Figure 6–5 summarizes this scenario.

The value of perfect information in this example involves only one state of nature, which means that with perfect information one simply chooses the alternative that has the highest payoff.

Problems treated under conditions of certainty have but one state of nature and a probability equal to 1 ($p_1 = 1$). The decision is made as soon as the strategy is found that will produce the best outcome. Usually this strategy is selected because it produces either a maximum or a minimum value for the result.

When there are several environments or states of nature, the results will vary according to which state of nature actually occurs. Note that in the preceding example with only one state of nature there was absolute certainty. Here, however, there is uncertainty. Suppose, for instance, that there are two states of nature possible, labeled x_1 and x_2, with a probability of .60 and .40 respectively. Figure 6–6 depicts this payoff matrix.

This means that if strategy A is chosen and state x_1 occurs, the payoff will be $100. If strategy A is chosen and state x_2 occurs, the payoff will be only $90. The expected value or average payoff is the sum of the payoff for each decision (the result of the conditional value) multiplied by the probability for each outcome. Thus, the expected value for strategy A is .60($100) + .40($90) = $60 + $36 = $96.

The value of information for more than one state of nature is the difference between the maximum expected value in the absence of additional information and the maximum expected value in the presence of additional information.

FIGURE 6–5   Payoff Matrixes with Imperfect and Perfect Information

Payoff Matrix with Imperfect Information		Payoff Matrix with Perfect Information	
Strategy	Payoff	Strategy	Payoff
A	$100	A	$100
B	$ 50	B	$150
C	$125	C	$ 50

What one should note here is that (1) the expected value of information has some upper limit, and (2) if the value of the information for the decision maker does not justify the expenditure for the additional information, there is no point in obtaining it. Or put in another way, if the cost of information is less than the value of perfect information, it pays to acquire that information.

INFORMATION THEORY

While the subject of information has been of interest for many years, it was only within the past several decades that mathematicians were able to treat the subject scientifically. In some quarters, researchers still hold fast to the dictum that unless the object of discussion can be quantified, it lacks suitable description. This was behind the attempt to define the concept of information more accurately and

FIGURE 6–6   Payoff Matrix with Two States of Nature and Expected Values

Strategies	States of nature	
	x_1	x_2
	Probabilities	
	.60	.40
A	$100	$ 90
B	$ 50	$ 20
C	$125	$100

Expected Value					
A .60($100)	+	.40($ 90)	=	$ 96
B .60($ 50)	+	.40($ 20)	=	$ 38
C .60($125)	+	.40($100)	=	$115

unambiguously through mathematical analysis. Indeed, the endeavor directed to a quantification of information was given a title reflecting the efforts— *A Mathematical Theory of Communication.*[4]

One should perhaps state here that the mathematical treatment of communication has application only to a specific set of circumstances and that in discussion of information theory the term *information* is used in a very specialized sense. The mathematical theory of information evolved over a number of years as communication engineers attempted to measure the amount of information that was communicated over telephones, telegraphs, and radios. This is not to say that the mathematical concepts or techniques lack relevance in human communication, for indeed there are similarities, but *direct* application is only to the equipment itself and not to the *users*. As will be noted, while attempts to apply the formal concepts to other disciplines have not been lacking, results have been slow in coming, and after two decades of experimentation the mathematical theory of communication is still dominated by and restricted to the field of telecommunications.

Interest is in the statistical aspect of information, which stems from the view that messages that have a high probability of occurrence contain little information, and therefore any mathematical definition of information should be based on statistical analysis—i.e., the probability of that particular message being chosen from a given set of messages.

According to the classical theory, information is viewed as an entity that is neither true nor false, significant nor insignificant, reliable nor unreliable, accepted nor rejected.[5] As such, it is concerned neither with meaning nor with effectiveness. This is so because, in the transmission of signs or words, it is the signs or physical signals that are transmitted and not their meaning. Thus, information theory is associated only with quantitative aspects, the *howmuchness* of the uncertainty or ignorance reduction.

A communication system will consist of the following five elements: (1) an information source, (2) an encoder/transmitter, (3) a channel, (4) a detector, and (5) a decoder. (See Figure 6–7.)

The information source selects the desired message out of a finite set of possible messages (verbal, written, and so on). The message is then transformed into a signal (encoded) and sent over the channel. The channel is the medium used for sending messages from the source to the receiver. The detector picks up the transmitted signal. The signal is finally decoded into a message.

When one wishes to communicate there must obviously be common agreement as to the language (symbols, phonemes, and so on) to be used. Specifically, both the sender and the receiver of information must agree on the set

---

[4] Claude E. Shannon and Warren Weaver, *A Mathematical Theory of Communication* (Urbana: University of Illinois Press, 1949).

[5] See T. F. Schouten, "Ignorance, Knowledge, and Information," in *Information Theory*, ed. Colin Cherry (New York: Academic Press, 1956), pp. 37–47.

FIGURE 6–7 Communication Model

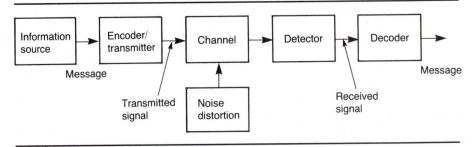

of symbols available in the language to be used. When the symbols or words that the sender selects are unknown to the receiver, no information is transmitted. This merely says that if I transmit a signal which has meaning for me but not for the intended receiver, obviously no information is transmitted. What is in my information bank is not in the receiver's. Thus, it is necessary that the sender and the receiver both know the set of possible messages from which a particular one will be selected.

It is precisely the restriction that both sender and receiver have the same information bank (source) that makes possible quantification of information. The larger the information bank of the sender and the receiver is, the greater the number of choices available and the more information needed to resolve the uncertainty. Put very simplistically, if a child's vocabulary is limited to 10 words, there is less uncertainty as to what word the child will say as compared to a grown-up whose vocabulary is less restricted. Thus, it can be seen that information, selection, and uncertainty are all interrelated.

Measuring the Amount of Information

One way to measure the amount of information in a statement is to enumerate the number of possible outcomes the statement eliminates. If only one outcome is possible, no information is required. For example, it is known that in the English language the letter q is always followed by the letter u. If we were attempting to spell any specific English word in which the letter q appears, no additional information is conveyed when we are told that the next letter is u. When there is zero uncertainty, no further information can be conveyed. Put simply, uncertainty decreases as information increases.

In measuring the amount of information, the unit used is the "bit," short for *bi*nary digi*t*. A bit is the smallest amount of information possible, and it represents a single selection between two alternatives, as between on and off, yes and no, open and closed, 0 and 1.

A central issue in information theory is the number of bits that are required (given a set number of alternatives) for the sender to communicate certain information. One example to illustrate this is to take your local telephone book and have a person choose a particular name from somewhere in the white pages. The question is, How many bits of information are required for you to choose that same name? or, to put it another way, How many bits are required for the sender to be certain that the receiver is getting the message? It was said that if only two alternatives were possible, one bit of information would be conveyed. Obviously there are many alternatives possible in this example, but still it is not difficult.

In Figure 6–8, the relationship between the number of alternative choices and the number of bits is depicted. The figure shows that, if the number of alternative choices is 4,000, then it will take 12 bits of information to completely sp
ecifiy the particular choice. If the telephone book has 500 pages, it would take no more than 9 bits of information to find out what page the particular name is on. This is simply done by continually halving the number of alternatives (decreasing uncertainty) by taking the following line of questioning.

Is the name in the front half of the book (pp. 1–250)? If the answer is yes, then the next question is, Is the name in the front fourth of the book (pp. 1–125)? It will take nine yes or no answers to specify the page number that the name is located on. The additional number of questions required will vary with the number of columns in the telephone book and the number of names vertically listed. If there are four columns, then two more additional questions are required: Is the name in the left half of the page (first two columns)? Likewise, most telephone books have approximately 100 names listed vertically, which means that it will take another seven bits of information to completely specify the name. Thus, given the above conditions, 18 bits of information are required to specify the particular name.

It was mentioned previously that at least two alternatives must be present before any information is conveyed. This is the simplest type of communication system and is termed the binary system, since for each bit two choices are required. It is for this reason that logarithms to the base 2 are used to measure the amount of information. The number of bits per alternative is then $\log_2 N$, where N is the number of alternative signs available in the entire repertoire of the sender and receiver. In the situation where only one sign (outcome) is possible ($N = 1$) it was stated earlier that no information can be conveyed. It takes at least two alternative signs to convey one bit of information. This is shown in the following manner: the amount of information $(H) = \log_2 1$ which is equal to 0. With two alternatives, such as yes and no, $N = 2$, and the number of bits (see Figure 6–8 for verification) equals one ($H = \log_2 2 = 1$). This is so because 2 raised to the first power is equal to 2. When there are four alternatives ($N = 4$) the number of bits is two.

Let us assume that each letter in the English alphabet has an equal chance of being selected. Let us also consider the blank space as another character, thus making the list of possible alternative signs employed in a message 27 (26 letters of the alphabet plus one space). The amount of information would be $H = \log_2 27$. A log table would show this to be 4.75 bits, or a quick examination of Figure 6–8 will show that for 27 alternative signs 5 bits are required. Now if we were to let n represent the number of actual signs in the message, then the number of bits in

FIGURE 6–8 The Relationship between the Number of Alternative Signs and the Number of Bits Needed to Move from Uncertainty to Certainty

Bits	0	1	2	3	4	5	6	7	8	9	10	11	12
Powers of 2	2^0	2^1	2^2	2^3	2^4	2^5	2^6	2^7	2^8	2^9	2^{10}	2^{11}	2^{12}
Number of alternative signs		1	2	4	8	16	32	64	128	256	512	1,024	2,048	4,096

the entire message (H_m) thus becomes $H_m = n \cdot log_2 N$, which is the information rate measured in bits.

In the following message, PLAY BALL, the number of bits would be $H_m = 9 \cdot log_2 27$. PLAY BALL is eight digits plus a space, which is nine. Thus, in this message 42.75 bits are required (9×4.75), or 43.

So far we have assumed that each outcome has an equal probability of being chosen. This, of course, is an oversimplification of the actual situation. Of the 26 letters in the English alphabet, not all have the same probability of appearing in English words. One would expect that the vowels would appear more frequently than some consonants. In fact, the letter *e* appears about 60 times more often than the letter *z*.[6] Consequently, the probability of the letter *z*'s occurrence is considerably less than that for the letter *e*. But note. When *z* does occur, its informational content will be greater than when *e* occurs because it does more to identify a word.

To take account of differences in the *probability* of messages, a message is assigned a probability *p* when it is selected from a predetermined set of $1/p$ messages. The *amount of information* that must be transmitted for that message is then $log_2 1/p$ or $-log_2 p$. If all *N* messages have an equal probability of being chosen, then the probability of any of these is $p = 1/N$ and then $-log_2 p = -log_2 N$. If the probabilities assigned to *N* messages are p_1, $p_2, \ldots, P_n$, then the amounts of information associated with each message are $-log_2 p_1, -log_2 p_2, \ldots, -log_2 p_n$.

Very seldom, however, is one concerned with a single message. Generally what is of interest is the capacity of the channel for generating messages and the average amount of information per message per channel. The average amount of such information from a particular source is generally given by the equation:

$$H = -(p_1 log_2 p_1 + p_2 log_2 p_2 + \ldots + p_n log_2 p_n)$$

or more simply

$$H = -\sum_{i=1}^{n} p_i log_2 p_i$$

---

[6] Schouten, "Ignorance, Knowledge, and Information," p. 36. The frequency probabilities associated with letters in the English messages are the feature of Edgar Allen Poe's classic story "The Gold Bug."

Channel Capacity

When information is transferred from one location to another, it is necessary to have a channel of some sort over which the information can travel. The measurement unit used when describing channel capacity is again the bit. The challenge is to devise efficient coding procedures that match the statistical characteristics of the information source and the channel. The upper limit of the amount of information that can be transmitted over a channel is termed the channel capacity.

For example, in a 100-words-per-minute teletype channel, it is possible to transmit 600 letters or space characters per minute, or 10 characters per second. Since the maximum information associated with one such character is 4.75 bits, the capacity of this channel is 47.5 bits per second.[7] Up to this point we have been concerned only with signals made up of discrete characters. Although dealing with continuous communication such as musical tones, video signals, speech waves, or color is somewhat more difficult and complicated, still it is not essentially different. Information theory is so general that it can accommodate any type of symbol.

INFORMATION THEORY AND ORGANIZATIONS

In recent years the many claims regarding the importance of information theory and its applicability to the theory of business organizations have bordered on the extravagant. Since 1949, when the classic information theory was first formulated by Shannon and Weaver, the literature has grown by leaps and bounds. More recently, the number of articles purporting to relate information theory to the business organization has noticeably increased. This association of information theory with business organizations undoubtedly arises from the frequent use of the word *information* in the business context. Here and there one speaks of management information systems, accounting information, the information explosion, information for decision making, information-control systems, and so on. In most of these instances, the term *information* is equated with mere data acquisition, with the quality of data, its flow through the system, its functional characteristics, and so on. However, as stated earlier, information theory as originally developed has little to do with these connotations.

It should not be forgotten that information theory was initially developed for application in telecommunications, where it is both possible and feasible to compute the amount of information that can be transmitted over a wire or a radio band. For determining channel capacity it has indeed been of significant benefit. However, its utility when applied to other disciplines has been of doubtful value and the not infrequent attempts to apply it to the business sector have been equally disappointing. In recent years some have endeavored to apply the formal theory

---

[7] Gordon Raisbeck, *Information Theory, An Introduction for Scientists and Engineers* (Cambridge, Mass.: MIT Press, 1963), p. 45.

to the fields of experimental psychology,[8] sociology,[9] decision making,[10] accounting,[11] and many other diverse situations.

Since information theory is concerned with reducing the uncertainty associated with many possible outcomes, its focus on the amount of information is understandable and its prescinding from the semantics problem is justifiable. However, in recent years scientists have elaborated on Shannon and Weaver's mathematical theory of communication that dealt primarily with quantitative technical problems in communications theory. They have also developed semantic information theories that are now being applied with more or less success to such diverse fields as linguistics, biology, psychology and sociology. One such recent theory is the Relativistic Information Theory of Guy M. Jumaire.[12] Although the results in settings other than engineering are indeed modest, it should be observed that other scientific concepts of today have also lacked a noble and auspicious origin. But for the present Rapoport states:

> However, one must admit that the gap between this sort of experimentation and questions concerning the "flow of information" through human channels is enormous. So far no theory exists, to our knowledge, which attributes any sort of unambiguous measure to this "flow." . . . If there is such a thing as semantic information, it is based on an entirely different kind of "repertoire," which itself may be different for each recipient. . . . It is misleading in a crucial sense to view "information" as something that can be poured into an empty vessel, like a fluid or even like energy.[13]

Since there is little direct application of information theory to business situations, does this mean that cybernetics cannot profitably be applied to these areas? The answer is obviously no. Just as it is necessary to have humans in many cybernetic systems to provide the feedback function, communication must also often rely on the human element to operate and control the system. Granted that the statistical aspects of information are inappropriate, this simply means that information must be treated from some

---

[8] Colin Cherry, *On Human Communication* (New York: Science Editions, 1961). Also F. C. Eric, "Information Theory in Psychology," in *A Study of Science,* Sigmund Koch, ed. (New York: McGraw-Hill, 1959).

[9] See Walter Buckley, *Sociology and Modern Systems Theory* (Englewood Cliffs, N.J.: Prentice-Hall, 1967).

[10] Russell L. Ackoff, "Toward a Behavioral Theory of Communication," in *Management Science,* vol. 4 (1957–58), pp. 218–34.

[11] Norton M. Bedford and Mohamed Onsi, "Measuring the Value of Information—an Information Theory Approach," *Management Services* (January-February 1966), pp. 15–22.

[12] Guy M. Jumaire, *Subjectivity, Information, Systems: Introduction to a Theory of Relativistic Cybernetics* (New York: Gordon and Breach, 1986).

[13] Anatol Rapoport, "The Promise and Pitfalls of Information Theory," *Behavioral Science* 1 (1965), p. 303.

other dimension to make it applicable to organizations. The remainder of this chapter will be concerned with human and organizational communication.

HUMAN AND ORGANIZATIONAL COMMUNICATION

Information has often been considered analogous to energy—one of the integrating links of the physical sciences. This does not hold, however, in the social sciences since some concepts of information theory do not apply there. For instance, in discussing human communication no mention is ever made of the principle of the conservation of information. Nor is the flow of information along organizational channels analogous to the flow of physical energy along physical channels. When physical energy flows from one point to another, it no longer can be found at its source but only at its destination. Information is essentially duplicated when it is shared with others in the organization. The information that a teacher imparts to the members of a class is not lost to the teacher in the communication process; rather it is replicated in each of the students, making the communicated information a clone, as it were, of the information of the teacher. Hence, unlike energy, information is not subject to the entropy principle.

Likewise, information and energy are basically different insofar as their subjective elements are concerned. When defining the amount of information in a message, we saw how probabilistic concepts had to be employed. When rigorously applying probability, one needs to know ahead of time all of the possible outcomes. Without this knowledge, the amount of information contained in a message remains undefined. Hence, one can hardly speak even analogously of the flow of information and the flow of energy.

In effect a conceptual gap exists between the concept of information theory in the physical sciences and the concept of information theory in the human organization. Despite this, one can still show that decision making in organizations does share many of the characteristics of information theory in the physical sciences. In a way individual members or groups of members in an organization resemble the components of a machine. This analogy holds true not because they are considered stripped of their human characteristics, but because these same human characteristics are irrelevant to communication theory as such.

Another analogy commonly found in the literature relates communication to "social cement," a substance that bonds the members together. Wiener, whose main concern with communication was a quantitative one, was nevertheless aware of other important aspects of organizational communication. He states:

> Communication is the cement that makes *organizations*. Communication alone enables a group to think together, to see together, and to act together. All sociology requires the understanding of communication.
>
> What is true for the unity of a group of people, is equally true for the individual integrity of each person. The various elements which make up each personality are in continual communication with each other and affect each other through control mechanisms which themselves have the nature of communication.

Certain aspects of the theory of communication have been considered by the engineer. While human and social communication are extremely complicated in comparison to the existing pattern of machine communication, they are subject to the same grammar; and this grammar has received the highest technical development when applied to the simpler content of the machine.[14]

Rightly suggested by the above quotes is that society's very existence is dependent upon communication. And yet, in spite of the intensive study of communication processes and the voluminous literature in existence, few substantive theories have evolved. There are several fundamental problems which serve to explain this situation. Thayer notes five conceptual difficulties which have impeded progress in the development of communication theory.[15]

Familiarity

The more familiar a concept, the more difficult it is to develop a sound empirical base for it. This fact can also be noted in the numerous definitions and connotations of the word *system* discussed throughout this text. The more popular a concept, typically the more ambiguity is possible. Another concept falling into this category is the very substance of this chapter — information. This word is used daily in a multiplicity of ways. For the accountant, the receipt of cost figures represents factual information; knowledge of a competitor's strategies may be construed as valuable information. Still, there are many other connotations given to the word. It is precisely this ambiguity surrounding the word *information* that has led to efforts to define it more rigorously.

Lack of Discipline

Thayer notes that a second difficulty in the development of a theory of communication is that it is disciplineless. No single discipline exists that purports to study communication in the main. He notes: "There are 'loose' professional associations of persons having some part interest in communication, of course, as well as academic programs built upon some special orientation; and there is undoubtedly an 'invisible college' of scholars whose scientific interests and pursuits with respect to communication do overlap to some degree. But there is nothing like the discipline foundation one sees in physics, for example."

This point is also made by Cherry: "At the time of writing, the various aspects of communication, as they are studied under the different disciplines, by no means form a unified study; there is a certain common ground which shows promise of fertility, nothing more.[16]

---

[14]Norbert Wiener, *Communication* (Cambridge, Mass.: MIT Press, 1955), p. 40.

[15]Lee Thayer, "Communication — *Sine Qua Non* of the Behavioral Sciences," *Vistas in Science* (1968), 48–51.

[16]Colin Cherry, *On Human Communication* (Cambridge, Mass.: MIT Press, 1967), p. 2.

Because communication is so basic to each discipline and is studied within its own disciplinary boundaries, the fragmented results do not extend beyond its walls and remain for the most part segmented, never adding up to more than the sum of the parts.

Scientific and Operational Approach

The haziness of the line separating theoretical aspects of communication from the pragmatic ones is still another difficulty. While research in other disciplines rarely alters social behavior, this is not always the case with communication. The point being made here is that much of the work designed for the scientific inquiry into communication inevitably becomes "rules or ways to communicate better," thus enhancing the practical knowledge of communicating. A clear delineation is required, since the operational aspects of communication differ from the scientific aspects.

Scientism and the Mystique of Technology

Another barrier, states Thayer, is the incompatability of our blind faith in scientism, on the one hand, and the nature of the communications phenomenon itself on the other. The caution expressed here is that employment of "scientific techniques" in this subject area does not ensure that worthwhile results will be forthcoming. It is quite likely that such approaches will reveal only that which is scientizable in the first instance. The application of new technology to old problems will not ipso facto solve these problems; indeed, the problems plaguing organizations today are the same ones that plagued them many years ago. The communication problems of today are identical to those centuries ago.

Basic Reconceptualization

The remaining difficulty noted is that once such an ubiquitous subject as communication becomes conceptualized, it is nigh impossible to reconceptualize it. The initial conceptualization is one of the largest obstacles to conceiving the subject in new and different ways. Thayer cites the oft-used formula $A{\rightarrow}B = X$ in which A communicates something to B with X result. Thayer maintains that the "thing" communicated is just as much a product of the receiver as it is of the sender, and in effect the message is coproduced. What is required is a basic reconceptualization of the underlying phenomena.

These, then, are some of the obstacles which stand in the way of a universal body of knowledge of communication. This is not to say, however, that there has not been any progress, but rather that the progress has been fragmented. Developments have come from many resolution levels—from the communications engineer who is basically concerned with the transmission of signals, to the behavioral scientist who is concerned with the behavioral aspects of communication.

LEVELS OF ANALYSIS OF COMMUNICATION

Although various authors present alternative approaches to the study of human communication, nearly all of them categorize them at differing levels. The particular scheme we wish to present is that of Thayer, who notes five levels of an analysis from which one can approach the study of communication:

a. the intrapersonal (the point of focus being one individual, and the dynamics of communications as such);

b. the interpersonal (the point of focus being a two or more person interactive system and its properties—the process of intercommunication and its concomitants);

c. the multi-person human enterprise level (the point of focus being the internal structure and functioning of multi-personal human enterprise);

d. the enterprise-environment level (the point of focus being the interface between human organizations and their environments); and

e. the technological level of analysis (the focus being upon the efficacy of those technologies—both hardware and software—which have evolved in the service of man's communication and intercommunication endeavors).[17]

The first level of analysis, that of the intrapersonal level, concerns itself with the individual's own physiological and mental processes. The individual acquires, processes, and consumes information about himself or herself and other events in the environment. The individual may acquire this information through reading, observation, speaking, writing, and so on. Emphasis at this level is on the inputting and processing of information by the individual, since it is here that communication occurs in the individual.

The next level of analysis is the interpersonal one, often termed intercommunication since a minimum of two people is required in the system. In this type of system, attention is focused on how individuals affect each other through intercommunication (influential level). Most of the standard texts on communication treat each of the above types in depth, and therefore attention will not be devoted to these.

The third analysis level is that of organizational communication, of particular interest to us. It is to this area that we will shortly direct our attention.

The enterprise-environment level is concerned with the ways in which the organization communicates with its environment. There have been a number of recent attempts to incorporate environmental variables into theories of organization, since the environment can be treated as information which either becomes routinely available to the firm or which the firm actively seeks. It is only through information that an organization can learn of and adapt to changes "out there" in the environment. Because this subject area is of critical importance in the study

---

[17]Thayer, "Communication," pp. 56–75.

of organization, a separate chapter is devoted to this environment-organization interaction.

The final level of analysis (technological) obviously refers to computerized management-information systems. While the mathematical theory of information of Shannon and Weaver can be included under this level, major attention must be devoted to the advances in technology which facilitate communication in organizations.

ORGANIZATIONAL COMMUNICATION AND STRUCTURE

That there is a relationship between the efficacy of the communication system of the organization and the structure of the organization is not to be denied. Horti[18] views the process of communication as "the dynamics of the organization structure" and proceeds to show that the communication system and organization structure are interdependent. He also states that the uniqueness of any organization is reflected in its structure, upon which the communication system is based. Thus, in order for a firm to remain viable, the communication system must be in balance with the organization structure.

Deutsch, in commenting on the relationship of communication and control, gives an opposing viewpoint:

> Communication and control are the decisive processes in organizations. Communication is what makes organizations cohere; control is what regulates their behavior. If we can map the pathways by which communication is communicated between different parts of an organization and by which it is applied to the behavior of the organization in relation to the outside world, we will have gone far towards understanding that organization. . . .[19]

The above suggests that organization structures follow the development of the communication system. Perhaps Deutsch was taking account of the informal channels of communication that exist and precipitate the informal organization structure. This is not to say that communication must follow the formal established channels, for, if this were the case, decision making would entail a long-drawn-out process. The existence of informal communication systems as well as informal power structures must be acknowledged, and indeed it is often through this vitalizing force that the organization succeeds. Although current organizational practice is hardly reason to state that the communication system should follow the organizational structure, this is what occurs. In the design of information systems, managers are asked what information is required for decision making in their particular functional and hierarchial positions. In other words, an implicit relationship exists between structure and information. Likewise, when a reorganization occurs in the organization, seldom if ever is this based on an analysis of

---

[18] Thomas Horti, "Organization Structure and Communication: Are They Separable?" *Systems and Procedures* (August 1968), pp. 6–10.

[19] Karl Deutsch, *The Nerves of Government* (New York: Free Press, 1966), chap. 5.

the existing communication network. Suffice to say that the organizational communication system and the organizational structure are intertwined.

The structure of the organization is important in social systems as well as in physical systems. The organization as a system embedded in a supersystem and as a system with subsystems of departments, groups, and individuals is a legitimate subject for inclusion in the study of general systems theory.[20]

Having established the legitimacy of examining human communications in organizations, we will give attention to a few of the more important concepts in the literature.

Communications has been studied by engineers, sociologists, information specialists, and management scientists. Communications has been defined as the "exchange of information between a sender and a receiver and the inference of meaning between the organizational participants."[21] This definition brings out both the "exchange of information" and the "inference of meaning." Effective communication takes place only when the receiver imputes the same meaning to the words or symbols that the sender intended.

Communication in organizations takes place at many levels and among many groups. It involves the transmission of messages among individuals and groups and its primary purpose is to inform. Figure 6–9 is merely a detailed model of the communication process presented earlier in this chapter. As noted there, communication is the process of encoding, sending, and decoding.

As regards patterns of organizational communication, one must admit that what we know about them came ultimately from laboratory experiments in small groups. These experiments have been conducted by many social science researchers to determine the effect of organizational structure on communications. The importance of these experiments was noted by Guetzkow and Simon.[22] According to them, communications essentially flow to individuals or to decision makers who process the information, arrive at a consensus, disseminate the decision to all concerned, and then implement the decision. Centralized organizational structures require that communications flow through a central person, as shown in the wheel in Figure 6–10. Decentralized structures which tend to employ the all-channel network (star), allow for freer communications between group members since anyone can communicate with anyone else; in the wheel network, one member can communicate with two neighbors.

The most efficient network appears to be the wheel since it permits communication of information without any delays as required in the all-channel and in the circle configurations. One conclusion reached by Guetzkow and Leavitt

[20] For a qualified discussion on this see Anatol Rapoport, *General System Theory: Essential Concepts and Applications* (Tunbridge, Wells, Kent: Abacus Press, 1986), pp. 156–63.

[21] Jane Gibson and Richard M. Hodgetts, *Organizational Communication: A Management Perspective* (Orlando: Academic Press, 1986), p. 4.

[22] H. Guetzkow and Herbert A. Simon, "The Impact of Certain Communication Nets upon Organizations' Performance in Task-Oriented Groups," in *Some Theories of Organizations* ed. A. H. Rubenstein and C. J. Haberstroh (Homewood, Ill.: Dorsey Press, 1966).

FIGURE 6–9   Detailed Analysis of the Communication Process

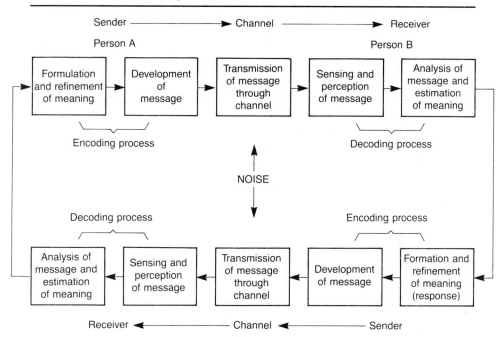

was that the differences in network were due not to the constraints on the communications channel but rather to the organizing efforts required by group members to gain consensus, to pool information, to forward it to one source, and to disseminate it to those concerned.

The results of the research on small group communications highlight the relationship between communication flows and performance. The following conclusions have been identified in a review of the research:

FIGURE 6–10   Wheel Structure

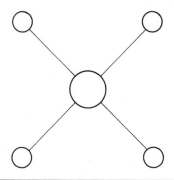

1. Centralized networks (the wheel) tend to develop single leaders who have a great deal of decision-making power. This person has access to more information and can therefore exercise a great deal of influence over group members by controlling the flow of critical information.

2. In decentralized structures, the wheel tends to produce the *least* satisfaction among group members; the open (star) communications pattern (all-channel) produces a *greater* degree of satisfaction among the group members; circle networks tend to produce the *greatest* degree of satisfaction among members.

3. Decentralized networks are more conducive to solving complex problems; centralized networks are better utilized when dealing with simple, routine tasks. Information regarding a routine task can be effectively collected by a central person and disseminated to group members only when needed.

4. Communications overload occurs more frequently in centralized networks (wheel). As the number of group members increases in a centralized network, the flow of information quickly overloads the central person's capacity to process the information.

5. Work groups will alter their communications networks in response to changes in the type of tasks they must complete. A centralized network may shift to an all-channel (star) pattern if a unique or unusual problem arises that needs special in-depth group analysis.[23]

ORGANIZATIONAL COMMUNICATION AND DECISION MAKING

Communication and decision making are inseparable, since the decision process must rely on information. Likewise, if decisions are to be carried out, they must be communicated to the people in the organization. Irrespective of how one approaches the subject of decision making — i.e., from a mathematical, statistical, psychological, or any other viewpoint — it always involves a choice among alternatives. Choosing implies that the alternatives are known, and this obviously involves communication. The range of alternatives is limited by the variety and amount of information available to the decision maker.

Decision making always takes place in an environment, which can have a weighty impact on the process. As mentioned previously and as will be discussed more thoroughly in a following chapter, an organization constantly adapts to changes in its environment. In its adaptation process, the organization makes a series of decisions based on events in the environment. In order to make such decisions, the firm must first engage in soliciting and/or receiving

---

[23] Peter P. Schoderbek, Richard A. Cosier, and John C. Aplin, *Management* (New York: Harcourt Brace Jovanovich, 1988), pp. 494–95.

information depicting the events. The entire decision-making process can be conceived of as a series of communication events. A decision is made based upon the receipt of communications from a variety of sources, including a memory bank of the organization, and is then communicated to others in the organization.

Simon defines three steps in the decision-making process: (1) the listing of all the alternative strategies, (2) the determination of all the consequences that follow upon each of these strategies, and (3) the comparative evaluation of these sets of consequences.[24] Suffice it to say that each of these involves information and communication.

COMMUNICATION PROBLEMS

1. The problem most cited in the communication literature is that of meaning, the semantic aspect of communication. A major point regarding meaning is that it is not inherent in messages and, therefore, it is not conveyed by the sender but rather is imposed by the receiver. The problem encountered, then, is to give to the message the precise meaning that the sender intended. The relationship between a word and the object that the word stands for exists only in the minds of the people using the word. That meaning rests with the individual is an important concept in all communication theory. It is through the commonality of experiences that meanings take on the same connotations. However, since persons' experiences vary widely, so does the meaning of objects and reference points. Boulding, in his oft-quoted work *The Image,* states that each experience encountered becomes a part of the individual's image of the world:

> The image is built up as a result of all past experience of the possessor of the image. Part of the image is the history of the image itself. At one stage the image, I suppose, consists of little else than an undifferentiated blur and movement. From the moment of birth, if not before, there is a constant stream of messages entering the organism from the senses. At first, these may merely be undifferentiated lights and noises. As the child grows, however, they gradually become distinguished into people and objects. The conscious image has begun. . . . Every time a message reaches him his image is likely to be changed in some degree by it, and as his image is changed his behavior patterns will be changed likewise. We must distinguish carefully between the image and the messages that reach it. The messages consist of information in the sense that they are structured experiences. *The meaning of a message is the change which it produces in the image.*[25]

Thus meaning is subject to change because of changes in our experiences. Even if two people were to experience the same situation, the meaning could not

---

[24] Herbert Simon, *Administrative Behavior* (New York: Macmillan, 1957), pp. 80–84.

[25] Kenneth Boulding, *The Image* (Ann Arbor: University of Michigan Press, Ann Arbor Paperbacks, 1956), pp. 6–7.

be the same for both of them since an experience is essentially a private affair. Of course, what we attempt to do is to utilize certain words that will stimulate recall of common experiences. But even though two people may share the identical stimuli, the response will be somewhat different. It is through words that mental associations with past experiences are recalled. Presumably if you have indulged in the contents of this text up to this point, your viewpoint of "systems" will have been altered. This alteration will be different for each individual, depending on one's particular experiences. In any case, the reader's "image" of systems will have changed.

2. A second problem noted in the literature is that of a participant's concern for the other participant's intentions. It is often heard that "it is our intentions in the use of language, not the language itself, which hinder or facilitate communication."[26] While such concerns can be useful in the communication process, they can also call forth behavior which tends to distort the intended meaning. Strategies are often encountered between superiors and subordinates which make it difficult, if not impossible, to assess true statements. We can all name an individual who we know tends to exaggerate, and in order to put related facts into a proper perspective (from our standpoint) we tend to discount some or much of such a person's statements. The individual aware of communication bias may in turn introduce a counterbias.

3. The interpersonal level of communication is not to be confused with the organizational communication system. Although individuals may be effective in their communications with other individuals in the organization, one should not mistakenly identify these interpersonal communications with a viable organization communication system. The danger is in thinking that other communication problems which may arise are soluble at the interpersonal level when in reality they are communication-systems problems.

4. Since organization structure and communication systems are interdependent, an elimination of the structural problems in the organization will also eliminate the communication problems. An inevitable fact is that the hierarchical structure in an organization must terminate in one focal point of authority. It does not follow that such a situation is problematical. Simon states: "Organization is innately hierarchical in structure; to the extent that one looks upon certain communicative consequences of that structure as 'problematical,' he is destined to deal with symptoms rather than causes."[27]

These are but a few of the organizational issues and problems encountered in the workaday world. Many other texts treat these and other communication issues in greater depth. Our main purpose here has been to elucidate several of the major problems encountered in organizations and to set the stage for the interaction of the organization and the manner that the organization adapts to changes via an "information" system.

---

[26] P. Meredith, *Instruments of Communication* (London: Pergamon Press, 1965), p. 36.

[27] Herbert Simon, *The New Science of Management Decisions* (New York: Harper & Row, 1960).

SUMMARY

The purpose of this chapter was twofold, first to present the concept of information theory and to discuss its main elements, and its area of applications; second, to discuss the more encompassing field of communication theory. The relationship between organizational structure and the communication system was treated, as well as the inseparability of decision making and the communication system.

Finally, some organization issues and problems were briefly mentioned.

KEY WORDS

Data   Bit
Information   Channel Capacity
Expected Value   Communications
Information Theory   Data Explosion
Intelligence   Scientism
Perfect Information

REVIEW QUESTIONS

1. Discuss how information is basically a statistical concept.
2. How many bits of information are required to specify the following messages? (Assume equal probabilities for each letter.)

   you passed the course

   do not quit

   be prepared

3. What factors limit the use of information theory in human communication?
4. In many textbooks on communication problems, a frequently cited one is the semantic (meaning) problem. Show how this is related to the communication model presented in this chapter.
5. Using an 8 × 8 matrix, ask a friend to specify a square (a number) within the matrix. After this has been done in the conventional way, reverse the procedure and show how you would do it.
6. International symbols are used in all Olympic games; international traffic signs are also increasingly used. Explain this in terms of communication theory.
7. Some people might argue that the field of communication is not contentless, but it is multidisciplinary. How would you harmonize these two statements?
8. List some gestures used by the following people that have a different meaning for the general populace.
   a. a baseball coach
   b. a football referee .
   c. a truck driver
   d. a student

9. How might the structure of an organization inhibit communication within it? Are there any organizational structures which are more suitable for communicating, or is communication an individualized problem of the person communicating?

10. Thayer notes that the more familiar a word or concept, the more ambiguous it is. Name five words or concepts that are very ambiguous. Give words or concepts in which there is little ambiguity.

11. Sasa Dekleva, a Yugoslavian basketball player once said: "Saying what you mean, and meaning what you say are as different as saying all Russian bears are the same." What do you think he meant?

Additional References

Barker, William H., II. "Information Theory and the Optimal Detection Search." *Operations Research* 125, no. 2 (March-April 1977), pp. 304–11.

den Hertog, J. Frisco. "Information and Control Systems: Roadblock of Bridge to Renewal?" *Data Base* 13, nos. 2 and 3 (Winter-Spring 1982), pp. 26–39.

Dow, John T. "A Metatheory for the Development of a Science of Information." *Journal of the American Society for Information Science* 28, no. 6 (November 1977), pp. 323–31.

Feldman, Martha S., and James G. March. "Information in Organizations as Signal and Symbol." *Administrative Science Quarterly* 26, no. 2 (June 1981), pp. 171–86.

Hilton, Ronald W., and Robert J. Swieringa. "Perception of Initial Uncertainty as a Determinant of Information Value." *Journal of Accounting Research* 19, no. 1 (Spring 1981), pp. 109–19.

Langlois, Richard N. "Systems Theory and the Meaning of Information." *Journal of the American Society for Information Science* 33, no. 6 (November 1982), pp. 395–99.

Lavalle, Irving H. "On Value and Strategic Role of Information in Semi-Normalized Decisions." *Operations Research* 28, no. 1 (January-February 1980), pp. 129–38.

McCarthy, Patrick S. "On the Use of Information Theory in Mode Choice Analysis." *Logistics & Transportation Review* 17, no. 3 (1981), pp. 327–42.

Mansfield, Una. "The Systems Movement: An Overview for Information Scientists." *Journal of the American Society for Information Science* 33, no. 6 (November 1982), pp. 375–82.

Schell, George P. "Establishing the Value of Information Systems." *Interfaces* 16, no. 3 (May-June 1986), pp. 82–89.

Schneider, Susan C. "Information Overload: Causes and Consequences," *Human Systems Management* 7 (1987), pp. 143–53.

Smith, Allan N., and Donald B. Medley. *Information Resources Management.* Cincinnati: Southwestern Publishing Co., 1987.

Thayer, Lee, ed. *Organizational Communications: Emerging Perspectives.* Norwood, N. J.: Ablex Publishing Co., 1986.

Tom, Paul L. *Managing Information as a Corporate Resource.* Glenview, Ill.: Scott, Foresman and Co., 1987.

CHAPTER 7

Organizational Environment

We have modified our environment so radically that we must modify ourselves in order to exist in this new environment.

Norbert Wiener

CHAPTER OUTLINE

INTRODUCTION

Recently, growing attention has been directed toward the relationship between phenomena of the external environment and phenomena internal to the organization. Organizations are viewed as transacting with environmental elements through the importing and exporting of people, material, energy, and information.

Stogdill maintains that an organization is in part a product of its physical and cultural environment, that it engages in an exchange with its environment, and that the viability of an organization is firmly rooted in the relationships that it maintains with its environment. Specifically, he notes that

> The physical environment and the nature of the resources available may place constraints upon the kinds of activities in which the organization can engage. The societal environment may prescribe the aims and structure of organization, as well as the right to organize. . . . The physical media of exchange will be determined in part by the social value placed on the available materials by the members of the larger society. . . . The survival of utilitarian organizations depends upon their ability to extract from the physical environment those materials necessary to sustain their operations.[1]

The need to incorporate environmental variables into the study of organizations is commonly accepted. But how one proceeds to operationalize these variables is open to much discussion.

Most definitions of the environment have been largely subjective. As some authors envision it, the environment is an arbitrary invention of organizations, since the firm itself chooses from a large variety of environmental dimensions those elements it defines as relevant. Two executives, or even two researchers for that matter, can arrive at different determinations of the environment for the same organization, even when viewing it from the same hierarchical level. That this criticism is valid hardly anyone will deny. Furthermore, this very same criticism can with equal ease and with equal force be applied to the topography of all organizational theory. Yet one would not willfully reject a study of organization simply because it lacked a conceptual framework acceptable to all. Consensual acceptance is no substitute for conceptual utility.

Underlying all organizational research are the implicit assumptions that (1) various dimensions of organizational environments *exist,* (2) that these various environments are more or less *identifiable* and researchable, and (3) that specific kinds of environment are *associated with* specific kinds of organizations. Without these basic assumptions, little progress would be possible in our understanding of the behavior of organizations.

Some of the studies attempting to clarify measures of environmental uncertainty have depended upon "subjective" data obtained from members of the organizations under study. While some researchers would term such data-

---

[1] Ralph M. Stogdill, "Dimensions of Organization Theory," in *Approaches to Organizational Design* ed. James D. Thompson (Pittsburgh, Pa.: University of Pittsburgh Press, 1966), pp. 40–41.

collection methods "weak" because they record subjective assessments of situations by organization members rather than the "objective" situation itself, the very same line of reasoning can be applied to researchers who "subjectively" determine which are the relevant properties of organizations and which are the relevant dimensions of the environment. Whose subjective perception of the environment is the valid one? Whose subjective perception of the environment coincides with the objective reality out there? Or is it even proper to ask such a question? Is not the *perceived* environment the environment that the decision maker reacts to in the final analysis?

Prescinding from the metaphysical implications of this philosophical problem, one notes that authors in examining the environment of an organization often treat it as if it were a *singular entity*. If the philosophical axiom that "function follows nature" is accepted, then one would also expect that a multidimensional organization would be immersed in multidimensional environments. That is why it is difficult to envision a singular specific environment for a firm such as General Electric, which has over 12,000 different products, or for Westinghouse with over 9,000. These eminently diversified corporations are structured around a vast array of products coming from divisions that report to a group vice president. Different divisions have their own separate production facilities and their own sales forces; they have their own suppliers; they have their own computers. Each has its own *task environment.*

Because of differing task environments, it is not always easy to answer the rather simple but direct question posed by Peter Drucker in many of his consultations: "What is your business?" With proliferating mergers, and acquisitions of companies with seemingly little in common regarding manufacturing, distribution, technical expertise, etc., the lines of a business are often quite blurred nowadays. It would be very hard for the president and other executives of ITT to say just what business they are in, since the parent company comprises over 200 companies, or for General Electric or Westinghouse to unequivocally state that they are in such-and-such a business. Complexity must be handled by complexity!

Now when one considers the various facets ascribed by organization scholars to the environment of divisions of large organizations — dimensions such as complexity, uncertainty, and change — one cannot help but notice the critical role of the assumptions underlying all organizational research. Different environments do exist. They certainly can be identified. Specific kinds of environments are associated with specific kinds of organizations and industries. The environment of the division of General Electric that produces nuclear reactors differs substantially from that of the more mature divisions producing home appliances. Their environments are different; they have been identified, hence are identifiable; and the environment of a division producing nuclear reactors would be quite similar to that of other companies producing nuclear reactors.

Because of the nonsingular nature of environments, one should not expect that all the environments of an organization change at one and the same time or change at a uniform rate. Environmental change, not unlike organizational change, can be partial only, or total. Therefore, in treating environments, one should be aware that the real environment can and often should be segmented. In Chapter 1 a diagram was presented of the organization/ environment interaction, and the variables spanning these areas were termed *boundary variables.* Some of them were partly in the system and partly in the environment. Over time they could move one way or the other. Labor, for example, is typically a systemic variable; however, during contract negotiations, it should be considered an environmental variable — something outside the company's control but significantly affecting the company's operations. Likewise, a new products technology that one has just developed is for the corporation a systems variable; for companies that do not share it, it is an environmental variable, something they may have to contend with. Some environmental elements may remain static, while other elements may be dynamic at particular instances of time. Indeed, these same elements may in time change places.

In short, the environment should not be conceived of in a unitary way. Because it may be multidimensional, the complexity, uncertainty, and change ascribed to it should not be viewed as being of one and the same fabric. If this is kept in mind, many of the obstacles to a clear understanding of organizational environment may easily be overcome.

ENVIRONMENTAL UNCERTAINTY

A major dimension of the environment is uncertainty. Environmental uncertainty appears to be a result of three conditions: (1) the lack of information regarding the environmental factors associated with a given decision-making situation, (2) the inability to assign probabilities with any degree of confidence with regard to how environmental factors will affect the success or failure associated with a particular organizational decision-making unit, and (3) the lack of information regarding the costs associated with an incorrect decision or action.[2]

Numerous authors point out that the certainty-uncertainty dimension is not really a dimension of the *environment;* rather, it is a characteristic of the decision maker's *perception* of information concerning the environment.

The fact that different decision makers perceive the environment differently obviously affects their decisions. When executives perceive information about the environment incorrectly, or when they fail to receive the information in sufficient time to act upon it, or when they fail to implement an appropriate strategy for lack of expertise, then their decision can have negative and harmful effects.

Jurkovich notes three situations in which the environment could be termed *nonroutine:* (1) when people complain that they cannot gain access to critical

---

[2] Robert B. Duncan, "Characteristics of Organizational Environments and Perceived Environmental Uncertainty," *Administrative Science Quarterly* 17, no. 3 (September 1972), 313–27.

information needed to make decisions, (2) there is doubt regarding the reliability of a significant portion of the information, and (3) the decision maker is uncertain regarding the set of information categories required.[3]

Although Emery and Trist[4] (to be discussed later in this chapter) are mainly concerned with change as the most important of the environmental constituents, still their analysis centers about uncertainty, which they treat as the dominant characteristic of a turbulent environment.

Likewise, Lawrence and Lorsch[5] (to be discussed later in this chapter) have contended that dissimilar environments call for different organizational behaviors. Their environmental questionnaire purported to measure uncertainty in three sectors of the environment, namely, the marketing, manufacturing, and research sectors. Responses were sought to questions of how the respondents perceived (1) the lack of clarity of information, (2) the general uncertainty of cause-and-effect relationships, and (3) the lack of definitive span of feedback about results related to each functional sector of the environment.

Duncan,[6] when operationalizing the uncertainty concept, adopted the first two components of Lawrence and Lorsch's analysis and added a third of his own. The first two components of operationalized uncertainty were (1) the lack of information about the environmental factors associated with a given decision-making situation, and (2) the lack of knowledge about how much the organization would lose if a specific decision was incorrect. His third component was the inability to assign probabilities with any degree of confidence with regard to how environmental factors will affect the success or failure of the decision-making unit in performing its function.

Duncan's model purports to integrate the change and complexity dimensions of organizational environments with that of uncertainty. Figure 7–1 shows that model. As one would expect in such a model, when change is rapid and there are a large number of diverse elements in the environment, there is high perceived uncertainty (Cell 4). When there are but few factors making up the environment, and these are quite similar and undergo few if any changes, then there is low perceived uncertainty (Cell 1).

In the chapter on system fundamentals, the complexity of a system was defined by the number of system components and the interrelationships among these components. There it was stated that, generally speaking, the more components the more complex the system, given an increase in diversity with an increase in numbers. Duncan, it appears, has dipped into the systems literature

---

[3] R. Jurkovich, "A Core Typology of Organizational Environment," *Administrative Science Quarterly* 19, no. 3 (September 1974), pp. 380–94.

[4] F. E. Emery and E. L. Trist, "The Causal Texture of Organizational Environments," *Human Relations* 18 (1965), pp. 21–32.

[5] P. Lawrence and J. Lorsch, *Organization and Environment: Managing Differentiation and Integration* (Cambridge, Mass.: Division of Research, Graduate School of Business Administration, Harvard University, 1967).

[6] Duncan, "Characteristics of Organizational Environments," p. 318.

FIGURE 7–1   Environmental State Dimensions and Predicted Perceived Uncertainty
Experienced by Individuals in Decision Units

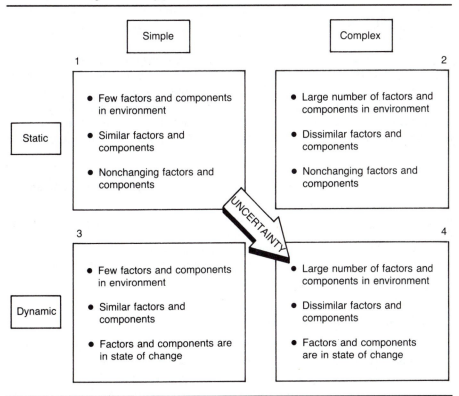

SOURCE: Adapted from R. B. Duncan, "Characteristics of Organizational Environments, and Perceived Environmental Uncertainty," *Administrative Science Quarterly* 17, no. 3 (September 1972) p. 320. Reprinted with permission.

to formulate his environmental model. The present authors feel that his model
does aid one in understanding environmental types. His discussion of change and
complexity in the environment is really a discussion of uncertainty, for the latter
follows the former. A rapidly changing environment or a very complex
environment will almost inevitably also be an uncertain environment.

According to James Thompson, "Uncertainty appears as the fundamental
problem for complex organizations and coping with uncertainty is the essence of
the administrative process."[7]

In recent years this has been demonstrated on a worldwide basis. The lack
of knowledge concerning the intentions of the OPEC countries regarding
increases in the cost of crude oil created uncertainty, not only for automobile

---

[7]James D. Thompson, *Organizations in Action: Social Science Bases of Administrative Theory* (New
York: McGraw-Hill, 1967), p. 159.

manufacturers throughout the world, but also for countless others who would be burdened with increased costs. And the earlier oil embargo by the Middle East oil producers made for an uncertain future for many governments. They did not know when the embargo would be lifted, if the price would be acceptable, if enough oil would be purchased, if a war would break out over the matter, and if retaliatory policies would be initiated by the oil-consuming nations. Because uncertainty renders decision making difficult and is disruptive of normal operations, organizations will devote more and more energy to reducing the level of uncertainty in the environment. Those organizations marked by mass-production technology will integrate vertically the better to control the sources of raw materials and channels of distribution. Organizations with an intensive technology, like the intensive care unit of a hospital, will try to acquire as much control as possible over the object worked upon, be it the patient or the X-ray machine. Organizations with a mediating technology — a technology that mediates, that connects the different users of a service or a product (like a telephone network) — will tend to diversify the better to avoid too great a dependence on any one market.

Thompson's 2×2 paradigm of organizational adaptation to the environment is given in Figure 7–2. Note that the unstable (changing) environments are characterized by uncertainty.

ENVIRONMENTAL CHANGE

As shown previously, the three major dimensions of the environment singled out by many researchers are complexity, uncertainty, and change. As with the uncertainty dimension, change too embraces a number of different aspects. Zey-Ferrell notes three different aspects of change: (1) the frequency of change in relevant environmental activities (rate), (2) the degree of difference involved in each change (variability), and (3) the degree of irregularity in the overall pattern of change (instability).[8] Change then may be due to an increased rate, to variability, or to the instability of the elements making up the relevant environment.

The reader may recall that rates and flows were discussed in systems dynamics, where the rate of change was an important systems variable. So too is it here. A high rate of change would require that the organization react quickly and often.

Jurkovich notes four types of environmental states: low stable change, high stable change, low unstable change, and high unstable change.[9] Note that Jurkovich adds the dimension of stability/instability to the element of change. In other words, not only is the rate of change important, but so too is the stability of the environment. For example, when the change rate is low and stable, the

---

[8] Mary Zey-Ferrell, *Dimensions of Organizations* (Santa Monica, Calif.: Goodyear Publishing, 1979), pp. 90–91.

[9] Jurkovich, "Core Typology of Organizational Environment," pp. 380–94.

FIGURE 7–2   Organizational Adaptation to Environment

	Stable Environment	*Unstable Environment*
Homogeneous Environment	I	III
Heterogeneous Environment	II	IV

Cell I   Typified by a 7-Eleven store near a gasoline station (convenience store).
Characterized by rules and categories for applying rules.

Cell II   Typified by a large department store.
Characterized by rules and categories for applying rules.

Cell III   Typified by EDP systems.
Characterized by uncertainty absorption (monitoring), contingency planning, decentralized decision making.

Cell IV   Typified by complex corporations like General Motors, IBM, etc.
Characterized by uncertainty absorption (monitoring), contingency planning, decentralized decision making.

SOURCE: James D. Thompson, *Organizations in Action: Social Science Bases of Administrative Theory* (New York: McGraw-Hill, 1967), p. 72.

organization has good control over its outcomes because it is able to anticipate the future and plan for it. When the change rate is low but unstable, the situation for the firm, he believes, is problematical — that is, although change is anticipated, its timing cannot be determined with any degree of certitude. High change rates and stable environments allow the organization to predict the rate of change, but they require quick response times. These rapid changes may not allow the organization to restructure its activities for goal attainment to take place; rather, a frequent restructuring of goals may result instead. In a situation of rapid change in an unstable environment, great unpredictability (uncertainty) erupts. Here is the situation where the organization must readily adapt its structure and processes to the environmental configuration if it is to survive. Organizations (as will be seen in the following chapter on scanning the environment) develop "busy" scanning systems that attempt to gather relevant information about these environmental changes. The information is sought out from the various segments of the environment either formally or informally. Such activity can involve anything from lobbying to setting up formal marketing research departments and even to outright industrial spying and stealing of patents and secrets. All of this scanning is undertaken to reduce the amount of uncertainty and to improve predictability.

Emery and Trist[10] deal primarily with the problem of change. They note that the environmental contexts in which organizations exist are themselves changing

[10] Emery and Trist, "Causal Texture of Organizational Environments," pp. 21–32.

under the impact of new technologies at an ever-increasing rate and toward increasing complexity. The better to understand the heretofore neglected processes in the environment that eventually become the determining conditions for the organization—the environment-organization exchange process—the concept of causal texture of the environment, introduced originally by Tolman and Brunswick,[11] is reintroduced for analytical purposes.

The two-way exchange between the environment and the organization can be depicted in the matrix shown in Figure 7–3. The Ls in the matrix indicate some potentially lawful connection; the subscript 1 refers to the organization and the subscript 2 to the environment. Thus L_{11} refers to processes solely within the organization; L_{12} and L_{21} refer to exchanges between the organization and environment; and L_{22} refers to processes solely within the environment—that is, its causal texture. This is the area of environmental interdependencies, just as L_{11} is the area of organizational interdependencies.

In considering these environmental interdependencies, two points ought to be noted. First, the laws connecting the parts of the environment to each other are often *incommensurate* with those connecting parts of the organization to each other, or even with those governing the organization-environment interchanges. Because of this, executives of organizations often face great dangers and experience serious difficulties arising from the rapid and gross increase in this area of relevant uncertainty. Second, environments of organizations differ in their causal texture regarding degrees of uncertainty, stability, clustering, etc.

Emery and Trist propose a fourfold typology—"ideal types" that approximate what exists in the real world of most organizations. These four ideal types of environment are:

1. The placid, unstructured randomized environment.
2. The placid, clustered environment.
3. The disturbed-reactive environment.
4. The turbulent fields.

The first type is the simplest environment, which does not require any specific goal structuring for organizational survival. Because of the random distribution of goals and *noxiants* ("goods" and "bads"), each member of the organization must act, to a large extent, according to his or her own perceptions. Here no distinction holds between strategy and tactics, the best strategy being simply trying to do one's best here and now. The trial-and-error method prevails.

The second type of placid environment is somewhat more complex than the first. It is characterized by clustering: its goals and noxiants are no longer randomly distributed; they somehow "hang together." They are serially related in much the same way that serial inputs are in systems. Clustering leads ultimately to goal orientation, to role definition, and to some degree of centralized coordination.

---

[11] E. C. Tolman and E. Brunswick, "The Organism and the Causal Texture of the Environment," *Psychological Review* 42 (1935), pp. 43–77.

FIGURE 7–3   The Environment-Organization Interaction Matrix

Inputs \\ Outputs	Organization	Environment
Organization	L_{11}	L_{12}
Environment	L_{21}	L_{22}

Here strategy emerges distinct from tactics, and survival is linked to what the organization knows of its environment. Organizations with Type 2 environments tend to grow in size, to become pyramidal (hierarchical), and to centralize their control and coordination.

The third type of environment is called the disturbed-reactive environment. It is no longer static like the first two but dynamic. It too is a clustered environment but one with many organizations similar to one's own. Each organization now must not only take account of its competitors in the field but must also realize that what it knows can also be known by others. Consequently, organizations with this type of environment tend to hinder the others from reaching similar or identical goals. They employ both strategies and tactics, and also "operations" — actions whereby one draws off one's competitors. Control becomes more decentralized, the better to deal with one's competitors. This third type of environment is characteristic of oligopolies in which the reaction of major market participants must be evaluated in policy decisions. The organization needs a considerable amount of long-range planning as well as short-run flexibility to respond rapidly to each and every threat to its market position or survival.

The fourth type of environment is characterized by complexity, by rapidity of change in the causal interconnections in the environment as well as in the field itself. The "ground" is, as it were, in motion. The emergence of turbulent conditions such as rapid product development and obsolescence can be seen in the environments of many formal organizations, particularly in the computer industry, where the turbulence is associated with an increasing degree of complexity worldwide and with rapid changes in external relationships to other firms and to other industries.

According to Emery and Trist, three trends have contributed to the emergence of forces in the turbulent fields:

1. The growth to meet Type 3 conditions of organizations and linked sets of organizations, so large that their actions are both strong enough and persistent enough to induce fundamental processes in the environment.

2. The ever-growing interdependence of the economic facet with the other facets of society. This means that for-profit organizations are becoming more and more enmeshed in governmental legislation and in public regulation.

3. The increased reliance on research and development to be able to

meet competitive challenges. This leads to a situation in which a change gradient is continuously present in the environmental field.

For organizations, these trends mean a gross increase in their area of relevant uncertainty. Consequently organizations find it nearly impossible to predict and to control the compounding consequences of their actions.

ENVIRONMENTAL COMPLEXITY

Just as a system increases in complexity with the addition of diverse components, so too does the environment. Duncan's complexity embraces but two determinants. He defines environmental complexity as the degree to which the elements in the focal environment are both (1) great in number and (2) dissimilar to one another (heterogeneity).[12] Child, too, before him had defined complexity by these same two components—range and diversity.[13] Duncan, however, arrived at an environmental complexity index by multiplying the number of decision factors by the number of environmental components or elements. This complexity index gives the decision maker an idea of the number of factors that have to be considered in making decisions respecting the organization-environment interface.

Whether or not this environmental complexity index serves a useful purpose is for practitioners (decision makers) to find out. It does, however, smack of Graicunas's now-abandoned quantitative approach to the determination of the maximum span of control. As we shall presently see, complexity may be a little more complex than Duncan's two determinants show.

Osborn and Hunt defined complexity in terms of three interrelated variables: (1) the amount of risk involved in organization-environment relations; (2) environmental dependency, or the degree to which an organization relies on elements in the environment for its own survival and growth; and (3) the degree of favorableness of interorganizational relationships.[14] Their study of 26 small service organizations focused on the relationship between environmental complexity and organizational effectiveness. Unfortunately, as Osborn and Hunt themselves point out, the rigidity of the structures of these quite similar service organizations was such that they probably provided a very poor measure of environmental risk, the first of the three complexity variables. In other words, the lack of heterogeneity in the organizations was probably associated with the lack of heterogeneity in the environment. Hence, the similarity or homogeneity of the risks encountered by the organization in its interface with the environment.

---

[12] Duncan, "Characteristics of Organizational Environments," p. 325.

[13] John Child, "Organizational Structure, Environment and Performance: The Role of Strategic Choice," *Sociology* 6, no. 1 (January 1972), pp. 2–22.

[14] Richard N. Osborn and James G. Hunt, "Environment and Organizational Effectiveness," *Administrative Science Quarterly* 19, no. 2 (June 1974), pp. 231–46.

Environments illustrative of varying degrees of complexity would be those of the computer and the appliance industries. While complexity itself may have numerous subdimensions, some indicators of complexity might be technological innovation, rates of change, number of factors that affect the industry, number of new product innovations, and the growth of the industry relative to that of the economy as a whole. There would be little doubt that if one developed objective measures of dynamic environments, the computer industry would be found to be operating in such an environment.

On the other hand, the authors suspect that the appliance industry is less complex on all of the above criteria, although this has not been verified. It would also be found to be operating in a more stable and less dynamic environment. A number of studies which dichotomize environments into dynamic versus stable employ some of the criteria mentioned above.

A number of authors highlight the point that a major dimension of complexity is heterogeneity. This appears so obvious that the point need not be pressed. Still, a word is in order the better to situate this concept in its proper context. True, the addition of similar elements in the environment does not necessarily add to complexity. For example, the addition of another competitor selling the same items as one's own organization does not of itself make the environment more complex. The same would hold for the addition of another outlet of the parent company at another location.

The concept of complexity, the reader will note, has already been treated in the discussion of system characteristics in Chapter Three. There it was pointed out that when treating complexity from a nonquantitative viewpoint (in contrast to the quantitative or statistical approach), complexity may be conceived as the property that results from the interaction of these four determinants: (1) the number of elements involved, (2) the attributes of these elements, (3) the number of interactions among these elements, and (4) the degree of organization among these elements. Far too often, complexity is predicated on just two of these determinants: (1) the number of elements and (3) the number of interactions among the elements. Determinants (2) and (4) are too often overlooked. Perhaps the reason for this is that these two latter determinants are *qualitative* while (1) and (3) are *quantitative.* But it is precisely these qualitative criteria that add immeasurably to complexity. And of these two, the one relating to the attributes of the elements should be singled out. Perhaps among the attributes one should specifically list heterogeneity, stability, and uncertainty. Such a listing would again show that the three main determinants of environments, namely, complexity, uncertainty, and change, are all interrelated. And unless the researcher includes these three, the treatment will be superficial.

SOME CONCEPTUAL AND OPERATIONAL PROBLEMS

That controversy exists in this field is to be expected, and, as with most fields of endeavor, there are those who challenge the appropriateness of the concepts and those who question the attempts at operationalizing the concepts. Not surprisingly, as in most social science areas, some of the criticism centers on measurement problems, but not all of it. Here we will briefly touch upon some of these problem areas.

1. There is no unanimity regarding the nature and extent of the environment's impact upon the organization. That the environment affects the organization no one will deny, but students of organization cannot agree among themselves on the nature, extent, and effects of the dependency nor on its causal direction. Besides, other variables may come into play which moderate the effect of this dependency. Interorganizational factors (e.g., dependency of one organization on another − a form of organizational-environmental dependency) also may account for the observed phenomena. At the present time, one can very likely find studies that support one's own position on this subject, whatever it may be.

It will be recalled that in the definition of system in Chapter One, six key concepts were involved and among these was that of environment. In explaining this concept two characteristics were highlighted: (1) the environment lies outside the system's control, more or less, and (2) the environment exerts considerable or significant influence on the system, thus determining, more or less, its behavior or performance. The third characteristic − the extent to which the system in turn influences or determines the environment − was not explicated. In the literature the focus appears to be more on how the environment influences the organization or how the environment serves as the locus of its inputs, outputs, or both. The control function of management, however, deals extensively with this aspect of the problem.

2. A second problem, that of the nonunitary nature of environments, has already been touched upon earlier in this chapter. It is precisely the failure to comprehend this concept, the failure to realize that there may be more than one segment to the organization's environment, that lies at the heart of understanding and applying systems. An organization's environment may be segmented, each segment having its own degree of complexity, change, and uncertainty. Failure to realize this can short-circuit an otherwise good research study.

3. A third problem centers around the lack of consensus regarding the relevant dimensions of environment. When examining the environment, the authors have chosen to include these three: complexity, uncertainty, and change. One may ask: What criteria were employed in arriving at these three? Why three? Why not four or five? And just how does one develop specific measures for each of these? Perhaps comments made in an earlier chapter are appropriate here. The more general the concept, the greater the extent of its applicability, the greater its usefulness. The more specific the concept, the fewer the types of organizations to which it can be applied. And so here: the fewer the concepts, the greater the range of applicability.

4. The measurement problem is naturally related to conceptual considerations. In a number of studies in this field, researchers have employed subjective measurements for the assessment of the uncertainty dimension of the environment. In empirical studies, subjective measurements are less desirable than objective ones, for they weakly and inadequately satisfy the requirements of scientific rigor. Yet not everything that masquerades under an objective guise is what it seems to be. Other methodologies may be equally open to the same criticism. Nor does the fact that one employs methodology-tight procedures ensure that the results will be worthwhile. A house is as good as its foundation, and a methodology based on a weak conceptual foundation will not long withstand the blasts of critics.

5. Because the environment has been conceptualized as being invested with three dimensions, it is not only impossible to ascertain the adequacy of the chosen dimensions, but it is also difficult to separate out their synergistic effects. Other factors could well interact in a completely different manner. For example, change itself may not be a critical factor when it occurs in a stable environment. Thus, no one questions that automobile manufacturers must consider the effect of the energy shortage on their business. Such change was imminent a number of years ago, and so the firms have had ample opportunity to plan for such changes. Likewise, competition arising from the introduction of foreign cars on the American market did not appear overnight. American manufacturers had plenty of time to draw up policies and to develop strategies for dealing with this.

SOME SELECTED RESEARCH RESULTS

Space limitations do not permit even a cursory discussion of the many and varied studies on organization-environment phenomena. As a result, only a few studies will be mentioned here. We have tried to choose some that will be sufficiently representative of the vast bulk of literature on this topic. As noted more than once in this chapter, the dimensions generally associated with environment are complexity, change, and uncertainty. Not too surprisingly, the focus is on how these various environmental dimensions are related to other organizational variables, often organizational effectiveness. Unfortunately, findings lack consistency even when the same environmental dimensions are used for study, sometimes because of variations in conceptualization, often because of diversity in operationalization. Nevertheless, a short review of some of the major studies follows. The reader is encouraged to consult the original works for an in-depth treatment of the studies.

Burns and Stalker[15] conducted a survey of 20 firms in the United Kingdom, drawn from various industries and operating in both relatively stable and dynamically changing environments. The specific characteristics that they

---

[15] Tom Burns and G. M. Stalker, *The Management of Innovation* (London: Tavistock Publications, 1961), p. 72.

examined were the rate of change in the scientific techniques and the markets of the selected industries.

In the early stages of their fieldwork, Burns and Stalker found two distinctly different sets of management methods and procedures that they came to classify as the *mechanistic* system and the *organic* system — a terminology reminiscent of Emile Durkheim's classic formulation. They found the mechanistic system to be apparently well suited to enterprises operating under relatively stable conditions, and the organic system to companies operating under changing conditions. The authors provided many detailed descriptions of characteristics associated with each of the two ideal types. These have been listed in adjacent columns for ready comparison in Figure 7–4.

A very nice summary of these same characteristics have been provided by Hage in his axiomatic treatment of organization theory.[16] In his summary Hage identified four organizational ends: adaptiveness, production, efficiency, and job satisfaction. He also identified four organizational means for achieving these ends: complexity, centralization, formalization, and stratification. Hage then attributed these means and ends to mechanistic and organic systems as shown in Figure 7–5. This clearly demonstrates basic differences between the two polar types of organization.

The Burns and Stalker study suggests that the effective firms in the more dynamic industries were more organic, that is, with wider spans of supervisory control, with a higher degree of task interdependence, with less attention to formal procedures, with more horizontal communication, and with more decisions made at the middle levels of the organization. In short, they were less bureaucratic, less structured. Organizations in the more stable industries, on the other hand, tended to be more mechanistic — with formal rules and procedures reached at higher levels of the organization, with narrower spans of supervisory control, with a high degree of task specialization, and with more vertical communication. In short, the mechanistic system was more bureaucratic, more structured.

According to Burns and Stalker, no one type of organization was more effective than the other: each was most effective in its own given environment. To sum up, the Burns and Stalker study showed that effective organizational units operating in relatively stable sectors of the environment were more highly structured than those in relatively dynamic sectors of the environment. It refuted the "one best design" concept of traditional management. This concept was replaced with one in which the most appropriate management system was contingent upon environmental and task demands facing the organization.

A later study is that of Paul R. Lawrence and Jay W. Lorsch.[17] Involved in this research effort were 10 organizations with different levels of economic performance in three distinctive industrial environments (the plastics industry, the food industry, and the container industry). For the study, top executives from each

---

[16] Jerald Hage, "An Axiomatic Theory of Organizations," in *Organizational Systems* ed. Koya Azumi and Jerald Hage (Lexington, Mass.: Heath, 1972).

[17] Lawrence and Lorsch, *Organization and Environment.*

FIGURE 7–4   Comparison of Mechanistic and Organic Systems of Organization

Mechanistic	Organic
1. Tasks are highly fractionated and specialized; little regard paid to clarifying relationship between tasks and organizational objectives.	1. Tasks are more interdependent; emphasis on relevance of tasks and organizational objectives.
2. Tasks tend to remain rigidly defined unless altered formally by top management.	2. Tasks are continually adjusted and redefined through interaction of organizational members.
3. Specific role definition (rights, obligations, and technical methods prescribed for each member).	3. Generalized role definition (members accept general responsibility for task accomplishment beyond individual role definition).
4. Hierarchic structure of control, authority, and communication. Sanctions derive from employment contract between employee and organization.	4. Network structure of control, authority, and communication. Sanctions derive more from community of interest than from contractual relationship.
5. Information relevant to situation and operations of the organization formally assumed to rest with chief executive.	5. Leader not assumed to be omniscient; knowledge centers indentified where located throughout the organization.
6. Communication primarily vertical between superior and subordinate.	6. Communication is both vertical and horizontal, depending upon where needed information resides.
7. Communication primarily takes form of instructions and decisions issued by superiors, of information and requests for decisions supplied by inferiors.	7. Communication primarily takes form of information and advice.
8. Insistence on loyalty to organization and obedience to superiors.	8. Commitment to organization's tasks and goals more highly valued than loyalty or obedience.
9. Importance and prestige attached to identification with organization and its members.	9. Importance and prestige attached to affiliations and expertise in external environment.

SOURCE: Adapted from T. Burns and G. M. Stalker, *The Management of Innovation* (London: Tavistock, 1964), pp. 119–22.

organization were interrogated by means of interviews and questionnaires. The focus of this study was on two structured characteristics of organizations—differentiation and integration.

By differentiation they understood "the difference in cognitive and emotional orientations among managers in different functional departments" (p. 11),

FIGURE 7–5   Characteristics of Mechanistic and Organic Management Systems

Mechanistic *(Emphasis on Production)*		*Organic* *(Emphasis on Adaptiveness)*	
Organizational means	Low	Complexity	High
	High	Centralization	Low
	High	Formalization	Low
	High	Stratification	Low
Organizational ends	Low	Adaptiveness	High
	High	Production	Low
	High	Efficiency	Low
	Low	Job Satisfaction	High

SOURCE: Jerald Hage, "An Axiomatic Theory of Organizations," in Koya Azumi and Jerald Hage, *Organizational Systems* (Lexington, Mass.: Heath, 1972), p. 272.

the "differences in attitude and behavior and not just the simple fact of segmentation and specialized knowledge" (p. 9). Differentiation includes, therefore, a psychological dimension, the *perceived* differences in attitudes and behavior of managers in different departments. The greater the psychological distance between managers in different departments, the greater the differentiation.

Integration, for the authors, signified "the quality of the state of collaboration that exists among departments that are required to achieve unity of effort by the demands of the environment" (p. 11). Thus, integration refers to both the quality of the interdepartmental relations and to the traditionally accepted processes by which such relationships are brought about.

Particularly appealing to Lawrence and Lorsch was the proposition that "different external conditions (environments) might require different organizational characteristics and behavior patterns within the effective organization" (p. 14). However, they were also interested in learning whether the certainty of information or knowledge about events in the environment is or is not an external dimension influencing the organizational variables.

The approach used by Lawrence and Lorsch was a comparative one. They selected for their study an industry dealing with rapid technological change and compared the organizations in this industry not only with one another but also with firms operating in more stable, less dynamic industries.

The Lawrence and Lorsch study found that in each industry the high-performing organizations came nearer to meeting the demands of their environment than did their less effective competitors, and that the most successful organizations tended to maintain states of differentiation and integration consistent with the nature of the environments and the interdependence of their parts. Their findings suggest a "contingency theory of organization," a term first coined by these researchers. They also found an important relationship among external variables, internal states of differentiation, integration, and the process of conflict resolution.

More specifically, they found that the state of differentiation in the effective organization was consistent with the diversity of the environmental parts, while the state of integration achieved was consistent with the environmental demand for interdependence. They also found that the states of differentiation and integration are inversely related. "The more differentiated an organization, the more difficult it is to achieve integration" (p. 157). However, a truly effective organization has integrating devices consistent with the diversity of the environment.

In summary, the Lawrence and Lorsch contingency theory of organizations seems to provide a framework for the major relationships that executives should consider as they design organizations to deal with specific environmental conditions.

William Dill's study of two Norwegian firms antedated the Lawrence and Lorsch contingency theory development by several years.[18] Though brief, the study is important for opening up to researchers the exploration and discovery of the impact of the environment upon the organization and on its functioning.

Dill's definition of environment (pp. 423–34) included both an internal and an external component. The internal component included such factors as the stress on formal rules, departmental independence in routine work, barriers to management interaction, and other short-run factors. The external environment contained the following subenvironments: customers, suppliers, competitors, and regulatory groups.

The findings that emerged from the study of the two firms (Alpha and Beta) were these: (1) Beta's management personnel were noticeably more autonomous in their decision making and (2) Beta's environment was considerably more differentiated, more heterogeneous than that of Alpha.

Osborn and Hunt[19] studied the effects of environmental complexity on organizational effectiveness in 26 small social service agencies and found that the amount of risk present in the external environment was unrelated to effectiveness. However both environmental dependency and interorganizational interaction were found to be positively related to effectiveness. In general, the results of this study did not support the findings of Lawrence and Lorsch. Likewise the more recent study by Pennings[20] in which he studied 40 branch offices of a brokerage firm and determined that the degree of association between organizational structure and environmental dimensions did not support the structural contingency model.

The scorecard is fairly even, with several studies supporting and several others not supporting the model. The adequacy of the model must await the results of further testing by researchers.

---

[18] William R. Dill, "Environment as an Influence on Managerial Autonomy," *Administrative Science Quarterly* 2 (1958), pp. 409–43.

[19] Osborn and Hunt, "Environment and Organizational Effectiveness."

[20] Johannes Pennings, "The Relevance of the Structural-Contingency Model of Organizational Effectiveness," *Administrative Science Quarterly* 20, no. 3 (September 1975), pp. 393–409.

SUMMARY

Granted that transactional interdependencies between the environment and the organization exist, the authors have attempted in this chapter to explore the nature of these relationships. Recognition was given to the fact that the environment, though often discussed as a unitary entity, is *actually* composed of subenvironmental factors.

Three environmental dimensions were noted here as important: complexity, change, and uncertainty. It was seen to be difficult to discuss these dimensions separately because of the carryover effect of the other dimensions. One could hardly attempt to examine complexity without recognizing uncertainty, and one could hardly study uncertainty without recognizing the change dimension. Still, it has been postulated that different types of environment call forth different organizational behaviors. Various models have been utilized for testing, and to date the results are still unclear.

Organizations engage in search activity (scanning) in order to acquire information about their environment. That different environments might call for different intensities of search activity is the subject of the next chapter.

KEY WORDS

Environmental Uncertainty   Environmental Complexity
Stable Environment   Heterogeneous Environment
Dynamic Environment   Organic System
Homogeneous Environment   Mechanistic System
Environmental Change

REVIEW QUESTIONS

1. Utilizing Figure 7–1, where would you classify the following?
   a. Railroads.
   b. Independent truckers.
   c. Meat packing.
   d. Electronics.
   e. Universities.
   f. Symphony orchestras.
   g. Aerospace companies.
   h. Mail system.
   i. Professional football.
   j. Fashion firms.
2. Firms in some industries have more impact on the environment than firms in other industries. List some firms which have a significant impact on the environment and some which have but little. Note whether or not there is any commonality of the firms in the different categories.

3. Are book publishing companies in the environment of universities or are universities in the environment of book publishing companies? Explain your answer.

4. Give several examples of a turbulent environment.

5. Look up the Fortune 100 list of largest firms from 20 years ago and compare it with the most recent one. List the number of firms that are no longer in existence and try to identify the type of environment in which they operated.

6. Iggy Prokopovich once said, "Your environment is what you make it, rather than the popular notion that you are the product of your environment." Explain.

7. Dizzy Dean is credited with saying, "In order to have a good pitcher, you gotta have a lot of lousy batters." What point was he unconsciously making?

8. Webster defines the environment as "the aggregate of all the external conditions and influences affecting the life and development of an organism." Can such a definition be operationalized? If your answer is no, then how does one study open systems in biology?

9. How would you define the dimensions of the environment for fish in a river, for deer in the woods, for ducks in the city park?

10. What factors would you include in the environment of your local barber or beautician on campus?

11. Suppose you were hired as an organization-environmental expert for a large industrial company. Write your job description, outlining your key areas of responsibility and major duties under each.

Case: The Professional Product Providers (PPP)

The PPP organization was founded for the express purpose of supplying products and services for the education of the handicapped. The organization was funded by both the federal and the state governments. It was charged with dispensing services to the various school districts that it served. While its services were limited to a certain number of school districts, other school districts not served by PPP did not necessarily provide alternative services for their own handicapped students. For this reason, PPP was asked to extend services to additional school districts.

PPP started out with a skeleton staff of 3, and in five years' time employed approximately 50 people. The staff consisted of professionals, semiprofessionals, and clerical personnel. The PPP central staff and field staff promoted special education at three different levels — the state department of public instruction, school districts of the state, and individual schools within the districts.

QUESTIONS

1. How would one go about measuring the environmental dimensions of stability and complexity in this organization?
2. Develop some measure of effectiveness for this organization.
3. What might be some goal statements of this organization?
4. What might be some measures of differentiation and integration for this organization?

Additional References

Aharoni, Yair; Zvi Maimom; and Eli Segev. "Performance and Autonomy in Organizations: Determining Dominant Environmental Components." *Management Science* 24, no. 9 (May 1978), pp. 949–59.

Brown, J. K. *This Business of Issues: Coping with Company's Environments.* New York: Conference Board, 1979.

Davis, K., and R. L. Blomstrom. *Business and Its Environment.* New York: McGraw-Hill, 1966.

Hambrick, D. C. "Environment, Strategy, and Power within Top Management Teams." *Administrative Science Quarterly* 26 (June 1981), pp. 253–71.

Kefalas, A.G. "Defining the External Business Environment." *Human Systems Management* 1, no. 3 (November 1980), pp. 253–60.

Kiesler, S., and L. Sproull. "Managerial Response to Changing Environments: Perspectives on Problem Sensing from Social Cognition." *Administrative Science Quarterly* 27 (December 1982), pp. 548–70.

Leifer, R., and A. Delbecq. "Organizational/Environmental Interchange: A Model of Boundary Spanning Activity." *Academy of Management Review* 3 (January 1970), pp. 40–50.

Lenz, R. T. "Environment, Strategy, Organization Structure and Performance: Patterns in One Industry." *Strategic Management Journal* 1 (1980), pp. 209–26.

Leontiades, M., and A. Tezel. "Planning Perception and Planning Results." *Strategic Management Journal* 1 (1980), pp. 65–75.

McAllister, Don M. *Evaluation in Environmental Planning: Assessing Environmental, Social, Economic, and Political Tradeoffs.* Cambridge, Mass.: MIT Press, 1980.

Meechan, C. J. "Problem of a Changing Business Environment." *Research Management* 22 (November 1979), pp. 35–38.

Meyer, Alan D. "Adapting to Environmental Jolts." *Administrative Science Quarterly* 27, no. 4 (December 1982), pp. 515–37.

Provan, Keith G.; Janice M. Beyer; and Carlos Krytbosch. "Environmental Linkages and Power in Resource-Dependence Relations between Organizations." *Administrative Science Quarterly* 25, no. 2 (June 1980), pp. 200–25.

Schoorman, F. D. and L. G. Hrebiniak. "Industry Differences in Environmental Uncertainty and Organizational Characteristics Related to Uncertainty." *Academy of Management Journal* 23, no. 4 (December 1980), pp. 750–59.

Snow, C. C., and D. C. Darran. "Organizational Adaptation to the Environment: A Review." *Proceedings of the Annual Meeting of the American Institute for Decision Sciences.* Edited by M. W. Hopfe and H. C. Schneider. Cincinnati, Ohio, 1975, pp. 278–80.

Snyder, N. H., and W. F. Glueck. "Can Environmental Volatility be Measured Objectively?" *Academy of Management Journal* 25, no. 1 (March 1982), pp. 185–91.

Strand, Rich. "A Systems Paradigm of Organizational Adaptations to the Social Environment." *Academy of Management Review* 8, no. 1 (January 1983), pp. 90–96.

Warriner, Charles K. *Organizations and their Environments: Essays in the Sociology of Organizations.* Greenwich, Conn.: JAI Press, 1984.

CHAPTER 8

The Environmental Scanning Process

There is one quality more important than "know-how" . . . this is "know-what" by which we determine not only how to accomplish our purposes, but what our purposes are to be.

Norbert Wiener

CHAPTER OUTLINE

INTRODUCTION

During the last decade a good deal of discussion has concerned the need to incorporate environmental variables more explicitly into the study of organizations. Organizational researchers admit that different environments impose unlike demands and provide varying opportunities. An organization learns of these demands and opportunities by gathering data about environmental events and by the subsequent analysis and evaluation of the data. This information is then utilized in organizational decision making for determining appropriate adjustments in strategies.

One can define the environment not as an objective "fact" but rather as an "image" in the entrepreneur's mind.[1] However, this can lead to misunderstanding.

Between the popular definition of environment as all those things that surround the organization — and the highly subjective — the executive's image — lies a middle ground incorporating both viewpoints.

Churchman defines environment as those factors which not only are outside the system's control but which determine in part how the system performs.[2] Things that are within the control of the organization are, according to Churchman, resources or means that the organization may use in whatever way it finds appropriate. These, therefore, do not constitute environment. Things that have no direct impact upon organizational performance are also not in the actual environment. These may become organizational environments only if there is a change in the organization's objectives or goals.

For present purposes, environmental information will be treated as information which becomes available to the organization or as that to which the organization acquires access through its scanning activity. Environmental information flows either are routinely communicated to the organization or are deliberately sought out.

This definition of environment coupled with the "image" concept provides a clearer idea of what "that over there" actually is. The mere enumeration of the things surrounding the organization does not provide the organization with any specific information about the environment; it only hints at potential sources of data that the organization should monitor. What constitutes external environmental information is analysis and evaluation of the properties of these sources of data, together with the executive's "image" of the environment, that is, the individual's *Weltanschauung*.

Most investigations of organizations and their environment rely heavily on the assumption that environmental demands or opportunities are presented to the organization in the form of "the problem" (constraints, threats, opportunities). However, once this assumption is challenged, one is then forced to ask how the organization becomes aware of these problems. How does it learn of impending threats? Organizational decisions are made as a result of the organization's ability

---

[1] E. T. Penrose, *The Theory of Growth and the Firm* (New York: John Wiley & Sons, 1959), p. 215.

[2] C. W. Churchman, *The Systems Approach* (New York: Delacorte Press, 1968), p. 36.

and willingness to scan its external environment for the purpose of identifying environmental problems. Scanning, then, becomes the first element to be investigated here.

SCANNING AND THE DECISION-MAKING PROCESS – A DEPENDENCY

At the outset, it will suffice to define scanning as the activity or the process of acquiring information for decision making. Simon describes the decision-making process (DMP) as comprising three principal phases, "(1) finding an occasion for making a decision, (2) finding possible courses of action, and (3) choosing among courses of action."[3]

The first phase of the decision-making process (DMP) – searching the environment for conditions calling for a decision – is termed by Simon the "intelligence activity." The second phase, designated the "design activity," refers to inventing, developing, and analyzing possible courses of action. The third phase, referred to as the "choice activity," consists of selecting a particular course of action from those available.[4]

Initially, one may be tempted to suggest that a relationship between scanning and decision making exists only for the first phase. Indeed, the relationship here is very explicit, for the terms "searching the environment" and "intelligence" imply gaining knowledge through active information-acquisition behavior. Intelligence has, in fact, been defined by some scholars as "data selected and structured such as to be relevant in a given context for a decision."[5]

A more careful examination, however, will reveal that the second phase of DMP, the design activity, is also dependent on scanning. One can hardly deny that inventing, developing, and analyzing possible courses of action are influenced by the kinds and amounts of information acquired through the scanning process. Similarly, whether or not a certain event or situation will be considered as a possible course of action will depend upon the degree of knowledge that the observer has about the event or situation. It will depend upon how well the decision maker is informed.

Finally, the dependence of the choice activity on scanning follows logically from the above. The less the environment is scanned, the fewer the possible courses of action, and therefore, the more limited the final choice.

If one were to add, as a fourth phase of DMP, the implementation and evaluation of the chosen courses of action (decision), then one would be forced to admit that the link between the final phase of DMP and scanning is quite significant. Implementation of a decision will require information about the

[3] H. A. Simon, *The Shape of Automation for Men and Management* (New York: Harper Torchbooks, The Academy Library, 1965), pp. 52–54.

[4] Simon, *Shape of Automation,* p. 53.

[5] O. H. Poensgen and Z. S. Zannetos, "The Information System: Data and Intelligence." Alfred P. Sloan School of Management, MIT, Working Paper no. 404-69, July 1969.

system that is prior to or concurrent with it (feedforward). The evaluation of a decision will require information for determining how effective the course of action was (feedback).

Each step in this process has as its inputs the outcome of activities in preceding steps. While DMP is depicted here as a chain of sequential activities, very often it takes the form of recurring chains with feedbacks.

If one accepts the proposition that "a person's judgment is no better than his or her information," one would expect to find the same amount of attention given to scanning as to decision making per se. Furthermore, if the four phases of DMP are of a sequential nature, and if the dependence of the intelligence activity on scanning is substantially high, then one would expect that every scientific treatment of DMP would be preceded by an extensive investigation of the scanning process.

A search of the literature, however, shows that this is not the case. Not only does most of the literature on decision making concentrate on the last two phases (the comparison of alternatives and making the choice), but in nearly all cases the scanning process has been assumed to have already taken place.

When empirical research on the subject includes some kind of controlled experimentation, the information necessary for making the choice is provided to the subjects in the form of an "information survival kit," called "data bank," "information base," or "information structure." As Lanzetta and Kanareff state:

> Empirical studies of decision-making have typically provided the decision-maker with an information base in terms of which a choice among alternatives must be made.... The information base includes specification of the alternatives, the possible consequences of a choice, the probabilistic data on the relationship between alternatives and outcomes. The information-acquisition processes preceding decision are assumed to have occurred, in essence, are simulated by the experimenter.[6]

As a result, most of our knowledge of scanning is far too incomplete and based upon implicit or explicit recommendations in the literature that deal only incidentally with information-acquisition behavior. It is our purpose to present a conceptual framework of the scanning process. In this framework it is assumed that factors outside the formal organization's boundaries do have a significant effect upon the functioning of the enterprise. Although opening the organization to its environment unduly complicates organizational behavior, still it is the only way one can get a feel of what the firm, the enterprise, the organization, is all about. Just because such a procedure necessitates an analysis of extremely complex interactions does not justify its omission. It is precisely when the organization interacts with its environment that its amazing complexity is revealed.

---

[6] J. T. Lanzetta and V. T. Kanareff, "Information Cost, Amount of Payoff, and Level of Aspiration as Determinants of Information Seeking in Decision Making," *Behavioral Science* 7 (November 1962), pp. 459–73. See also C. W. Churchman, "Operations Research as a Profession," *Management Science* 17, no. 2 (October 1970), pp. B37–53.

INFORMATION ACQUISITION IN OPEN SYSTEMS

In general, the inputs of an open system can take the form of either *energy* and/or *information.* As Thayer states,

> The basic processes of an organization, which are the basic processes of all open living systems are:
>
> 1. *Importing* from the environment certain raw materials and resources for conversion into products or services which are *exported* for consumption by the same or other parts of the environment; and
>
> 2. *Acquiring* data from the environment, and from its internal parts, to be "consumed" in problem-definition and decisioning in the service of its attempts to alter its intended-states-of-affairs, its internal structure or function, or some aspect or domain of its environment.[7]

The second process, that of acquisition of data from the environment, constitutes what has here been called the "scanning process."

Any goal-seeking system must be related to the outside environment through two kinds of channels: the *afferent* (a scanning system), through which it receives information about the environment; and the *efferent* (decision system), through which it acts on the environment.[8] The relationships between the organization and the environment are shown in Figure 8–1. For logical completeness, a third subsystem has been added to Simon's two subsystems. This third subsystem is the intelligence or internal organizing system. The addition of this system reflects the author's understanding of the role of information in the decision-making process. Organizational actions which are the outputs of the decision subsystem are not based upon outputs of the scanning subsystem, but rather upon the outputs of the internal organizing subsystem, which evaluates the scanning subsystem. Data received by the scanning system are eventually fed into the decision system, where they are utilized for problem-solving purposes. Contrary to popular belief (at least in the business literature), evaluated data do not constitute information unless and until they enter the decision system. What the scanning system receives from the environment are raw sensory data about some aspects of the external environment.

The next step in this sequence is a selective conversion of the scanning data into a form suitable for consumption. It is only through this conversion process that data become information.

It is this information which serves as a basis for decision making. The proportion of the available data to be converted into consumable information will be determined by the problem at hand. This will be indicated by the discrepancy between the intended state (X) and the actual state (Y) of the system; of in terms of Figure 8–1, the *existence, direction,* and *magnitude* of the (XY) interval.

---

[7] L. Thayer, *Communication and Communication Systems* (Homewood, Ill.: Richard D. Irwin, 1968), p. 101. ©1978 by Richard D. Irwin, Inc.

[8] H. A. Simon, *The Sciences of the Artificial* (Cambridge, Mass.: MIT Press, 1969), p. 66.

FIGURE 8–1   The Environment-Organization (EO) Interaction System

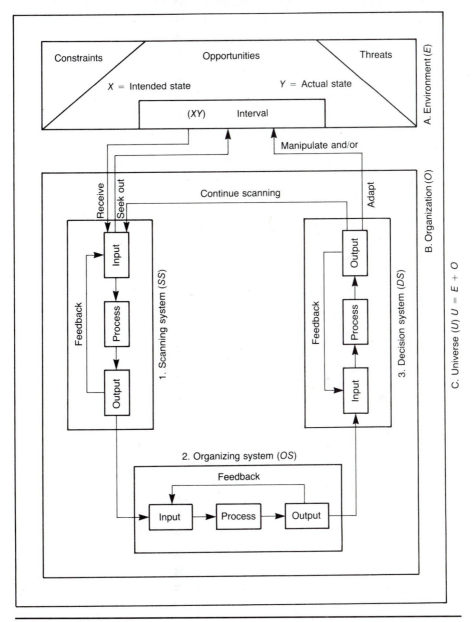

SOURCE: A. G. Kefalas, "Scanning the External Business Environment," Ph.D. dissertation, The University of Iowa, 1971, p. 56.

Thus, what actually goes into the decision system are data, not information. Information is formed in the mind of the problem solver or decision maker as an outcome of a comparison between the problem and the data. Adrian McDonough has presented this in Figure 8–2.[9] Figure 8–1 implies that the observer or designer of the system can distinguish conceptually between incoming and outgoing flows. In fact, it is quite likely that individuals within the organization may be engaged in any or all three of the activities at various times.

Thus far the environment-organization interaction system has been treated conceptually as a hierarchical system — one composed of interrelated subsystems, each subordinate to the one above it.[10] At the bottom of the rank order is some lowest level of elementary subsystem to which no other is subordinate. This "boxes-within-boxes" way of looking at complex phenomena is not a mere partitioning of elements but one formed by the relationships of the several parts with one another and with the whole.

EO interaction is regulated by the decision-making process. More precisely, the decision maker coproduces the future of the system along with the environment, which he or she does not control. The decision-making process in its entirety (i.e., sensory system + organizing system + decision system) becomes a subsystem within the total system of the EO interaction. Through these three subsystems, the organization tries to build associations between the environment and itself so that the EO system can be brought into a harmonious functional relationship.

In short, the scanning subsystem gathers data on the deviations between the actual state and the intended state (the XY interval). The scanning subsystem assesses the existence, direction, and magnitude of the XY interval. The decision subsystem minimizes the existing deviation between the actual state and the intended state.

SPHERES OF EXTERNAL INFORMATION[11]

One can assume that an infinite amount of information exists in the environment. Some of this is available if sought out, some is unavailable even if pursued with limitless resources, while some will be unknown and therefore inaccessible.

In Figure 8–3A the environment of the firm, represented by the area E, is the one from which executives of organizations seek to acquire information. Circle A represents the area with information accessible to the firm. All of this information is generally not sought out because of either economic considerations or other constraints. Circle B represents the total area of external information that an

---

[9] McDonough, *Information Economics,* p. 71.

[10] Simon, *Shape of Automation,* p. 87.

[11] The following discussion is from Kenyu Nishi, "A Study of the Information-Acquisition Process in Japanese Computer and Information Processing Service Industries," Ph.D. dissertation, University of Iowa, December 1976, pp. 88–91.

FIGURE 8–2   Information

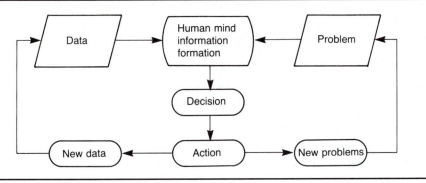

SOURCE: A. M. McDonough, *Information and Management Systems* (New York: McGraw-Hill, 1963).

FIGURE 8–3   A Conceptual Relationship among Different Spheres of External Information

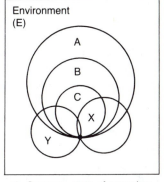 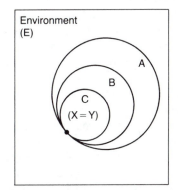

A. Common state of scanning   B. Ideal type of scanning

Box E: External environment.
Circle A: Accessible information area.
Circle B: Total area of external information obtained.
Circle C: Information area obtained and subjectively recognized to be strategic.
Circle X: Information area that is subjectively assumed to be strategic.
Circle Y: Ideal-type strategic-information area recognized objectively as being strategic.

executive obtains: it includes both strategic and nonstrategic elements. Circle C represents the amount of information that an executive actually obtains and which is recognized at the same time as being strategic. (This is subjectively determined.) Circle X represents the information area that an executive wants to acquire and which is assumed to be strategic. Circle Y represents the ideal-type strategic information area that is recognized as being objectively strategic.

In Figure 8–3A, a sizable gap exists between Circle X and Circle Y. In other words, a gap exists between the ideal-type objectively recognized strategic information and the information that the executive assumes to be strategic. This discrepancy arises from the different scanning behaviors of individual executives or of the organization as a whole. Similarly, a gap also exists between Circle X and Circle C. Here again, the gap arises between the information an executive wants to acquire (Circle X) and the strategic information he or she actually obtains (Circle C).

Figure 8–3B represents an ideal model of the scanning process, where Circle X, Circle Y, and Circle C are all in the same sphere. In this model, there is no discrepancy between the ideal-type strategic information (Circle Y), the information that an executive thinks is strategic and wants to acquire (Circle X), and the information that an executive actually obtains and recognizes as being strategic (Circle C).

Perhaps an illustration of the scanning behavior of management would be helpful here. A management executive scans the environment for external information with strategic value. The executive realizes that the strategic information desired often is not available. (This is probably the actual situation nearly everywhere.) This means that management will have to make decisions under conditions of uncertainty. Moreover, of all the information a manager receives, only some is needed, relevant, and laden with strategic value. In real life, the typical executive may not always realize what information is necessary and relevant, and what has strategic value. Sometimes the manager has strategic information in hand but does not recognize it as strategic. Much depends upon the perceptual differences of individual executives.

However, when a manager recognizes the strategic nature of the incoming information, the perceived Circle X will not be exactly the same as before: a new type of Circle X will develop. Although the ideal-type Circle Y is not conceptualized, scanning behavior will change and will come closer to the sphere of Circle Y—that is, Circle X moves toward Circle Y. Not only does Circle X move towards Circle Y, but also Circle A and Circle B grow larger and move to the position where all circles nest in one large circle. Figure 8–3B shows this ideal situation. In other words, individual executives (or an organization as a whole) learn of scanning techniques and cover increasing spheres of external information, and this fact in turn reduces the gaps among the different spheres (Circles X, Y, and C) so that the behavior approaches the ideal model. This approach to the ideal-type model is a way of reducing uncertainty. In this lies the real value of scanning for information—its reduction of uncertainty in a decision-making situation.

THE ENVIRONMENT-ORGANIZATION INFORMATIONAL FLOWS

In a previous chapter, the two-way exchange between the organization and the environment was represented by a two-by-two matrix:

Inputs \ Outputs	Organization	Environment
Organization	L_{11}	L_{12}
Environment	L_{21}	L_{22}

Here the L_{12} and the L_{21} transactions are of interest because they represent the exchanges between the environment and the organization. Since these represent information flows, this relationship may be more meaningfully depicted as in Figure 8–4.

The box labeled *Environment* represents the various components of the total environment from which information may routinely be acquired or from which information may be sought. The box labeled *Organization* lists the sources of internal information available to the organization. Top management relies primarily on information external to the organization. First-line management relies primarily on information internal to the organization. How information is acquired will be dealt with in the next section.

THE NATURE OF THE SCANNING SYSTEM

The term strategic problem refers to the basic feature of every open system: *maintenance and regulation of flow of information between the system and its environment.* Adaptability of the system to environmental demands and opportunities constitutes a *conditio sine qua non* for survival of the open system. It is in this sense that the term strategic is used.

Scanning Intensity

One might hypothesize a priori that at the organizational level the degree and intensity of scanning will depend, among other things, upon

1. The availability of organizational economic resources.
2. The perceived nature of the relationship between the organization and its environment.
3. The frequency and magnitude of changes in the states of the environment.

For the sake of simplicity, the degree and intensity of scanning will be considered as being dependent on the second and third factors only.

Suppose there are only two kinds of relationship possible between the enterprise and its environment: (1) symbiotic and (2) synergistic. A symbiotic relationship is of the functionally necessary type: the relationship between the two systems is necessary for survival of both systems. A synergistic relationship, on the other hand, is not functionally necessary, but its existence enables the two systems

FIGURE 8–4   Flow Relationships of External Information between Environment and
Organization

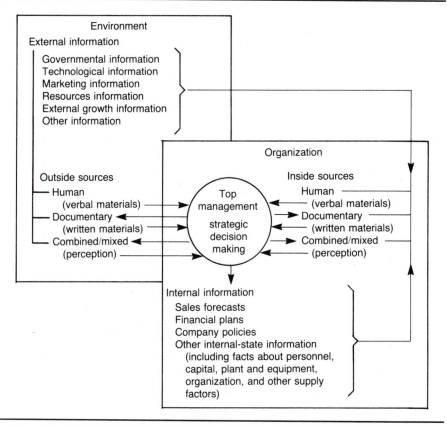

SOURCE: Kenyu Nishi, "A Study of the Information-Acquisition Process in Japanese Computer and Information Processing Industries," Ph.D. dissertation, The University of Iowa, December 1976, p. 99.

to achieve a performance that is greater than the sum of the two individual performances taken separately.

Disregard of a symbiotic relationship increases the probability that the relationship will eventually get out of control, leading temporarily to undue exploitation of each system and ultimately to disintegration of both systems. In the EO interaction system, certainly the ultimate loser will be the organization, for that is the system that has to adapt.

Disregard of a synergistic relationship also increases the probability that the relationship will eventually get out of control, but the consequences here are much less severe than in the previous case. For what is at stake is not the whole relationship but rather the additional increment in the system's ability to survive, attributable to the synergistic effect.

In an open system the degree of environmental scanning will depend upon the importance of the relationship between the two. This, in turn, will be influenced by the amount of interaction. The test for the intensity of interaction is the same as the test for the dependence of a system on certain parameters. To test whether a parameter is effective, one observes the system's behavior on two occasions when the parameter has different values.

The relationship between scanning and environmental states would be considerably more complicated were one to take into consideration the third factor mentioned above, that is, the frequency and magnitude of changes in the environment.

Were one to classify a slowly changing environment as relatively stable and a frequently changing environment as relatively dynamic, then one would expect to find the degree of scanning to be higher in the dynamic environment than in the stable. This is so because the variety inherent in a frequently changing environment can only be handled or controlled through equal variety in the system designed to monitor it. Since "information kills variety," dynamic environments call for "busier" scanning systems.

Figure 8–5 combines both determinants of scanning intensity. The numbers in the cells represent scanning intensity in ascending order.

The environmental sector that is perceived as being symbiotically related to the organization and that has a high frequency and magnitude of changes requires the most monitoring. The opposite is true of the symbiotic/stable combination.

MODES OF SCANNING

Scanning was defined as the process whereby the organization acquires information for decision making. This certainly must include human activity. As with all human activity, scanning is subject to all biological, psychological, social, cultural, and economic laws governing human behavior.

Biologically, the process of information acquisition is today fairly well understood. From the psychological and, to some extent, the cultural and social viewpoints, scanning is considered as part of the process of thinking and problem solving.

From the economic viewpoint the question of information is said to be subject to the law of efficiency, which states that the cost of acquiring information should not exceed the benefits to be derived from the acquired information.[12] Despite the obvious soundness of this principle, empirical research has thus far failed to provide any substantial evidence confirming its operationality.

---

[12] J. March and H. A. Simon, *Organizations* (New York: John Wiley & Sons, 1958); H. Simon, *Models of Man* (New York: John Wiley & Sons, 1957); J. Marschak, "Towards an Economic Theory of Organization and Information," in *Decision Processes,* ed., R. M. Thrall et al. (New York: John Wiley & Sons, 1954); J. March, *Handbook of Organizations* (Chicago: Rand McNally, 1965).

FIGURE 8–5   Determinants of Scanning Intensity

Degree of Change of the Environment \ Nature of the Relationship	Symbiotic	Synergistic
Stable	1	2
Dynamic	4	3

SOURCE: A. G. Kefalas, "Scanning the External Business Environment," Ph.D. dissertation, The University of Iowa, 1971, p. 62.

As with most aspects of human behavior, scanning covers a broad continuum of possibilities that merge imperceptibly into one another. For purposes of analysis, however, it may be necessary to establish some recognizable, even if arbitrary, reference points within the continuum.

Since the organization and the environment are viewed here as parts of the same system, the different modi operandi of scanning utilized by the organization will depend upon the states of the variables defining the system organization (L_{11}), and upon the states of the parameters determining the exchange activity between the organization and its environment (L_{12} and L_{21}).

Although the two sets of determinants of the modes of scanning are not independent of each other, only the second set (L_{12} and L_{21}) will be examined here. As Ruesch puts it, "In our modern technological society, environmental change is so rapid that modern man's way of adaptation consists of holding the internal surroundings stable."[13] One may perhaps safely hypothesize that since "the outer environment determines the conditions for goal-attainment," and since data about the degree of goal attainment are gathered by the scanning system (Figure 8–1), *the mode of scanning is for the most part determined by the external environmental stimuli.*

In general, one can distinguish between two basic methods of scanning: *surveillance* and *search.*[14] The term *surveillance* refers to "a watch over an interest." The term is similar to what is termed "current awareness," the function of which is to give the information seeker some *general* knowledge.

Search, on the other hand, aims at finding a *particular* piece of information for solving a problem. The meaning of this term is familiar enough and has been dealt with in a number of works.

For present purposes, the difference between the two basic modes of scanning may be considered to be one of degree and not of kind. The difference lies in the degree of involvement of the scanner and in the formalization of the

---

[13] Jurgen Ruesch, "Technology and Social Communication," in *Communication: Theory and Research*, ed. Lee Thayer (Springfield, Ill.: Charles C. Thomas, 1967), p. 466.

[14] F. J. Aguilar, *Scanning the Business Environment* (New York: Macmillan, 1967).

scanning procedures as measured by the degree of commitment of scarce resources in time.

Also the degree of involvement of a receptor will be influenced by the degree and frequency of the environmental changes that it is designed to monitor. Depicted in Figure 8–6 is the continuum that begins with surveillance at the extreme left end and terminates with search at the extreme right. The degree of involvement in time spent also runs in the same direction.

The two basic scanning modes can be further subdivided into viewing and monitoring for surveillance, and investigation and research for search. Figure 8–7 shows the complete "scanning tree," consisting of the two basic branches and the four smaller but more detailed subbranches.

All scanning modes can be viewed as processes for:

1. Seeking a problem solution.
2. Gathering data about problem structure that will ultimately be used in discovering a problem.
3. Increasing one's awareness or familiarity with an environment.
4. Making an information decision.[15]

The scanner assigns different values to each branch of the scanning tree. The assigned values obviously are not "true" values but rather estimates of the gain to be expected from further scanning along the same branch of the tree. Figure 8–8 presents the four modes of scanning and their relative values or utility as (1), (2), (3), and (4).

From Figure 8–8, it can be seen that surveillance (viewing + monitoring) has high value as an information-gathering process rather than as a means for finding solutions to problems. This relationship is reversed in the search (investigation + research) situation. Most research is indeed aimed at finding a satisfactory solution to a specific problem. Research activity, like all human problem-solving activities, is a varying mixture of trial, error, and selectivity. The selectivity derives from various rules of thumb that suggest which pattern should be tried first and which leads are promising.

The Mechanics of the Scanning System

Organizational information-acquisition activity cannot be investigated apart from its function in the survival of the system. Organizations, as goal-seeking or goal-guided open systems, depend for their survival upon their ability to adapt to

---

[15]The term information decision refers to the decision whether to (1) make the decision with the existing information or (2) acquire more information. The term could approximately be equated with "continuation or termination of search." See, for example, J.C. Grayson, Jr., *Decisions under Uncertainty: Drilling Decisions by Oil and Gas Operators* (Cambridge, Mass.: Harvard Business School, Division of Research, 1960), chap. 11.

FIGURE 8–6 Modes of Scanning

	Surveillance	*Search*
	Involvement in Time, Etc. Low	Involvement in Time, Etc. High

FIGURE 8–7 The Scanning Tree

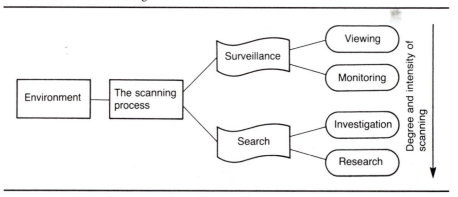

SOURCE: A. G. Kefalas, "Scanning the External Business Environment," Ph.D. dissertation, The University of Iowa, 1971.

FIGURE 8–8 Scanning Modes and Their Values or Uses

Values or Uses *Scanning Continuum (modes)*	*(1)* *Seeking a Problem Solution*	*(2)* *Gathering Data about Problem Structure*	*(3)* *Increasing One's Awareness*	*(4)* *Making an Information Decision*	*Involvement, Time*
Surveillance: *Viewing*	*Low*	*High*	*High*	*Low*	*Low*
Monitoring	*Relatively low*	*Relatively high*	*Relatively high*	*Relatively low*	*Relatively low*
Search: *Investigation*	*Relatively high*	*Relatively low*	*Relatively low*	*Relatively high*	*Relatively high*
Research	*High*	*Low*	*Low*	*High*	*High*

SOURCE: A. G. Kefalas, "Scanning the External Business Environment," Ph.D. dissertation, The University of Iowa, 1971, p. 68.

environmental states. Assessments of the states of the environment are made through the receptor or scanning system.

Figure 8–9 on the following page is a simplification of Figure 8–1 illustrating the system's dependence upon scanning for its adaptability and ultimate goal attainment.

From the environment, the scanning system acquires data about the XY interval ideally in terms of

1. Existence of the XY interval (i.e., $X - Y > 0$).
2. Direction (i.e., $X - Y = \pm$)
3. Magnitude (i.e., $X - Y = c$).[16]

Signals about the XY interval can be viewed as feedback emanating from the environment. It will be convenient to view these as negative feedbacks which may be either "direct" or "real" or "anticipatory."[17] Thus, assuming that the XY interval can vary from Y_1 to Y_2 for the actual state and from X_1 to X_2 for the intended state, then it follows that the scanning system is confronted with a 3×3 matrix of signals which it has to watch. For each of these nine signals, the scanning system will have to note and transmit to the organizing systems (1) existence, (2) magnitude, and (3) direction of X and Y movements. In addition, the system will have to monitor each state movement separately. Thus, there are four additional feedbacks that have to be taken into account: $X_0 - X_1, X_0 - X_2, Y_0 - Y_1, Y_0 - Y_2$.

Feedback information about these changes in the external environment of the simple system with *one* known intended state and *one* known actual state is conveyed to the organizing system by way of the mismatch signal. The capability of the scanning system to transmit in the mismatch signal not only data about the existence of the XY interval but also its direction and magnitude will affect the number of trials the decision system will have to make to adapt itself suitably to the external environment.[18]

Obviously, the organization will try to minimize the number of trials needed for perfect adaptation and survival. However, keeping the XY interval as small as possible will depend on the organization's willingness and ability to maintain a scanning system whose mismatch signal can indicate not only the existence but also the direction and magnitude of the XY interval. Gains derived from minimizing the number of trials would be proportional to the cost for such a scanning system.

Also, one can assume that the organization will seek not an optimal but a satisfactory solution, that is, a solution that satisfies all the constraints. Such a solution will consist of a flexible multistage scanning system like the one depicted in Figure 8–7.

---

[16] D. MacKay, "Towards an Information-Flow Model of Human Behavior," in *Modern Systems Research for Behavioral Scientist* ed. W. Buckley (Chicago: Aldine Publishing, 1968), pp. 359 ff.

[17] N. Wiener, *Cybernetics* (Cambridge, Mass.: MIT Press, 1967), p. 95 ff.

[18] MacKay, "Towards an Information-Flow Model," p. 362.

FIGURE 8–9 A Goal-Guided Open System

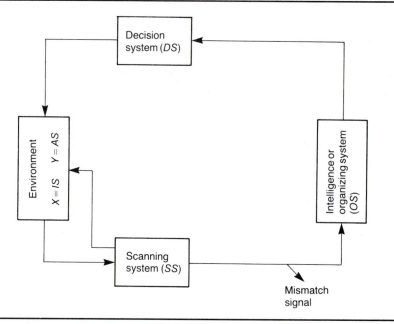

SOURCE: A. G. Kefalas, "Scanning the External Business Environment," Ph.D. dissertation, The University of Iowa, 1971, p. 69.

Modes of Scanning and the Mismatch Signal

The four modes of scanning can be associated with different *degrees of completeness of the mismatch signal* in terms of the *existence* of a mismatch ($X - Y > 0$), *direction* ($X - Y$ at $t = 0$ is greater or less than $X - Y$ at $t = 1$), and *magnitude* ($X - Y = c$ where c is a number greater than zero).

Figure 8–10 presents the expected relationships between the four modes of scanning and completeness of the mismatch signal. Viewing, for example—the weakest form of scanning—does virtually nothing in terms of producing a complete mismatch signal and therefore contributes almost nothing to the reduction of the number of trials needed to achieve suitable adaptation. This mode, however, does provide information about the existence of the *XY* interval and can therefore relieve the higher modes of searching for this kind of information. It is only because of this contribution that the value assigned to viewing is positive. A system engaged in viewing should be considered a search-directing rather than a search-performing system.

On the the other end there is research. The output of the research phase of the scanning process ought to be complete in the sense that statements about the *XY* interval should also indicate its direction and magnitude. Here the decrease in the number of trials will be the greatest. It should be emphasized, however, that research is the end phase of a multistage process and not a totally independent activity.

FIGURE 8–10   Modes of Scanning and Completeness of Mismatch Signal

Scanning Continuum (modes) / Completeness of the Mismatch Signal	Existence of XY Interval $(X - Y \neq 0)$	Direction of XY Interval $(Y - Y)_0 < (X - Y)_1$	Magnitude of Interval $(X - Y = c)$	Number of Trials
Surveillance: Viewing	Maybe	No	No	Very large
Monitoring	Yes	Maybe	No	Large
Search: Investigation	Yes	Yes	Maybe	Relatively small
Research	Yes	Yes	Yes	Very small

SOURCE: A.G. Kefalas, "Scanning the External Business Environment," Ph.D. dissertation, The University of Iowa, 1971, p. 73.

Since all four phases of scanning process belong to the same overall system, values assigned to lower steps can be regarded as savings of effort for the higher steps. Ideally, search should begin with assessing the direction of the XY interval, since its existence should have already been detected through one of the surveillance modes.

The direction and magnitude of the XY interval will depend on two factors: (1) the organization's internal capability to correctly formulate the intended state X and to actually produce the state Y that approximates the desired state, and (2) the nature of the environment.

Although the first factor definitely pertains to information acquisition, it is exclusively a matter of goal setting, goal seeking, and goal attaining. The second factor more directly touches upon the nature of the environment.

If the magnitude of the feedback is proportional to the intensity of the stimulus picked up from the environment, then one should expect a large deviation to be accompanied by a correspondingly strong feedback.

The particular mode of scanning, however, will depend on the direction of the XY interval (i.e., its sign). When $X_1 - Y_1 > X_0 - Y_0$, that is, when $FB_{t=1} > FB_{t=0}$, the negative feedback indicates an unfavorable development in the L_{12} or L_{21} relationship. The discrepancy between the goal and the achievement is now larger than before. This certainly constitutes a problem, and, following Cyert and March's thinking, one would expect search to be the predominant mode of scanning. According to *these* authors, search is always stimulated by a problem; and therefore *all organizational search is "problemistic search."*

When $X_2 - Y_2 < X_0 - Y_0$, that is, when $FB_{t=2} < FB_{t=0}$, then the negative feedback indicates an improvement in the L_{12} or L_{21} relationship. The discrepancy between the goal and the actual state of the system is now less than before. In this case one would expect surveillance to be the predominant form of information-acquisition behavior. In J. D. Thompson's thinking, surveillance is *the* scanning mode for the discovery of opportunities. According to him then, *all surveillance* by an organization is *"opportunistic surveillance."*

Although both movements—movement away from the intended state and movement toward it—can be perceived as problems, the "not-reaching-the-goal" situation (i.e., $X_1 - Y_1 > X_0 - Y_0$) is much more of a problem than the "doing-better-than-expected" situation (i.e., $X_2 - Y_2 < X_0 - Y_0$).

The particular scanning mode will also be affected by the environmental source that generates the feedback signal. If the environment feedback affects a fairly large number of the system's variables, the system's ability to control any discrepancy is diminished, and the amount of information and the number of trials needed to reach the desired (terminal) state are increased. In this kind of dynamic environment one needs fairly close monitoring. Again, the sign of the feedback signal will determine the mode of scanning.

SOME SCANNING PROPOSITIONS

There have been several major studies concerned with how business executives obtain environmental data that can affect strategic decision making.[19] Although the studies varied in methodology, still each was directed toward assessing several of the dimensions articulated in this chapter.

The propositions presented here have been derived both from a review of the related literature and from research findings of several of the studies. Even though previous studies on scanning have developed the rationale for the propositions here being presented, a short review of their derivation and a word on their importance will not be out of place.

SELECTED ENVIRONMENTAL PROPOSITIONS

Executive Scanning Style Related to the State of External Environment

Proposition One. Managers working for organizations in a relatively dynamic environment are expected to spend a greater amount of their time

---

[19] The five major studies explicating propositions concerning the environment are: Aguilar, *Scanning the Business Environment;* Warren J. Keegan, "Scanning the International Business Environment: A Study of the Information-Acquisition Process," Ph.D. dissertation, Harvard University, 1967; Asterios G. Kefalas, "Scanning the External Business Environment," Ph.D. dissertation, The University of Iowa, 1971; Asterios G. Kefalas and Peter P. Schoderbek, "Scanning the Business Environment: Some Empirical Results," *Decision Sciences* 4, no. 1 (January 1973), pp. 63–74; and Nishi, "A Study of the Information-Acquisition Process in Japanese Computer and Information Processing Industries," Ph.D. dissertation, The University of Iowa (Iowa City, IA. December 1976).

acquiring external information than are managers working for organizations in a relatively stable environment.

Proposition Two.   All managers are expected to spend a greater proportion of their time acquiring external information about the technology and marketing sectors than about the remaining sectors (government, resources, external growth, and other sectors) of the external government. This relationship is more likely to be true for the relatively dynamic environment than for the relatively stable environment.

Executive Scanning Style Related to the Hierarchical Levels of Management

Proposition Three.   Managers at higher levels of responsiblity in the overall industrial organization are expected to spend a greater amount of their time acquiring external information than managers at lower levels of responsibility.

Executive Scanning Style Related to the Functional Specialty

Proposition Four.   Executives are expected to acquire more external information related to their own functional areas than to other individual functional areas. An exception to this rule is likely to be marketing, which is expected to dominate most of the executives' scanning styles, regardless of their functional specialties.

Executive Scanning Style Related to the Environmental State

Proposition Five.   Managers overall are expected to believe that human and documentary sources, considered individually or together, are more important and are more frequently utilized than combined/mixed sources. This relationship is expected to be equally true for both environments.

Executive Scanning Style Related to the Hierarchical Levels of Management

Proposition Six.   Executives in higher levels of responsiblity are expected to utilize human sources of external information more than are managers in lower levels of responsibility. This relationship is expected to be equally true for both industries—the relatively dynamic and the relatively stable.

Executive Scanning Style Related to the Environmental State

Proposition Seven.   Managers are expected to rely to a greater degree on surveillance-oriented modes than on search-oriented modes in acquiring external information. This relationship is expected to hold for both industrial environments—the relatively dynamic and the relatively stable.

Executive Scanning Style Related to the Hierarchical Levels of Management

Proposition Eight. Managers in higher levels of responsiblity are expected to utilize more surveillance-oriented modes than search-oriented modes. This relationship is expected to be more characteristic of the relatively stable environment than of the relatively dynamic environment.

Proposition One is derived from a number of sources. The major independent variable of the Lawrence-Lorsch contingency theory of organizations is environmental uncertainty.[20] Thompson notes that "technologies and environment are basic sources of uncertainties for organizations."[21] Emery and Trist categorized environments by their causal texture, both as regards degrees of uncertainty and other important aspects.[22] Firms in dynamic environments where more uncertainty abounds require more information about changes in order to adapt.

Proposition Two emanates from Aguilar's (1967) finding that the "market tidings" sector was mentioned three times as often as any other sector. This suggested the need for a formal department in the organization for scanning the market sector. This may well be the reason for present-day market research departments.

Proposition Three is supported by much of the literature, even though Aguilar's study did not find this to be true. Top managers are said to employ, in their strategic decision making, information that is external to the organization. This need has been termed "information crisis" by many writers.

Lawrence and Lorsch's 1967 study serves as the foundation for Proposition Four dealing with functional specialties. They state:

> As organizations deal with their external environments, they become segmented into units, each of which has as its major task the problem of dealing with a part of the conditions outside the firm. This is the result of the fact that any one group of managers has a limited span of surveillance. Each one has the capacity to deal with only a portion of the environment.[23]

Both Propositions Five and Six deal with the sources of external information and have their roots in the findings of other researchers. It is expected that human

---

[20] P. R. Lawrence and J. W. Lorsch, *Organization and Environment: Managing Differentiation and Integration* (Cambridge, Mass.: Division of Research, Graduate School of Business Administration, Harvard University, 1967).

[21] J.D. Thompson, *Organizations in Action: Social Science Bases of Administrative Theory* (New York: McGraw-Hill, 1967).

[22] E. Emery and E. L. Trist, "The Causal Texture of Organizational Environments," *Human Relations,* 18 (1965), pp. 21–32.

[23] Lawrence and Lorsch, *Organization and Enviroment,* p. 8.

and documentary sources would be more important than other sources of information (Proposition Five) and that top executives would utilize human sources more than lower levels of management would (Proposition Six). Aguilar found that personal sources greatly exceeded impersonal ones in importance (71 percent vs. 29 percent), thus indicating the relatively high reliance that managers place on their own personal communication network.[24]

The different modes of scanning in the external environment (Propositions Seven and Eight) find support in Aguilar and Cyert and March.[25]

SUMMARY

This chapter presented a conceptual framework of the firm and its environment. The scanning process was viewed as the process of linking the organization to its environment.

The enterprise, viewed here as an open system, was related to its environment through two mechanisms commonly found in all open systems: the afferent or sensory system, and the efferent or motor system. These two systems are here called the scanning and decision systems, respectively.

Enterprises are man-made systems, often referred to as "artificials," whose actions are less integrative than biological systems. For this reason, to these basic systems a third was added. This system has been referred to here as the intelligence or internal organizing system.

These three subsystems are viewed as mechanisms enabling the organization ultimately to choose the correct action to diminish the difference between an intended state and an actual state. The achievement of this goal will depend, in part at least, on the state of the variables defining the system (organization) as well as upon certain environmental parameters significantly affecting the organization.

The fate of the organization and of the more comprehensive EO system will depend on keeping both the *effective variables* (L_{11} and L_{22}) and the *effective parameters* (L_{12} and L_{21}) within certain desired limits. Survival is a quality inherent in both subsystems taken together and not in either of the two systems taken separately.

The function of the scanning system is the assessment of the relation between the intended state and the actual state. Data on these states are communicated via a mismatch signal to the internal organizing system for evaluation, and from there to the decision system for the necessary information formation. By means of these three systems the organization tries to build associations between the states of the environment and the organization actions that will bring the EO system into a harmonious relationship.

---

[24] Aguilar, *Scanning the Business Environment*, p. 68.
[25] Aguilar, *Scanning the Business Environment,* pp. 18–24; R. M. Cyert and J. G. March, *A Behavioral Theory of the Firm,* (Englewood Cliffs, N.J.: Prentice-Hall, 1963), pp. 120–27.

Different environments will call for different scanning modes. In a stable environment in which the differences between the actual and the intended states are not very large and do not vary often, the amount of scanning required will be relatively small. However, scanning here will be well organized and considerably formalized. A dynamic environment, on the other hand, requires considerably more scanning, although the degree of formalization will be much less than in the case of the stable environment.

Specific modes of scanning the environment are determined by the magnitude and by the direction of the discrepancy between the goal and its realization.

Environmental sectors that in the past revealed an opportunistic type of feedback are expected to evoke more surveillance than search. Problem sectors, on the other hand, are expected to trigger problemistic search.

In the following chapter we will examine the ways in which organizational effectiveness is determined.

KEY WORDS

Scanning

Surveillance

Mismatch Signal

Scanning Modes

Decision-Making Process

Search

External Information

Information Acquisition

REVIEW QUESTIONS

1. In the previous chapter an organization was said to interact with its environment. Just how does this occur? Is it possible for a firm to have more than one environment?

2. If you were hired as a "scanning manager" in an organization, what do you suppose your job would be? Write a job description for your new job.

3. Most firms would acknowledge that they do informal scanning but do not have any department solely concerned with this activity. In reality, many firms do scan on a formal basis. How is this so, and what department in a large organization might be noted as doing most of the scanning? Could this vary with the type of organization? Give some examples.

4. Utilizing the organization-environment interaction matrix presented in the chapter, sketch out the matrix of an organization which you are familiar with.

5. Discuss the environment for your university. Be careful to note things which are in the system but perhaps controllable at a higher level of the organization.

6. The transactional dependency of L_{22} implies that some elements of the environment are affecting other elements of the environment. Give some examples of this.

7. Speculate on what some of the determinants of scanning would be for an insurance company. Contrast this with a firm in the computer industry.

8. How do data differ from information, and why is this distinction necessary? What does "information formation" mean?

9. Name several industries which you think have a dynamic environment and several which have a stable environment. Does it make any sense to talk about a stable company in a dynamic environment or a dynamic company in a stable environment? Discuss fully.

10. Draw up a list of questions which attempt to discriminate between the various modes of scanning, administer this questionnaire to your students, and then determine whether they can distinguish between the various modes of scanning.

APPENDIX: A RESEARCH METHODOLOGY FOR ASSESSING THE SCANNING PROCESS

Several of the authors of the studies mentioned in this chapter employed the same basic research procedures, usually through a two-part process. The first part of the study was concerned with an environmental classification of the firms chosen. In both of the previously cited studies of Kefalas and Nishi, two sets of criteria were used to classify firms in either the relatively dynamic or relatively stable group. Figure 8–11 presents the classification model which includes both objective and subjective criteria. While the selected criteria are not all precise measures, they are nonetheless representative of the best measures available to the researchers. The objective measures chosen correspond to those used by Lawrence and Lorsch in their classic study and deal with environmental uncertainty. The main idea running through the items is that an industry confronted with a dynamic environment will have greater uncertainty than an industry in a stable environment.

Part two of the study concerned the collection of data on the executives' scanning styles by means of a questionnaire, a copy of which follows. Three measures of environmental uncertainty were used: (1) clarity of information on environmental sectors, that is the degree to which executives felt informed of events occurring outside the company; (2) clarity of relationships with respect to events occurring outside the organization and their impacts upon the organization; and (3) the time span of definitive feedback relating to actions taken by the organization in regard to any environmental sector. The empirical findings of the propositions advanced and a thorough coverage of the methodology can be found in the recent study of Nishi.[26]

---

[26] Nishi, "A Study of the Information-Acquisition Process in Japanese Computer and Information Processing Industries."

FIGURE 8–11 An Environmental Classification Model

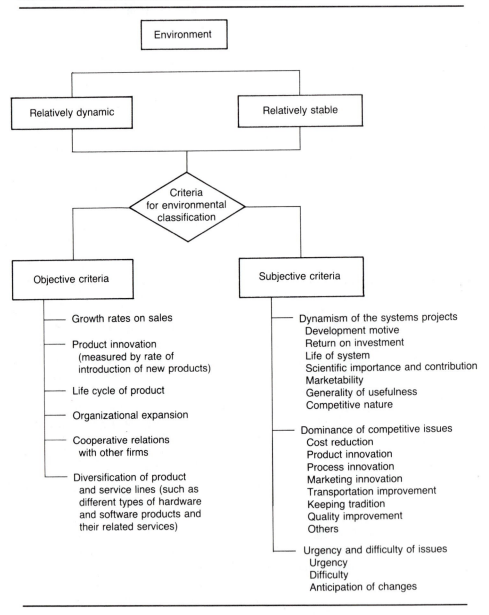

QUESTIONNAIRE ABOUT SCANNING THE EXTERNAL ENVIRONMENT –
Management Information–Acquisition Process
Overall Instructions for the Questionnaire

Please answer all questions. They cover the information needed to complete a study concerning the scanning process in your organization. The entire questionnaire consists of:

    I. Background Information about the Respondent.
    II. Questions about Company Strategies or Competitive Issues.
    III. Questions about Environmental Certainty.
    IV. Questions about Scanning for Strategic Information.

This questionnaire is to be used for an empirical research study only and will never be used for any type of personal investigation or any other purpose. Please feel free to express your scanning activities in acquiring external information for strategic decisions and planning in your company organization.

The information received will be strictly confidential. Please do not sign or attach your name on any part of the questionnaire.

Thank you very much for your time and cooperation.

I. BACKGROUND INFORMATION ABOUT THE RESPONDENT

1. Please indicate your job title: _____
2. Please check your position (i.e., management level) in your organization: Top ____, Middle ____, or Lower ____
3. If you are in the middle-management level, please check the level at which you most frequently participate in decision making:
   Upper level ____, Lower level ____
4. Please check your functional responsibility in your organization:

    a) General management   _____
    b) Planning   _____
    c) Finance/Accounting   _____
    d) Marketing/Sales   _____
    e) Technical/R&D   _____
    f) Other (specify):   _____

II. QUESTIONS ABOUT THE COMPANY STRATEGIES OR COMPETITIVE ISSUES

(If you have already covered this section when you had an interview with the researcher, please skip the following items 1 through 4.)

1. Please identify your company's major strategies or competitive issues and RANK them in order of importance. You can choose them from the following items:

                                      Rank

    (1) Cost reduction   _____
    (2) Product innovation   _____

(3) Process innovation _____
(4) Marketing innovation _____
(5) Transportation improvement _____
(6) Keeping tradition _____
(7) Quality improvement _____
(8) Other (specify): _____

2. As a whole, to what extent is the issues category considered urgent, and how difficult is it to achieve an effective solution? Please indicate the degree on the 7-point scale and circle it.

	Least						Most
	1	2	3	4	5	6	7
(1) Urgency							
	1	2	3	4	5	6	7
(2) Difficulty							

3. Do you expect any drastic change in your company strategies or competitive issues within a year?
Yes _____, No _____
4. If yes, please circle an appropriate point on the 7-point scale as a consequence of the drastic change.

	Least						Most
	1	2	3	4	5	6	7
(1) Urgency							
	1	2	3	4	5	6	7
(2) Difficulty							

III. QUESTIONS ABOUT THE ENVIRONMENTAL CERTAINTY

Before answering questions provided below, please read the separate material, SUBJECT AREAS OF EXTERNAL INFORMATION, and identify the definition of each subject area of external information.

1. *Clarity of Information*: Please circle a point on the 7-point scale which best describes the degree to which you feel informed about the developments occurring in each area.

	Least Informed						Most Informed
	1	2	3	4	5	6	7
Government sector							
	1	2	3	4	5	6	7
Technology sector							
	1	2	3	4	5	6	7
Marketing sector							
	1	2	3	4	5	6	7
Resources sector							

	Least Informed						Most Informed
	1	2	3	4	5	6	7
External growth sector							
	1	2	3	4	5	6	7
Other sectors							

2. *Clarity of Cause-Effect Relationships*: Please circle a point on the 7-point scale, which best describes the degree of clarity of cause-effect relationships between the developments in each area of the external environment and their impact upon your organization.

	Least Clear						Most Clear
	1	2	3	4	5	6	7
Government sector							
	1	2	3	4	5	6	7
Technology sector							
	1	2	3	4	5	6	7
Marketing sector							
	1	2	3	4	5	6	7
Resources sector							
	1	2	3	4	5	6	7
External growth sector							
	1	2	3	4	5	6	7
Other sectors							

3. *Time Span of Feedback*: Please circle a point, on the 7-point scale, which best describes the length of time in which information feedback about results of actions taken by your organization may come back, for each sector.

	More than 3 Yrs.	2–3 Yrs.	One Year	Six Mo.	One Mo.	One Week	One Day
	1	2	3	4	5	6	7
Government sector							
	1	2	3	4	5	6	7
Technology sector							
	1	2	3	4	5	6	7
Marketing sector							
	1	2	3	4	5	6	7
Resources sector							
	1	2	3	4	5	6	7
External growth sector							
	1	2	3	4	5	6	7
Other sectors							

IV. QUESTIONS ABOUT SCANNING FOR STRATEGIC INFORMATION
Questions about Time Span for External Information

1. *Amount of time spent acquiring external information:* Please circle the approximate amount of time you spend in a day acquiring external information of strategic value.

Hours

0.5	1	2	3	4	5	6	7	8

2. *Scanning time distribution*: Please indicate the percentage distribution of your scanning time for the sector information listed below. Total percentage should add up to 100 percent.

Government-sector information	_____
Technology-sector information	_____
Marketing-sector information	_____
Resources-sector information	_____
External growth-sector information	_____
Other-sectors information	_____
	100%

Questions about Sources of External Information

Please indicate sources of information you usually use, by marking a (√) in Column I. In Column II, please indicate its degree of importance, by circling on the 7-point scale. In Column III, please circle a relative degree of frequency — i.e., how often you utilize the sources that you marked.

III.   *Frequency*

A. Human sources *Outside*	I. Use	II. *Importance* Least Most	(Less) Yearly Semiann. Qtrly. Monthly Weekly Daily
Government officials	____	1 2 3 4 5 6 7	1 2 3 4 5 6 7
Investigators of overseas business affairs	____	1 2 3 4 5 6 7	1 2 3 4 5 6 7
Corporation lawyers, advisors, consultants, service agents, etc.	____	1 2 3 4 5 6 7	1 2 3 4 5 6 7
Members of the Board of Directors (outside members)	____	1 2 3 4 5 6 7	1 2 3 4 5 6 7
Scholars and men of experience	____	1 2 3 4 5 6 7	1 2 3 4 5 6 7
Colleagues and friends	____	1 2 3 4 5 6 7	1 2 3 4 5 6 7
Customers and consumers	____	1 2 3 4 5 6 7	1 2 3 4 5 6 7
Suppliers	____	1 2 3 4 5 6 7	1 2 3 4 5 6 7
Distributors	____	1 2 3 4 5 6 7	1 2 3 4 5 6 7
Competitors	____	1 2 3 4 5 6 7	1 2 3 4 5 6 7
Others (not classified elsewhere)	____	1 2 3 4 5 6 7	1 2 3 4 5 6 7

	I. Use	II. Importance Least ... Most	III. Frequency (Less) Yearly Semiann. Qtrly. Monthly Weekly Daily
A. Human sources			
Inside			
Direct superiors	____	1 2 3 4 5 6 7	1 2 3 4 5 6 7
Other superiors	____	1 2 3 4 5 6 7	1 2 3 4 5 6 7
Peers	____	1 2 3 4 5 6 7	1 2 3 4 5 6 7
Subordinates	____	1 2 3 4 5 6 7	1 2 3 4 5 6 7
Sales personnel	____	1 2 3 4 5 6 7	1 2 3 4 5 6 7
Staff	____	1 2 3 4 5 6 7	1 2 3 4 5 6 7
Librarians (working for the company)	____	1 2 3 4 5 6 7	1 2 3 4 5 6 7
Others (not classified elsewhere)	____	1 2 3 4 5 6 7	1 2 3 4 5 6 7
B. Documentary sources			
Outside			
Governmental publications	____	1 2 3 4 5 6 7	1 2 3 4 5 6 7
Newspapers and magazines in general	____	1 2 3 4 5 6 7	1 2 3 4 5 6 7
Industrial newspapers and magazines	____	1 2 3 4 5 6 7	1 2 3 4 5 6 7
Scientific and technical magazines	____	1 2 3 4 5 6 7	1 2 3 4 5 6 7
Scientific research papers	____	1 2 3 4 5 6 7	1 2 3 4 5 6 7
Indices and abstracts	____	1 2 3 4 5 6 7	1 2 3 4 5 6 7
Unpublished papers and research materials	____	1 2 3 4 5 6 7	1 2 3 4 5 6 7
Research publications	____	1 2 3 4 5 6 7	1 2 3 4 5 6 7
Reports from service agencies	____	1 2 3 4 5 6 7	1 2 3 4 5 6 7
Material from data banks	____	1 2 3 4 5 6 7	1 2 3 4 5 6 7
Letters and reports from agents stationed in overseas countries or from cooperating foreign businesses	____	1 2 3 4 5 6 7	1 2 3 4 5 6 7
Competitors' reports on scientific research and development	____	1 2 3 4 5 6 7	1 2 3 4 5 6 7
Professional books and references	____	1 2 3 4 5 6 7	1 2 3 4 5 6 7
Texts and handbooks	____	1 2 3 4 5 6 7	1 2 3 4 5 6 7
Others (not classified elsewhere)	____	1 2 3 4 5 6 7	1 2 3 4 5 6 7
Inside			
Scientific research reports (inside)	____	1 2 3 4 5 6 7	1 2 3 4 5 6 7
Data file materials	____	1 2 3 4 5 6 7	1 2 3 4 5 6 7

III.   *Frequency*

B. Documentary sources	I. Use	II. *Importance*		III. Frequency (Less) Yearly Semiann. Qtrly. Monthly Weekly Daily
Inside		Least Most		
House organs and newsletters	____	1 2 3 4 5 6 7		1 2 3 4 5 6 7
Memos circulated inside the company	____	1 2 3 4 5 6 7		1 2 3 4 5 6 7
Reports from branch offices or remote business offices	____	1 2 3 4 5 6 7		1 2 3 4 5 6 7
Others (not classified elsewhere)	____	1 2 3 4 5 6 7		1 2 3 4 5 6 7
C. Combined mixed sources				
Outside				
Conferences and symposiums	____	1 2 3 4 5 6 7		1 2 3 4 5 6 7
Industrial meetings, lecture meetings, seminars, etc.	____	1 2 3 4 5 6 7		1 2 3 4 5 6 7
Visiting overseas firms	____	1 2 3 4 5 6 7		1 2 3 4 5 6 7
Field trips and observations	____	1 2 3 4 5 6 7		1 2 3 4 5 6 7
Business and trade shows	____	1 2 3 4 5 6 7		1 2 3 4 5 6 7
T.V. and radio	____	1 2 3 4 5 6 7		1 2 3 4 5 6 7
Films, slides, tapes, etc.	____	1 2 3 4 5 6 7		1 2 3 4 5 6 7
Formal meetings (outside)	____	1 2 3 4 5 6 7		1 2 3 4 5 6 7
Ad hoc informal meetings and discussions (outside)	____	1 2 3 4 5 6 7		1 2 3 4 5 6 7
Labor negotiations (outside)	____	1 2 3 4 5 6 7		1 2 3 4 5 6 7
Others (not classified elsewhere)	____	1 2 3 4 5 6 7		1 2 3 4 5 6 7
Inside				
Research presentation and seminars (inside)	____	1 2 3 4 5 6 7		1 2 3 4 5 6 7
Company product shows	____	1 2 3 4 5 6 7		1 2 3 4 5 6 7
Films, slides, tapes, etc. (inside)	____	1 2 3 4 5 6 7		1 2 3 4 5 6 7
Visiting and observing company facilities	____	1 2 3 4 5 6 7		1 2 3 4 5 6 7
Formal meetings (inside)	____	1 2 3 4 5 6 7		1 2 3 4 5 6 7
Ad hoc informal meetings (inside)	____	1 2 3 4 5 6 7		1 2 3 4 5 6 7
Labor negotiations (inside)	____	1 2 3 4 5 6 7		1 2 3 4 5 6 7
Others (not classified elsewhere)	____	1 2 3 4 5 6 7		1 2 3 4 5 6 7

Questions about Modes of Scanning:
   Please read the following explanations in order to understand the various ways of acquiring external information. Then answer the question at the end.

1. *Viewing* is a relatively low-attention-level way of keeping in touch with the environment through generally oriented exposure to information which might be relevant for management. In viewing, you have no specific purpose in mind for possible exploration. You can use viewing to acquire background information, or information which enables you to better understand specific job-related matters, and to pick up warning signals of matters which may become significant and which may therefore indicate areas which should be scanned more intensively.

2. *Monitoring* is one level higher on a scanning continuum than the viewing level, and is concerned with focused attention (not involving active search) with a more or less clearly defined information subject area or source area of information. In this mode you are sensitive to specific kinds of information and are ready to assess its value or importance as such information is encountered.

3. *Investigation* is the next level and refers to a relatively narrow, structured effort to search out specific information for a specific purpose. In this mode, you actively work to acquire specific information. Reading appropriate materials, letting people know of your interest in order to encourage communication, keeping your eyes on the environment to check on the results of some current policy or activity or to uncover new information on any one of many issues known to be of interest, are all examples of the investigation mode.

4. *Research* is the highest of modes on the scanning process continuum. It is a formally structured effort of searching specific information for a particular purpose. Much of the activity performed in a research and development (R&D) unit would be included in this category. Market research is another example of environmental research.

The relationship of these four modes is diagrammatically illustrated. The continuum starts with "viewing" at the left end and terminates with "research" at the right end. The intensity of scanning increases as the activities move from surveillance to research. Associated with the increasing intensity of scanning is the degree of involvement (time and effort) of the scanner.

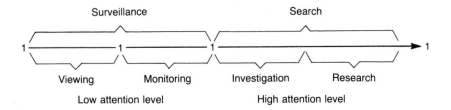

Please indicate the percentage of information you acquired within the last few weeks while using each of the four modes of scanning. Percentages should add up to 100.

1. Viewing   _____
2. Monitoring   _____
3. Investigation _____
4. Research   _____
   Total   100%

Additional References

Boulton, William, et al. "Strategic Planning: Determining the Impact of Environmental Characteristics and Uncertainty." *Academy of Management Journal* 25, no. 3 (September 1982), pp. 500–509.

Godiwalla, Y. M., et al. "Environmental Scanning – Does it Help the Chief Executive?" *Long Range Planning* 13 (October 1980), pp. 87–99.

Hambrick, Donald C. "Environmental Scanning, Organizational Strategy, and Executives' Roles: A Study in Three Industries." Ph.D. dissertation, Pennsylvania State University, 1979.

_____. "Environment, Strategy, and Power within Top Management Teams." *Administrative Science Quarterly* 26, no. 2 (June 1981), pp. 253–75.

_____. "Specialization of Environmental Scanning Activities among Upper Level Executives." *The Journal of Management Studies* 18 (July 1981), pp. 299–320.

Jain, S. "Environmental Scanning in U. S. Corporations." *Long Range Planning"* 17, no. 2 (1984), pp. 117–28.

Kast, F. "Scanning the Future Environment: Social Indicators." *California Management Review* 23 (Fall 1980), pp. 22–32.

Kefalas, A. G. "Scanning the External Business Environment." Ph.D. dissertation, The University of Iowa, 1971.

Kiesler, S., and L. Sproull. "Managerial Response to Changing Environments: Perspectives on Problem Sensing from Social Cognition." *Administrative Science Quarterly* 27, no. 4 (December 1982), pp. 548–70.

Lenz, R. T., and J. L. Engledow. "Environmental Analysis Units and Strategic Decision-Making: A Field Study of Selected Leading-Edge Corporations." *Strategic Management Journal* 7, no. 1 (1986), pp. 69–89.

_____. "Environmental Analysis: The Applicability of Current Theory." *Strategic Management Journal* 7, no. 4 (1986), pp. 329–46.

Nanus, B. "QUEST – Quick Environmental Scanning Technique." *Long Range Planning* 15 (April 1982), pp. 39–45.

Nishi, Kenyu. *Management Scanning Process.* Tokyo: Saikon Publishing, 1979.

Nishi, Kenyu; Charles Schoderbek; and Peter Schoderbek. "Scanning the Organizational Environment: Some Empirical Results." *Human Systems Management* 3 (1982), pp. 233–45.

O'Connell, Jeremiah J., and John W. Zimmerman. "Scanning the International Environment." *California Management Review* 22, no. 2 (Winter 1979), pp. 15–23.

Preble, J. F. "Corporate Use of Environmental Scanning." *University of Michigan Business Review* 30 (September 1978), pp. 12–17.

Stubbart, C. "Are Environmental Scanning Units Effective?" *Long Range Planning* 15 (June 1982), pp. 139–45.

Thomas, P. S. "Environmental Scanning – The State of the Art." *Long Range Planning* 13 (February 1980), pp. 20–28.

What Lies Ahead – a New Look. An Environmental Scan Report. (Alexandria, Va.: 1983).

CHAPTER 9

Organizational Effectiveness

There is nothing more difficult to carry out, nor more doubtful of success, nor more dangerous to handle, than to initiate a new order of things. For the reformer has enemies in all who profit by the old order, and lukewarm defenders in all those who would profit by the new order.

Machiavelli

CHAPTER OUTLINE

INTRODUCTION

One topic, currently much discussed, though with few tangible and incontrovertible results, is organizational effectiveness. Scores of management scientists, sociologists, psychologists, and practitioners of other scientific disciplines have devoted much thought, research, and critical analysis to their own and others' theories, yet have advanced little that is significantly different from and superior to previous findings. Nevertheless, one finds in current journals so-called newer and better approaches to the topic of organizational effectiveness. Perhaps before discussing the various approaches that have been advanced, it might be worthwhile to consider why this condition persists.

One very common and logical reason for intellectual confusion within a particular field of study is the diversity of concepts employed by practitioners. Since concepts, propositions, and theories are the basic components of scientific inquiry, differing concepts lead to different propositions, and sets of different propositions lead ultimately to different systematizations of propositions that constitute theory as such.

Now, concepts are basically an individual's way of perceiving phenomena. In the conceptualization process one abstracts from or selects from reality certain essential characteristics. Because the concepts formed neither affirm nor deny anything of reality, they are neither true nor false. They merely enable one to make statements about reality; they are not the statements or propositions themselves. Hence, Professor X's conceptualization of effectiveness is just as "valid" as Professor Y's. Consequently, concepts must be evaluated, not in the light of their inherent truth or falsity, which they cannot possess, but in terms of their usefulness for the advancement of scientific knowledge. For this purpose, concepts should have certain characteristics.[1] They should be clear (free from vagueness and ambiguity), wide-ranging or broad in scope (highly abstract, since the greater the abstraction the greater the scope or range of phenomena to which they can apply), and should possess systematic import (the more the concept is incorporated into propositions and theories, the more useful the concept).

The way in which organizational effectiveness is conceptualized can vary considerably from one author to the next. One may see no great distinction between effectiveness and efficiency—at least nothing to get excited about. Another may dichotomize the concepts. To some, efficiency means optimizing the yield from available resources, while effectiveness has to do with allocating one's resources and efforts to the best ends. As Peter Drucker succinctly phrases it: "Efficiency is concerned with doing things right. Effectiveness is doing the right things."[2] Sometimes one is regarded as being inclusive of the other. For most theorists, efficiency is defined as but one aspect of the broader concept of

---

[1] Bernard Phillips, *Social Research: Strategy and Tactics* (New York: Macmillan, 1966), pp. 32–35.

[2] Peter Drucker, *Management: Tasks, Responsibilities, Practices* (New York: Harper & Row, 1973), p. 45.

organizational effectiveness. Effectiveness, in other words, includes efficiency. On the other hand, some researchers have made efficiency the more inclusive concept of the two.[3]

However, this is not the only way of dichotomizing these concepts. Chester Barnard viewed the subject from still another perspective. Activities of an individual in an organization aimed at achieving ends sought by the organization he defined as effective, while the behavior of an individual in an organization directed to the satisfaction of personal goals he considered efficient. Thus, activities that take place in organizations should, from this viewpoint, be judged by two distinct critera: effectiveness (the degree to which organizational goals are attained at least cost) and efficiency (the individual's personal satisfaction derived from the activity).[4]

In the early stages of any scientific discipline one may find an abundance of vague, ambiguous, and ill-defined terms. Individual practitioners tend to coin their own terms and to avoid as much as possible those of others. This tendency to multiply words, to mint one's own coinage, retards scientific progress by making research difficult to interpret or to replicate.

A second cause of intellectual confusion is extreme operationalism. Concepts are defined not only theoretically in a general way but also operationally in very specific ways. Even though different researchers may have the same theoretical definition of a concept, the definitions they use for research purposes may vary substantially. Operational definitions are essential, since without them it would be impossible to carry out meaningful research, but when clarity and ease of objective measurement occupy the center of the stage in concept formation, the concept may have little or no meaning apart from the detailed operations used in the specific study. Clarity, precision of definition, and especially ease of measurement are but means to achieve the goals of explanation and prediction, not the goals in themselves.

When one examines the managerial literature on effectiveness, one is struck by the many and different criteria used to assess effectiveness. The notion that there is only one criterion has generally been abandoned. However, some of the criteria currently employed overlap, and others bear only tangentially on the core variables. John Campbell has rendered a distinct service to researchers by reviewing the vast literature and setting forth the various indicators of organizational effectiveness in use, and spelling out the object lessons to be learned from past research.[5]

A third reason for the confusion underlies the previously cited ones. Since concepts are but ways of perceiving phenomena, differing perspectives on

---

[3] See Selwyn Becker and Duncan Neuhauser, *The Efficient Organization* (New York: Elsevier, 1975), p. 46.

[4] William B. Wolf, *The Basic Barnard* (Ithaca, N.Y.: Cornell University Press, 1974), p. 69.

[5] John P. Campbell, "Contributions Research Can Make in Understanding Organizational Effectiveness," *Organization and Administrative Sciences*, 7, nos. 1–2 (Spring-Summer 1976), pp. 29–45.

organizational effectiveness must lie at the root of the problem. Why these different perspectives? The answer seems to be that the two fundamentally different meanings associated with organizational effectiveness depend on how one views the organization – from inside or from outside.[6]

The perspective that views the organization from within is a typically managerial one. Are the invested organizational resources being used efficiently and productively? This perspective tends to measure organizational effectiveness by return on investment.

In the other radically different perspective, the organization is viewed from the outside – in its relationships to the larger society. From this perspective, a cost-benefit analysis would be the most appropriate. Dubin calls these two viewpoints the fundamental dilemma – the efficient resource utilization perspective and the social utility perspective. Both perspectives cannot be maximized at the same time, for they are poles apart. When effectiveness is high from one viewpoint, it will not be from the other. This basic distinction pervades our whole economy. Dubin shows how pervasive it is by examining three broad areas of business organization: economic activities in the marketplace, personnel management, and organizational structure. He spells out four steps in an operating strategy that should dispel any undue optimism that a single approach to organizational effectiveness can be universally applied.

Step 1. Accept the fact that a choice must be made between the internal efficiency and the social utility perspectives. It's inevitable!

Step 2. Within a given business organization, the preferred choice will not always be that which is consistent with the view of an insider. In this area, consistency is not a virtue!

Step 3. Be cognizant of the fact that the measurement of organizational effectiveness is very different in the two opposing viewpoints. Each has its own appropriate criteria.

Step 4. Recognize that, under some circumstances (like industrial pollution), both opposing views of organizational effectiveness will have to be employed at the same time, and in the ensuing compromise keep clearly in mind how much of one is being traded off for the other. In this way, appropriate amounts of both operating efficiency and social utility can be attained at the same time.

Further insight into the problem may be gained by considering the problem in its historical setting. This is what Donald Schon has attempted to do.[7] He developed what he believed were the four stages in the study of organizational effectiveness.

---

[6] For example, see Robert Dubin, "Organizational Effectiveness: Some Dilemmas of Perspective," *Organization and Administrative Sciences* 7, nos. 1–2 (Spring-Summer 1976), pp. 7–13.

[7] Donald A. Schon, "Deutero-Learning in Organizations: Learning for Increased Effectiveness," *Organizational Dynamics* 4, no. 1 (Summer 1975), pp. 2–16.

In the first stage (from World War I to World War II), the goals of sound organization — that is, quality of product, good leadership, and optimum performance techniques — were utilized as static models. The problem of organization was seen as how to achieve and maintain a level of performance that approximated these models.

In the second stage (from World War II until the late 1950s) emphasis shifted from static models to dynamic ones; the development of new products and services came to the fore. The creativity model replaced the optimum performance model.

In the third stage (from the late 1950s until the mid-1960s) interest shifted from individual and group performance to the performance of the organization as a whole. This was a period of intense investigation of the organization as such.

In the final stage (from the mid-1960s to the present) effectiveness, according to Schon, has been linked to the organization's ability to cope with change. It is seen as depending on the ability and willingness of an organization to continually redesign itself in response to changing values and a changing context for learning. Structural and environmental variables come to the fore.

Perhaps what Katz and Kahn wrote over two decades ago as they surveyed the literature is as valid today as it was then:

> The literature is studded with references to efficiency, productivity, absence, turnover, and profitability — all of these offered implicitly or explicitly, separately or in combination, as definitions of organizational effectiveness. Most of what has been written on the meaning of these criteria and on their interrelatedness, however, is judgmental and open to question.[8]

This same attitude is expressed by W. Richard Scott in his article surveying the effectiveness of organizational effectiveness studies. His critique still rings true today.

> After reviewing a good deal of the literature on organizational effectiveness and its determinants, I have reached the conclusion that this topic is one about which we know less and less. There is disagreement about what properties or dimensions are encompassed by the concept of effectiveness. There is disagreement about who does or should set the criteria to be employed in assessing effectiveness. There is disagreement about what indicators are to be used in measuring effectiveness. And there is a disagreement about what features of organizations should be examined in accounting for observed differences in effectiveness.[9]

Of the various approaches to the understanding and measurement of organizational effectiveness, three stand out in the literature: the goal approach,

---

[8] Daniel Katz and Robert L. Kahn, *The Social Psychology of Organizations* (New York: John Wiley and Sons, 1966), p. 149.

[9] W. Richard Scott, "Effectiveness of Organizational Effectiveness Studies," in *New Perspectives on Organizational Effectiveness,* ed. Paul S. Goodman, Johannes M. Pennings, and associates (San Francisco: Jossey-Bass Publishers, 1977), pp. 63–95. Quotation appears on pp. 63f.

the systems-resource approach, and the financial viability approach. These will now be treated in this sequence.

GOAL APPROACH

For many years this has been the traditional method for studying organizational effectiveness. In this method effectiveness is measured by the degree to which the organization achieves its goals or objectives.

Since there are various types and levels of goals, these should be clearly defined and analyzed prior to their use in evaluating organizational effectiveness.

In a general way, an organizational goal is "a desired state of affairs which the organization attempts to utilize"[10] or a desired set of aims or tasks. Initially, one must realize that the state or aims or tasks desired must be desired by someone. Organizations as such do not have desires; men and women who run them or who work in them do. Consequently, in the goal approach (true also for the systems-resource approach) one must from the very start decide on whether to follow a normative or a descriptive path for determining goal attainment criteria.

In the normative path one defines the goals that must be attained, the resources that must be acquired, or the tasks that must be performed if the organization is to be successful. Of the various problems inherent to this approach, one is paramount: Who will specify the goals of the organization — the researcher or the organizational decision makers? A researcher who does so runs the risk of using biased personal values that may be totally unrelated to the stated or actual goals of the organization. If the decision makers do so, these organizational goals may be only the official ones and not the ones the organization is actually trying to reach.

The descriptive path attempts to bypass the personally broad value premises of the researcher for empirically based data. It is no longer a question of what the organization is striving after, but what types of goals characterize organizations that are successful. Here too lurks a built-in limitation: How does one define a successful organization objectively, free of the value judgments of the researcher?

According to the proponents of the goal approach, the following five assumptions seem to underlie its use:

1. Organizational effectiveness is defined in terms of the degree of goal attainment.[11]

2. The organizational goals are explicitly set by the dominant interest groups or decision makers, using rational bases for their decision making.

3. Organizations generally have mutliple and conflicting goals.

---

[10] Amitai Etzioni, *Modern Organizations* (Englewood Cliffs, N.J.: Prentice-Hall, 1964), p. 6.

[11] James L. Price, "The Study of Organizational Effectiveness," *Sociological Quarterly* 13, no. 1 (Winter 1972), pp. 3–15.

4. In organizations, goal optimization is to be preferred to goal maximization. Since organizations may have multiple and often conflicting goals, goal maximization may be more of a theoretical exercise than a realizable policy.

5. Every organization is related to its environment, from which it receives resources and to which it contributes products or services. Organizational goal attainment is, therefore, contingent upon the organization maintaining suitable relationships with its environment.

Goals can be conceptualized on more than just one level—societal, organizational, and individual. In the early stages of the problem, attention was focused on individual and group goals. However, one must be careful not to identify individual goals with organizational ones. These goals are not only conceptually distinct but often can be operatively so. Individual goals can work at cross purposes with organizational goals just as they can coincide with and reinforce them. Today, effective managers try to integrate individual goals with those of the organization, and also with those embedded in the culture of society. Societal goals, especially in this present decade, can no longer be ignored.

Besides recognizing goals on various micro and macro levels, goal-approach theorists also distinguish between official, operative, and operational goals. Following Charles Perrow's lead,[12] theorists define *official* goals as the often formally stated ones put forth in the organizational charter, annual reports, public statements, and other authoritative pronouncements. They are usually somewhat vague and ambiguous, formal, broad in scope, and of the "apple pie and motherhood" variety. For outsiders, especially government, law, and the establishment, they legitimate the activities of the organization. For the insider, they also serve to symbolically differentiate the organization from others.[13]

Operative goals "designate the ends sought through the actual operating policies of the organization; they tell us what the organization is actually trying to do, regardless of what the officials say are the aims.[14] Operative goals may coincide with the officially documented ones, or they may not.

Operational goals are operative ones for which criteria for evaluation already exist. They are operative goals operationally defined. Since quantitative variables are more easily measurable than qualitative ones, regardless of relevance, it is not surprising that the quantitative criteria are the ones most often operationalized by researchers.

Of the three types of goals mentioned, the operative are most pertinent to an organization's effectiveness. Because official goals are the more normal, more general, and more motivational (hence vague), they can be spelled out more easily

---

[12] Charles Perrow, "The Analysis of Goals in Complex Organizations," *American Sociological Review* 26, no. 6 (December 1961), pp. 854–66. In *Organizational Analysis: A Sociological View* (Belmont, Calif: Wadsworth Publishing, 1970), Perrow devotes chapter 5 (pp. 133–74) to an exposition and illustration of his five types of goals: societal, output, system, product, and derived.

[13] Official goals can also be used as motivational factors. See Scott, "Effectiveness," pp. 64–65.

[14] Perrow, "Analysis of Goals," p. 855.

than the actual (operative) goals of the organization. However, the success or failure of an organization will hinge on the actual goal(s) pursued, or the direction in which the organization is headed. The actual measurement of organizational effectiveness is usually carried out either by asking individual members of the organization to identify the goals toward which the organization is tending, or by inferring from the behavior of individual members the goals the organization actually has.

The typical organizational condition seems to be one where goal conflict exists among different groups of members. Members form coalitions (alliances) because they believe their own personal goals will be realized through the activities of the coalition. As a result, the goals an organization appears to be actually pursuing will be a function of past goals and past practices (generally taken for granted), present goals of the coalition currently "ruling" the organization, and the future desired states for the organization.[15] Researchers favoring this view tend to interview members of the dominant coalition(s).

Some authors raise the question whether organizations actually have goals. Cyert and March[16] maintain (1) that individuals alone and not collectivities of individuals have goals, and (2) that to define a theory of organizational decision making, one apparently needs at the organizational level something analogous to individual goals.

While not all students of organization will readily subscribe to the foregoing statements, it appears that in practice they need these premises to discuss the subject of conflict — conflict between individual and organizational goals, between stated and implied goals, and so on.

Cyert and March suggest that the problem may be dealt with by viewing the organization as a *coalition* of individuals.

> Let us view the organization as a coalition. It is a coalition of individuals, some of them organized into subcoalitions. In a business organization the coalition members may include managers, workers, stockholders, suppliers, customers, lawyers, tax collectors, regulatory agencies, etc. In the governmental organization the members include administrators, workers, appointive officials, elective officials, legislators, judges, clientele, interest group leaders, etc. In the voluntary charitable organization there are paid functionaries, volunteers, donors, donees, etc.[17]

Not only do the different coalitions have different and often conflicting goals, but the goals are continually changing. Some of the changes are induced by changes in the external environment; others may be modified by internal coalitions, such as unions. Even when goals remain the same, their applications

---

[15] Richard M. Cyert and James B. March, *A Behavioral Theory of the Firm* (Englewood Cliffs, N.J.: Prentice-Hall, 1963), chap. 3.

[16] Cyert and March, *Behavioral Theory of the Firm*, p. 26.

[17] Cyert and March, *Behavioral Theory of the Firm*, p. 27.

may change because of changes in the environment, in the organization, or in both. For example, the goals of a university are essentially unchanging, yet many factors—the needs of students, new technology, changing patterns of funding, and others—can modify the application of the goals from year to year.

Coalitions also change. Union representatives, governmental officials, management personnel, stockholders, and directors routinely leave their coalitions. With a different mix of members, different preferences may be expected to prevail in the ordering of goals.

Not only is goal formulation affected by changing coalitions; other forces are also at work. In discussing these, the present authors will adopt the classic approach to the subject: the definition of organizational goals as the goals of top management or of the stockholders. Goal completion is then attained through the inducement of rewards and the introduction of appropriate internal controls. Insistence on goal consistency, of course, brings with it certain limitations, such as unresolved conflict within the organization. For any organization to survive, however, there must be some degree of goal compatibility among the various coalitions.

Cyert and March list five goals of a business—all of which must be realized. Although they make no formal attempt to order them according to their importance, an implicit ordering occurs as a result of the circumstances affecting the goals. The five areas noted by Cyert and March are the areas of production, inventory, sales, market share, and profit. According to them, changes in the goals reflect changes in the structure of the existing coalitions (stockholders, managers, employees, and others). On the other hand, Peter Drucker[18] holds that if a business organization is to remain viable, results must be achieved in eight different areas: market standing, innovation, productivity, physical and financial resources, profitability, manager performance and development, worker performance and attitude, and public responsibility. These eight areas are generally recognized as being nontransferable, i.e., not to be identified with other organizational types or applied to certain types of private (not-for-profit) organizations. Drucker's eight goals are representative of systems goals and demonstrate quite convincingly their multiplicity. From this it is easy to see why the various coalitions making up an organization can have different goals and objectives.

Some of the problems surrounding the area of organizational effectiveness may stem from the heavy emphasis on quantitative measures for assessment. Richard Steers, in his analysis of 17 multivariate studies of organizational effectiveness, discovered 15 different operational definitions.[19] Only one of these

---

[18] Peter Drucker, *The Practice of Management* (New York: Harper & Brothers, 1954), p. 63.

[19] Richard M. Steers, "Problems in the Measurement of Organizational Effectiveness," *Administrative Science Quarterly* 20, no. 4 (December 1975), pp. 546–58.

TABLE 9–1   Frequency of Occurrence of Evaluation Criteria in 17 Models of
Organizational Effectiveness

Evaluation Criteria (15 Operational Definitions)	Times Mentioned (N = 17)
Adaptability-flexibility	10
Productivity	6
Satisfaction	5
Profitability	3
Resource acquisition	3
Absence of strain	2
Control over environment	2
Development	2
Efficiency	2
Employee retention	2
Growth	2
Integration	2
Open communications	2
Survival	2
All other criteria	1

SOURCE: Richard M. Steers, "Problems in the Measurement of Organizational Effectiveness," *Administrative Science Quarterly,* 20, no. 4 (December 1975), p. 549.

was found in more than half of the studies. This was the adaptability-flexibility aspect, followed not too closely by productivity and employee job satisfaction (see Table 9–1).

In a somewhat similar and detailed review, John Campbell presented a synthesized list of 30 possible indicators of organizational effectiveness—criteria measures indicated by the empirical literature he had surveyed.[20] Among the more commonly employed indicators were overall performance as measured by employee or supervisory ratings, productivity measured by output data, and employee job satisfaction as evidenced by the self-reported questionnaire data. A somewhat condensed summary of these measures is given in Figure 9–1.

Another view of goal formulation, equally important to researchers of organizational effectiveness, is that of Herbert Simon.[21] He suggests that a useful way of discovering the operating goals of an organization is to identify the constraints under which the decision maker(s) operate. Many of these constraints have their origin in the diverse goals the organization pursues just to keep the

---

[20] John Campbell, "On the Nature of Organizational Effectiveness," in *New Perspectives on Organizational Effectiveness* ed. Paul S. Goodman and others (San Francisco: Jossey-Bass, 1978), pp. 13–55. The summary appears on pp. 36–39.

[21] Herbert Simon, "On the Concept of Organizational Goal," *Administrative Science Quarterly* 9, no. 1 (June 1964), pp. 1–22. This article, with minor revisions, appears as chapter 12 in Herbert A. Simon, *Administrative Behavior,* 3d ed. (New York: Free Press, 1977).

FIGURE 9–1   Synthesized List of Possible Indicators of Organizational Effectiveness

1. *Overall effectiveness.* The general evaluation that takes into account as many criteria facets as possible. It is visually measured by combining archival performance records or by obtaining overall ratings or judgments from persons thought to be knowledgeable about the organization.

2. *Productivity.* Usually defined as the quantity or volume of the major product or service that the organization provides. It can be measured at three levels: individual, group, and total organization via either archival records or ratings or both.

3. *Efficiency.* A ratio that reflects a comparison of some aspect of unit performance to the costs incurred for that performance.

4. *Profit.* The amount of revenue from sales left after all costs and obligations are met. Percent return on investment or percent return on total sales are sometimes used as alternative definitions.

5. *Quality.* The quality of the primary service or product provided by the organization may take many operational forms, which are largely determined by the kind of product or service provided by the organization. They are too numerous to mention here.

6. *Accidents.* The frequency of on-the-job accidents resulting in lost time. Campbell and others (1974) found only two examples of accident rates being used as a measure of organizational effectiveness.

7. *Growth.* Represented by an increase in such variables as total manpower, plant capacity, assets, sales, profits, market share, and number of innovations. It implies a comparison of an organization's present state with its own past state.

8. *Absenteeism.* The usual definition stipulates unexcused absences but even within this constraint there are a number of alternative definitions (for example, total time absence versus frequency of occurrence).

9. *Turnover.* Some measure of the relative number of voluntary terminations, which is almost always assessed via archival records. They yield a surprising number of variations, and few studies use directly comparable measures.

10. *Job satisfaction.* Has been conceptualized in many ways but perhaps the modal view might define it as the individual's satisfaction with the amount of various job outcomes he or she is receiving. Whether a particular amount of some outcome (for example, promotional opportunities) is "satisfying" is in time a function of the importance of that outcome to the individual and the equity comparisons the individual makes with others.

11. *Motivation.* In general, the strength of the predisposition of an individual to engage in goal-directed action or activity on the job. It is not a feeling of relative satisfaction with various job outcomes but is more akin to a readiness or willingness to work at accomplishing the job's goals. As an organizational index, it must be summed across people.

12. *Morale.* It is often difficult to define or even understand how organizational theorists and researchers are using this concept. The modal definition seems to view morale as a group phenomenon involving extra effort, goal communality, commitment, and feelings of belonging. Groups have some degree of morale, whereas individuals have some degree of motivation (and satisfaction).

FIGURE 9–1 *(continued)*

13. *Control.* The degree of, and distribution of, management control that exists within an organization for influencing and directing the behavior of organization members.

14. *Conflict/Cohesion.* Defined at the cohesion end by an organization in which the members like one another, work well together, communicate fully and openly, and coordinate their work efforts. At the other end lies the organization with verbal and physical clashes, poor coordination, and ineffective communication.

15. *Flexibility/Adaptation* (Adaptation/Innovation). Refers to the ability of an organization to change its standard operating procedures in response to environmental changes. Many people have written about this dimension, but relatively few have made attempts to measure it.

16. *Planning and Goal Setting.* The degree to which an organization systematically plans its future steps and engages in explicit goal-setting behavior.

17. *Goal Consensus.* Distinct from actual commitment to the organization's goals, consensus refers to the degree to which all individuals perceive the same goals for the organization.

18. *Internalization of organizational goals.* Refers to the acceptance of the organization's goals. It includes the belief that the organization's goals are right and proper. It is *not* the extent to which goals are clear or agreed upon by the organization members (goal clarity and goal consensus, respectively).

19. *Role and norm congruence.* The degree to which the members of an organization are in agreement on such things as desirable supervisory attitudes, performance expectations, morale, role requirements, and so on.

20. *Managerial interpersonal skills.* The level of skill with which managers deal with superiors, subordinates, and peers in terms of giving support, facilitating constructive interaction, and generating enthusiasm for meeting goals and achieving excellent performance. It includes such things as consideration, employee centeredness, and so on.

21. *Managerial task skills.* The overall level of skills with which the organization's managers, commanding officers, or group leaders perform work-centered tasks, tasks centered on work to be done, and not the skills employed when interacting with other organizational members.

22. *Information management and communication.* Completeness, efficiency, and accuracy in analysis and distribution of information critical to organizational effectiveness.

23. *Readiness.* An overall judgment concerning the probability that the organization could successfully perform some specified task if asked to do so. Work on measuring this variable has been largely confined to military settings.

24. *Utilization of environment.* The extent to which the organization successfully interacts with its environment and acquires scarce and valued resources necessary to its effective operation.

25. *Evaluations by external entities.* Evaluations of the organization, or unit, by the individuals and organizations in its environment with which it interacts. Loyalty to, confidence in, and support given the organization by such groups as suppliers, customers, stockholders, enforcement agencies, and the general public would fall under this label.

FIGURE 9–1   *(continued)*

26. *Stability.* The maintenance of structure, function, and resources through time, and more particularly, through periods of stress.

27. *Value of human resources.* A composite criterion which refers to the total value or total worth of the individual members, in an accounting or balance sheet sense, to the organization.

28. *Participation and shared influence.* The degree to which individuals in the organization participate in making the decisions that directly affect them.

29. *Training and development emphasis.* The amount of effort the organization devotes to developing its human resources.

30. *Achievement emphasis.* An analog to the individual need for achievement, referring to the degree to which the organization appears to place a high *value* on achieving major new goals.

SOURCE: J. P. Campbell, "On the Nature of Organizational Effectiveness," in *New Perspectives on Organizational Effectiveness,* ed. P. Goodman, J. M. Pennings, and associates (San Francisco: Jossey Bass, 1978), p. 36–39.

various pressure groups at peace with one another. Some of these constraints may be imposed by formal rules or regulations or by informal norms or by "suggestions" of those higher up in the organizational hierarchy or by significant persons with power in the environment. Some may also reflect the personal desires, preferences, or norms of the decision maker. Consequently, to discover the constraints under which a decision maker makes a decision, one must know something of the environment under which the organization operates; the past practices and precedents of the organization; the preferences of the chief decision maker's colleagues; the head's personal preferences, needs, and aspirations; perception of what the organization's clientele is looking for; perception of what society at large will tolerate; view of the attitudes of major stockholders; and so on.

Because of these difficulties, some theorists have argued that the goal approach is vitiated by an inability to identify organizational goals.[22] Others, like John Price, have drawn up guidelines to aid one in defining organizational goals. Price's formulation can be condensed to these essentials:

The focus of research should be on:

1. The major decision makers of the organization.

2. Organizational goals as opposed to private individual ones.

3. Operative goals as opposed to official goals.

4. Intentions and activities (what the participants think the organization is trying to accomplish and what organizational members are observed doing).[23]

---

[22] Ephraim Yuchtman and Stanley Seashore, "A System Resource Approach to Organizational Effectiveness," *American Sociological Review* 32, no. 1 (December 1967), p. 892.

[23] Price, "Study of Organizational Effectiveness," pp. 5–6.

Once defined, operative goals should be weighted relative to their importance, if this is possible. Not all organizational goals are equally important. Some rank higher in the hierarchical ladder than others, or there may be no clear-cut hierarchy at all. This hierarchical ranking — priority — of operative goals can be ascertained either by asking decision makers to so rank them or by observing how much of what resources (personnel, production, research and development, etc.) is allocated to the achievement of various organizational goals.

Despite all the conceptual and practical problems that researchers of organizational effectiveness experience, it still remains true that corporate managers do continue to initiate, implement, and evaluate strategic and tactical policies, do set long-range, intermediate, and short-range goals toward which they strive, do often employ management by objectives whereby specific operative and operationalized goals are set and assessment made respecting the degree of attainment of these objectives. Despite the plethora of criterion measures, managers at different levels of organizational complexity and of different types of corporate enterprise do manage to select those criteria that they have found enable them to control day-to-day operations in their specific organization at that specific time. Despite the many difficulties of measurement and the achievement of total organizational effectiveness, management is still able to identify the less effective groups or subgroups of workers and to do something at improvement in practical ways. Organizations not only continue to survive but also to perform essential services and functions, despite the conceptual and methodological confusion rampant in the field of organizational effectiveness. One should no more imagine that organizations will cease being effective because of the seemingly impossible task of universally defining and applying effectiveness measures than that the institution of marriage will cease because of the amazing lack of consensus in defining and measuring what love really is!

Impermanence of Goals

Since the attainment of goals is the yardstick by which organizational performance is generally appraised, it should be obvious that in a changing environment the goals of an organization will be dynamic by nature. Organizations must adapt to their environment, for what is "out there" represents both a threat to survival for the firm unable or unwilling to change, and an opportunity for growth for the organization in tune with its environment. Because goal setting embraces the task of defining the relationships between the organization and its environment, a change in these relationships requires an alteration of the goals themselves.

Generally, the more uncertainty in the environment, the more dynamic the goals or the greater the need for goal reappraisal. Firms in dynamic industries need to reappraise their goals more frequently than those in stable industries. According to Thompson and McEwen,[24] reappraisal of goals appears to become

---

[24] James D. Thompson and William J. McEwen, "Organizational Goals and Environment: Goal Setting as an Interaction Process," *American Sociological Review* 23, no. 1 (February 1958), pp. 23–31.

more difficult as the product of the organization becomes less tangible and more difficult to measure. They cite the federal government's goal of maintaining favorable relations with a foreign country, and also the example of the university. The product of the university, as manifested in the performance of its graduates, is generally vague and imprecise.

Organizations, imbedded in specific and ever-changing environments, continually need to alter their goals if they are to survive and grow. In organizations whose product or service is clearly definable and amenable to measurement, there is generally rapid feedback on goal attainment and a subsequent reappraisal of goals. In organizations whose goals are less tangible and less amenable to measurement, it is more difficult to determine their acceptability; feedback indicating that the goals are unacceptable is both longer in coming and less effective. Many governmental goals that are stated in high-sounding terms lack quantifiability; they are less often challenged since no one is able to determine whether they are being met.

To sum up, the approach of Cyert and March that views organizations as coalitions of individuals is perhaps the most useful for understanding the way organizations really function. The configuration of coalitions making up an organization is often directly related to the suboptimization of goals, and ultimately to organizational effectiveness. Examples abound — from collective bargaining agreements between labor and management to vertical integration in industry whereby coalitions in one organization form coalitions with other organizations, the better to control the environment. Thus an automobile company purchases a steel mill; canners buy farms; and paper manufacturers acquire large areas of forests. Coalitions may also take place with other organizations having a common purpose. For the development of the Concorde airplane, France and the United Kingdom shared costs and technology. Federal agencies often develop goals in conjunction with state agencies. Municipalities may share common transportation and water facilities.

Ultimately, organizations consist of coalitions, and therefore organizational effectiveness will depend upon the success of these coalitions in attaining the goals determined by a particular coalition within the organization.

THE SYSTEMS-RESOURCE APPROACH

As could be expected, the systems-resource approach is a systems approach in which organizational effectiveness is defined in terms of how well the system integrates all its component parts and how well it is able to cope with the changing environment from which it obtains its resources (inputs) and to which it contributes its products or services (outputs). In other words, effectiveness is defined in terms of inherent consistency (integration of system parts into a working whole) and of organizational congruence with the environment (utilization of environment for input-output processes).

As the reader is well aware by now, every systems approach is of necessity teleological and regulatory. The telos is the goal, end, or objective toward which a system is tending. As explained in Chapter 1, systems embody interacting components, and the interaction results in some *final* state or goal or equilibrium position where the activities are conducive to goal attainment. Furthermore, this final state (telos) doesn't just happen. It is not the result of random activities of interrelated and interdependent components. The interacting components must be regulated in some fashion so that the system objectives will ultimately be realized. In human organizations this implies the setting of objectives and the determining of the activities that will result in goal fulfillment. This constitutes planning. Control implies that the original design for action will be adhered to and that all untoward deviations from the plan will be noted and corrected. For effective control, feedback is required.

The systems resource approach to organizational effectiveness is often linked to Yuchtman and Seashore. Theirs is an open-systems model in which a continuous interchange of energy and/or information takes place between the system and its environment. Because the entities being exchanged are neither limitless nor abundant but limited and scarce, they have value. However, this value is determined, not by the specific *ends* or goals of the organization, but by their being generalized *means* of organizational activity. With this in mind, Yuchtman and Seashore have defined organizational effectiveness as "the ability of the organization, in either absolute or relative terms, to exploit its environment in the acquisition of scarce and valued resources.[25] Accordingly, an organization will be most effective when it maximizes its bargaining position and optimizes its resource procurement.

Mott, among others who adopt the same approach, defines organizational effectiveness as the ability of an organization to mobilize its centers of power for action—production and adaptation.[26] The mobilization concept implies adaptation of the organization both to internal production problems and to external environmental ones.

The systems resource approach, according to Campbell, assumes that the demands placed on any organization are so dynamic and so complex that one cannot define in any meaningful way its real goals. The organization has as its first and most general goal that of survival and nondepletion of its resources. When assessing an organization's effectiveness theoretically, one should begin therefore by asking if the organization is internally consistent with itself, whether its resources are being used judiciously. From the practical viewpoint, one should not even bother to ascertain what goals the organization is pursuing; rather, one should inquire about conflict among work groups, communication, morale, racial tension, job satisfaction, absenteeism, accident rates, skills of managers and supervisors, and so on. The tasks of the organization would not concern the

---

[25] Yuchtman and Seashore, "Systems Resource Approach," p. 898.

[26] Paul E. Mott, *The Characteristics of Effective Organizations* (New York: Harper & Row, 1972), p. 17.

researcher, who would instead focus on its survival probabilities and its system strength.

The differences between the systems-resource approach and the goal approach would tend to dissipate if the researcher were to take the next logical steps. In Campbell's words:

> If the goal-oriented analyst attempts to diagnose why an organization scores the way its does on the criteria, he soon will be led back to system-type variables. . . . If the natural systems analyst wonders how various systems characteristics affect task performance, he very soon will be trying to decide which tasks are the important ones on which to assess performance. Unfortunately, in real life these second steps are not taken. The goal-oriented analyst tends not to look into the black box, and the natural systems oriented analyst does not like to worry about actual task performance unless he is pressed.[27]

It seems clear to the present authors that the systems-resource approach is an inclusive one — inclusive of goal definition, goal attainment, and goal measurement. The goal approach and the systems-resource approach are not very different in reality. For instance, Yuchtman and Seashore define organizational effectiveness as the ability of the organization to exploit its environment in the acquisition of scarce and valued resources. The operative definition that they offer involves the acquisition of a bargaining position with regard to obtaining these scarce and valued resources. The way they operationalize this definition is interesting.

> Seashore and Yuchtman used data from seventy-five insurance sales agencies located in different communities throughout the U.S. The analysis of the data yielded ten factors that were stable over time: business volume, production cost, new member productivity, youthfulness of members, business mix, manpower growth, management emphasis, maintenance cost, member productivity, and market penetration. Seashore and Yuchtman noted that factors such as business volume and penetration of the market could be considered goals, but member productivity and youthfulness of members certainly cannot. They concluded that most all of the factors associated with performance can be considered as goals, but these factors can also be regarded as important resources gleaned from the environment. Seashore and Yuchtman concentrated on these variables as means to ends rather than as ends in themselves. Thus, the goals model and system resources model use the same variables as indexes of effectiveness but call them ends and means, respectively.[28]

James Price has faulted the Seashore and Yuchtman study on the following three counts:

1. The idea of optimization is stressed throughout the study, but nowhere in their study of 75 independently owned and managed life

---

[27] Campbell, "Contributions Research Can Make," p. 32.

[28] Mary Zey-Ferrell, *Dimensions of Organizations: Environment, Context, Structure, Process, and Performance* (Santa Monica, Calif.: Goodyear Publishing, 1979), p. 347.

insurance agencies do they develop measures of optimization. None of their performance factors or their 23 indicator variables refers to optimization.

2. General measures of organizational effectiveness seem not to have concerned them. Thirteen of the 23 indicator variables, such as renewal premiums collected (dollars), numbers of lives covered per 1,000 insurables, etc., are clearly limited to insurance companies. Six of the remaining 10 are probably also limited to insurance companies — for example, managers' personal commissions and the percentage of business in employee trust.

3. Different concepts refer to the same phenomenon, thus violating the basic rule of mutual exclusiveness. Eight of the 23 measures refer to efficiency and 5 refer to size, so that 13 of the 23 measures refer to concepts other than effectiveness.[29]

To sum up, it is difficult to avoid using goals either explicitly or implicitly when assessing organizational effectiveness. Even a systems-resource approach involves operative goals, whether one calls them by that name or not.

THE FINANCIAL VIABILITY APPROACH

A third view of effectiveness that is different from, but not incompatible with, the goal or the resource approach is financial viability.[30] This expression simply denotes the ability of an organization to pay its bills on time and acquire a surplus. Price notes four reasons for the inclusion of financial viability as a criterion of effectiveness.

1. It is relatively easy to measure. The goal approach requires the decision maker to spell out the organizational goals — a task that can be quite difficult to do for nonprofit and service organizations. Also, the systems-resource approach is hard to operationalize in some areas. Although the financial viability approach also poses some problems, these are less than those found in the two approaches discussed previously.

2. Financial viability appears to be strongly and positively correlated with traditional views of effectiveness. Price argues that both the goal and the resource approach yield the same results as the financial viability approach. Assuming that valid measures of goal achievement and resource acquisition exist, the comparison of these with those for the organization's financial viability appears to be positive. That is, those organizations that achieve their goals typically have financial viability; those organizations that are successful in acquiring resources are also financially successful. Price registers the point

---

[29] Price, "Study of Organizational Effectiveness," pp. 8–10.

[30] James L. Price and Charles W. Mueller, *Handbook of Organizational Measurement* (Marshfield, Mass.: Pitman, 1986).

that financial viability would not be very important *unless* it was strongly and positively correlated with goal achievement and resource acquisition.

3. Financial viability allows for the creation of a theoretical model of determinants of effectiveness. The point Price makes here is that since various constituents of organizations have different goals, it is difficult to collapse these goals into a single measure of effectiveness. It simply is not feasible to combine the different strands. Financial viability is a measure of effectiveness that transcends many facets of management.

4. Financial viability is applicable to social systems other than organizations. Financial viability can equally be used as a measure of effectiveness for entities such as families, communities, and society in general. While the resource approach is also applicable to these other entities, the applicability of the goal approach to these other entities is somewhat more questionable.

Price does not claim to introduce a new conceptualization of effectiveness but rather amends the traditional views by using a more simple measure that appears to have wider applicability in a greater variety of settings.

The financial viability viewpoint is not without criticism, however. It runs counter to currently accepted conceptualizations of effectiveness. Once a concept's meaning becomes standardized, there is resistance to any reconceptualizing of it by others, especially by other scholars. This would be especially true if some particular unit were to come off poorly by a reconceptualization. Professional staff in many not-for-profit organizations such as hospitals and social service agencies would undoubtedly resist any new emphasis on financial viability at the expense of "quality care" measures. Price counterargues that he is not denying the goal approach. He merely wishes to make the point that goal achievement is not the same as effectiveness. Many scholars including the present authors would contend that Price is on shaky ground with this argument and that the goal approach is pervasive in its use in all types of organizations. Indeed profit organizations rely almost exclusively on goal achievement even if it is only financial goal achievement. Traditionally, this has been the most critical aspect of organizational effectiveness as well as of survival.

A second argument anticipated by Price against the acceptance of the financial viability approach is that such conceptualization renders impossible standardized measurement. For example, Price states that it would be difficult to compare the effectiveness of a hospital to a manufacturing organization. Although Price views this as a serious argument, these authors view it with less concern. The point is that financial analysis does indeed allow for a comparison of a hospital with a manufacturing organization. If it is a for-profit hospital such as those managed by the Humana company, the rate of return on investment is directly comparable to that of a manufacturing firm. Even in not-for-profit hospitals, a direct comparison is feasible. Many of these latter type hospitals use a somewhat different accounting terminology. Instead of *net income* it is often termed "excess of revenue over expenses;" instead of *retained earnings* it is termed "fund balance" or "reserve for future expansion." Instead of *profit* it is termed "surplus." Using rate of return on investment as a measure of effectiveness allows for a comparison of even the most dissimilar operations. This singular measure includes all aspects of performance – labor, the use of capital, engineering, etc. Some organizations produce thousands of products; others are involved in hundreds of services. With rate of return on investment or with other financial measures, it is possible for ITT to compare the operations of over 200 different profit centers that produce vastly different products. Financial viability measures are easily understood and are strictly objective. Two of the financial viability measures noted by Price include the aforementioned return on investment (ROI) and return on equity (ROE). Needless to say, there are numerous other financial ratios suitable for use.

One must be careful, however, when making comparisons that only like items are compared. For example, Price suggests that profit ratios could be used to compare all voluntary hospitals located in the same area. One obviously must be certain that the hospitals provide like services. If one is primarily a "birthing" or a neonatal hospital and another a cancer center, a comparison of cost/day/patient of the two would be meaningless. Likewise, a comparison of the cost per student in two different state universities would give vastly different results if one had a medical school and the other did not.

Whether or not financial measures will become more common as measures of organizational effectiveness only time will tell. Suffice it to say that for the past one hundred years they have been accepted both theoretically and practically by the business community.

SUMMARY

The literature on organizational effectiveness is still in a preliminary state with no theories or even definitions acceptable to all. Definitions are about as numerous as the number of authors who have investigated the field. Effectiveness, like the concept of environment, can be viewed as either unidimensional or as multidimensional, with the posture that one takes being dependent upon one's view of the organization.

One view prominent in the literature links effectiveness with goal attainment. Other authors take the position that effectiveness is the degree to which the organization can maintain the integration of its parts. In the latter case, adaptation and survival become measures of effectiveness.

While neither the goal approach nor the systems approach appears invulnerable to the thrusts of opponents, there seems to be more acceptance (judging from the literature) of the goal approach. With this in mind, the authors have addressed some of the issues regarding goals. Goals are formulated by various coalitions, each having potential self-interest at heart. The lack of goal congruency leads to conflict and suboptimization of goals. Goals are transient in that they change with changes in the various internal and external coalitions.

A third approach to organizational effectiveness is that of financial viability—the ability of an organization to pay its bills on time and to acquire a surplus. Reasons for and against this as a general criterion of effectiveness were stated at some length. However, only time will tell whether this becomes as common a measure as the goal and systems-resource approaches.

KEY WORDS

Effectiveness	Operational Goals
Goal Approach	Official Goals
Efficiency	Goal Conflict
Operationalism of Concepts	Goal Impermanence
Coalitions	Systems-Resource Approach
Operative Goals	Financial Viability Approach

REVIEW QUESTIONS

1. How might different organizational structures (product versus function) affect measures of effectiveness?

2. "In social service agencies process is more important than results." Relate your comment on this quotation to organizational effectiveness.

3. With several other students, develop some measures of effectiveness for the following:
   - *a.* A student in college.
   - *b.* A library.
   - *c.* A barber shop.
   - *d.* A hunter or fisherman.
   - *e.* A hospital
   - *f.* A co-op natural foods store.

4. One of the objectives of the federal government is so-called full employment. In view of other governmental goals, what are some of the shortcomings of such a policy?

5. In the text, Mott defines organization effectiveness as "the ability of an organization to mobilize its centers of power for action—production and adaptation." Critically assess this definition of effectiveness.

6. What are the measures of effectiveness for
   - *a.* Your instructor.
   - *b.* The department head.
   - *c.* The dean of the college.
   - *d.* The president of the university.
   - *e.* The board of regents.
   - *f.* The athletic director.
   - *g.* The football coach.
   - *h.* The librarian.
   - *i.* The computer center director.
   - *j.* The greenhouse keeper.
   - *k.* The power-plant supervisor.
   - *l.* The hospital chaplain.

7. Interview a person who works for the Internal Revenue Service of the federal government and develop measures of effectiveness for her or him.

8. What do you suppose are the effectiveness dimensions of a correctional institution?

9. Which of the effectiveness indicators listed by Campbell do you consider to be weak or meaningless? To which would you give high priority? Why?

10. Name an organization that uses
    a. Morale as a measure of effectiveness.
    b. Managerial interpersonal skills as a measure of effectiveness.
    c. Motivation as a measure of effectiveness.
    d. Efficiency as a measure of effectiveness.

11. Talk to one of the campus police and try to come up with some measures of effectiveness for these officers.

12. What typical problems occur when goals are subjectively evaluated?

Additional References

Angle, Harold L., and James C. Perry. "An Empirical Assessment of Organizational Commitment and Organizational Effectiveness." *Administrative Science Quaterly* 26, no. 1 (March 1981), pp. 1–14.

Cameron, Kim. "Critical Questions in Assessing Organizational Effectiveness." *Organizational Dynamics,* Autumn 1980, pp. 66–80.

————. "The Relationship between Faculty Unionism and Organizational Effectiveness." *Academy of Management Journal* 25, no. 1 (March 1982), pp. 6–24.

Cameron, Kim, and David A. Whetten, eds. *Organizational Effectiveness: A Comparison of Multiple Models.* New York: Academic Press, 1983.

Connolly, T; E. J. Conlon; and S. J. Deutsch. "Organizational Effectiveness: A Multiple Constituency Approach," *Academy of Management Review* 5 (1980), pp. 211–17.

Habayeb, A. R. *Systems Effectiveness.* New York: Pergamon Press, 1987.

Hitt, Michael A., and R. Dennis Middlemist. "A Methodology to Develop Criteria and Criteria Weightings for Assessing Subunit Effectiveness in Organizations." *Academy of Management Journal* 22, no. 2 (June 1979), pp. 356–74.

Jobson, J. D., and Rodney Schneck. "Constituent View on Organizational Effectiveness: Evidence from Police Organizations." *Academy of Management Journal* 25, no. 1 (March 1982), pp. 25–46.

Katz, Daniel; Robert L. Kahn; and J. Stacy Adams, eds. *The Study of Organizations.* San Francisco: Jossey-Bass Publishing, 1980.

Kaufman, Roger A., and Susan Thomas. *Evaluation Without Fear.* New York: New Viewpoints, 1980.

Macy, B. A., and P. H. Mirvis. "A Methodology for Assessment of Quality of Work Life and Organizational Effectiveness in Behavioral-Economic Terms." *Adminstrative Science Quarterly* 21 (1976), pp. 212–22.

Molnar, J. J., and D. C. Rodgers. "Organizational Effectiveness: An Empirical Comparison of the Goal and Systems Resource Approach." *Sociological Review* 17 (1976), pp. 401–13.

Molz, Rick. "How Leaders Use Goals." *Long Range Planning,* 20:5 Pct. 1987, pp. 91–101.

Price, James L. *Organizational Effectiveness.* Homewood, Ill.: Richard D. Irwin, 1968.

Price, James L, and Charles W. Mueller. *Handbook of Organizational Measurement.* Marshfield, Mass.: Pitman, 1986.

Shea, G. P., and R. A. Guzzo. "Group Effectiveness: What Really Matters?" *Sloan Management Review* 28, no. 3 (Spring 1987), pp. 25–31.

Spray S. Lee, ed. *Organizational Effectiveness: Theory — Research — Utilization.* Kent, Ohio: Kent State University Press, 1976.

Steers, R. *Organizational Effectiveness: A Behavioral View.* Pacific Palisades, Calif.: Goodyear Publishing, 1977.

Zammuto, Raymond F. *Assessing Organizational Effectiveness: Systems, Change, Adaptation, and Strategy.* Albany: State University of New York Press, 1982.

CHAPTER 10

Goals and Problems

To paraphrase a familiar epigram: "If you allow me to determine the constraints, I don't care who selects the optimization criterion."

Herbert A. Simon

CHAPTER OUTLINE

INTRODUCTION

In the previous chapter goals were treated as one of the approaches for measuring effectiveness. The subject of goals is so pivotal to the understanding of organizations that a separate chapter is being devoted to this topic.

The classical view of the firm assumes that the "economic man" always acts in a perfectly rational manner, thus maximizing his utility (profit) as a producer. In this view the firm possesses no consciousness and no human behavioral characteristic other than rationality. However, in reality a firm is composed of people who interact socially with one another and who have many different goals.

Once the assumption of the economic man is relaxed, the concept of goals or objectives can be introduced and treated in a much more comprehensive manner. Goals are categorized by different authors in many different ways. In the last chapter we considered official goals, operative goals, and operational goals. Perrow,[1] however, provides another classification scheme: social goals, output goals, system goals, product characteristic goals, and derived goals.

Social goals are those that pertain to the satisfaction of social needs. Governments and corporations are directly involved in providing goods and services that will satisfy many of society's needs.

Output goals are those that pertain to satisfying the consumers for whom organizations provide their goods and services. These change with a change in consumers. Many of the goods and services traditionally provided by the government for its citizens are now provided by private businesses for their consumers or clients.

Systems goals are those that pertain to the structure and process of the system's components. These typically fit into the traditional general systems theory model. Systems goals basically include survival and adaptation to new environmental variables as well as to the rate growth, rate of profit, etc.

Product characteristics goals are those that the organization would like to see realized in its goods and services. These obviously change over time. For the last decade or so, many manufacturing firms have been very much concerned with safety as one of their product characteristic goals.

Derived goals are those that pertain to the impact of an organization as a social entity on its environment. These goals encompass those often subsumed under an organization's social responsibilities.

OPTIMIZATION VERSUS SUBOPTIMIZATION

A system is concerned with converting inputs into outputs. Certain models, as discussed earlier, allow evaluation of this transformation process by showing when a particular variable such as profit is optimized. Examples of the

---

[1] Charles Perrow, *Organizational Analysis: A Sociological View* (Belmont, Calif.: Wadsworth Publishing, 1970), pp. 133–74.

optimization process are inventory and linear programming models. One should note at the onset that optimization applies only to models and not to real life, and generally it applies only to components of systems rather than to total systems. This last statement must be qualified, since so much depends upon how the system under study is defined. If the total system is defined in a simplistic way, then perhaps one can model the whole system. On the other hand, by following the path that leads to ever larger and more complex systems in which the present system is imbedded, we would eventually arrive at the universe as the total system. Modeling this would be an impossible and futile exercise.

The term *optimization* implies making the object under consideration as fully perfect or functional as possible. It is the most favorable or most attainable end possible under specified conditions. There are many mathematical models for optimizing—value analysis, linear programming, the transportation method, goal programming, dynamic programming, queuing theory, network analysis, and others. It is not the intent of the authors to describe in detail these many techniques since they are available in any operations research manual; however, a short description here of the above should suffice.

Value Analysis.   Value analysis is finding the best value among alternatives, given a number of variables. For example, the inventory model that considers the number of items used per year, ordering costs, and the cost of carrying inventory illustrates this technique.

Linear Programming.   Linear programming involves the maximization or minimization of an objective function, given certain constraints. Linear means straight; it means that the power of the variable is of the first degree and that the output is directly proportional to the input. Linear programming models are generally used for determining the best allocation of resources.

Transportation Method.   The transportation method is a form of linear programming that minimizes transportation costs in the routing of products from several different sources to different destinations. Such models assume a limited supply of homogeneous products and a fixed demand.

Goal Programming.   Goal programming is a special version of linear programming useful for solving problems where resources are limited and all of the subgoals cannot be attained simultaneously. Goals must be prioritized and a determination made whether or not overaccomplishment or underaccomplishment is to be allowed.

Dynamic Programming.   Dynamic programming is a form of linear programming that looks at solutions over a time interval. It derives the name dynamic from the actual procedure used to arrive at the optimum value and not from the type of problem involved, as in other techniques. Whenever a problem involves different stages and the result at one stage affects the subsequent stage, dynamic programming is probably the most efficient technique to use.

Queueing (Waiting Line) Theory.   Just as production firms are concerned with moving products efficiently through a plant, service firms are concerned with moving people efficiently through a service center (grocery checkouts, toll booths, copying centers, etc.). All waiting line problems have the same structure: some input arrives at some facility for either servicing or processing. The objective of queueing theory is to balance the cost of having enough personnel to serve the customer against the cost of losing customers due to an inadequate number of personnel.

Network Analysis.   Network analysis deals with a flow diagram of work activities sequenced to show interdependencies and interrelationships. Its basic objective is to provide progress reports of a project, and the expected effect on the total system of any slippages, should they occur.

In optimization models the following are features common to them all:

1. The problem is defined.
2. The model is formulated, i.e., relationships are identified.
3. The objective function is stated.
4. Rules or procedures (algorithms) are followed in solving the problem.

A fundamental feature of optimization is that one and only one objective function can be optimized at a time. This is precisely why organizations or their departments try, at some time or other, to optimize almost every functional objective. When the objectives of subsystems are optimized, this optimization may do very little for the total system. Subsystems must be integrated so that they can contribute to the optimization of the overall system objectives and not just of their own. Examples abound of the optimization of subsystem goals in organizations. Paying huge dividends conflicts with the growth goal requiring capital investment; granting huge salaries and benefits to employees conflicts with the profit objective; keeping high inventories that make for better customer service conflicts with cost control; pursuing zero defects programs typically adds costs to quality control. In his many writings Churchman has repeatedly made the point that the choice of the *proper objective,* even when given adequate information, is neither an easy nor an obvious one.

SUBOPTIMIZATION

Suboptimization is simply the process of settling for less than the best—the suboptimum—and, in the real world, it is more common than optimization. One reason for this is costs. Costs require trade-offs, and herein lies the problem!

Environmental problems are not specific to any one country—they are worldwide, and so too are the problems associated with AIDS, drugs, and cancer. Earlier in this text we presented a systems approach to the drug problem but at a low resolution level. Now this same example can serve to illustrate the problems encountered when suboptimization enters into the discussion.

If law enforcement of our international borders were dramatically increased (maximizing the enforcement objective), little would be changed unless our judicial system also changed. If our judges in the state and federal systems do not adjudicate the cases submitted by law enforcement officials, all our law enforcement efforts would be in vain. Also if a nationwide educational program concerning drugs were set in motion for our school children, the overall effect would still be small unless related measures were also taken—like making drugs inaccessible for them (not readily available from pushers operating in school playgrounds) and making them very expensive to procure—much more costly than at present in the states. The drug problem, we all somehow sense, must be the concern of innumerable subsystems—the governmental (law enforcement, the courts, and foreign policy), the medical, the educational, the civic, the religious, etc. If we read the data correctly, according to some authorities, it is this very approach—the optimization of individual *subsystems*—that has failed to bring meaningful progress to this problem area. There is little doubt in the minds of many citizens that the government could solve the problem if it were willing to sacrifice other goals for this one. However, choosing societal goals is much more complex a task than choosing organizational goals. Even within relatively restricted boundaries the choices are seemingly inexhaustible.

Take medical technology, for instance. Should the government allocate its allotted resources to research AIDS, Alzheimer's syndrome, cancer, pulmonary afflictions, heart diseases, brain disorders, degenerative conditions of the elderly, crippling maladies among the very young, or scores of other worthwhile health projects? Limited resources of time, money, and personnel must ultimately dictate solutions that are less than the best. Such allocations of resources would amount to *suboptimal* allocations.

When we began this chapter, we pointed out that only system components can be optimized and not the total system as such. Programs that provide for optimization and simulation, while not new in the modeling world, have nevertheless been slow in coming because of implementation problems. One such program of interest is DYSMOD (Dynamic System Modelling Optimiser and Developer.[2] Here an objective function is defined within the simulation model which sums total model behavior. Within the model, parameters with a range of numerical values are identified for optimization. The algorithm used starts with a simulation run that calculates the value of the objective function, given the initial

---

[2] R. Keloharju and E. F. Wolstenholme, "The Basic Concepts of Systems Dynamics Optimization," *Systems Practice,* March 1988, pp. 65–86.

parameters. Subsequent runs alter the parameters one at a time and use the objective function as the overall measure of performance. Iterations are continued until the value of the objective function stabilizes. This may require a hundred or more iterations.

In this model any existing parameter (constants and table functions) may be selected for optimization. In practice, the variables chosen would be those supported either from theory or from the data already collected.

Leaving models alone for a moment, we find that in the real world people are typically unable to identify all of the feasible alternatives, let alone evaluate what will make for an optimal solution. That is why optimization is possible only in the context of models in which the parameters are givens. In real-world situations the critical issue lies in determining what precisely is the problem. Unfortunately, few guidelines are available for problem identification since most of the literature concentrates on problem solving. A glance at the contents of most management texts will confirm this statement.

Optimization is feasible only when the problem is couched in the form of an optimization model. As noted above, optimization of real-world problems is seldom, if ever, achieved because it is nigh impossible to identify or to calculate the optimum. This necessarily leads to suboptimization of what is generally termed the "next best solution." In so doing we identify an objective function that we hope to attain. The systems theorist, van Gigch, believes that if one accepts the notion that optimization is the way to go, then one ought to follow certain principles. The nine principles that van Gigch[3] has collated for this purpose certainly merit our serious attention.

Principle 1:   When dealing with a hierarchic system, lower-system objectives should be in agreement with higher-level objectives. This principle can also be stated as: Lower-system criteria must agree with higher-system criteria.

This principle reaffirms the notion that decisions made at lower resolution levels should support the overall system. However, viewed from another perspective, this principle could be extended to suggest that the higher-system goals should also satisfy as much as possible the lower-system goals. Underlying this expanded principle would be the assumption that achievement of lower-system goals leads to optimization of higher-level goals, and the pursuit of higher-level goals results in completion of lower-order goals. Although this is definitely to be hoped for, regrettably it may not be the case. The trick is to choose the right goals at the higher level that satisfy more of the lower-level ones. The implementation of subsystem goals, all of which are suboptimal, ought to result in the optimization of higher-level ones. The rub here, however, is that executives do not even try to choose a single goal that is an aggregate of many lower goals. It may be feasible, for instance, for the plant manager to aggregate goals at the

---

[3] J. van Gigch, *Applied General Systems Theory* (New York: Harper & Row, 1978), pp. 368–70.

plant level of an organization, but it is not always feasible for the executive to aggregate into a single profit goal such organizational goals as service, production, marketing, and growth. If, however, lower-level goals were to be prioritized and assigned values, then it might be possible to get an overall value of the system goal. Of course, a major problem then is determining who sets the subjective values.

Different stakeholders have different utility functions. Besides the commonly accepted stakeholders which include stockholders, creditors, debtors, employees, customers, and suppliers, Ackoff[4] also includes the government, the public, and especially people not yet born. For the latter his point is that through the destruction and pollution of the environment and the exhaustion of limited natural resources, fewer choices will be available for future generations. To deprive future generations of choices is to deprive them of their basic rights. Ackoff believes that one could account for such stakeholders' values by the use of *future impact assessment studies.* Incorporating these into systems analysis would certainly make for greater complexity.

Principle 2:   Suboptimizations should be scored and ranked according to the extent to which they increase the utility of each subsystem without, in so doing, reducing the utility of any other systems and the overall system.

The significant point here is that it may be possible to improve the overall performance of a system by a tradeoff among lower-systems. A win-win situation may exist when two subsystems can both increase their utility by a tradeoff, or a variant may exist that increases the utility of one system more than it decreases the utility of another. An example of the latter would be a decrease in sales calls to small customers that would lead to a proportionate increase in calls to large customers with a disproportionate increase in sales. Principle 2 is known as Pareto Optimality.

Principle 3:   The danger of costly suboptimizations will be reduced as the scope of the system considered is enlarged. The influence of spillover effects and of externalities will be lessened as they are "internalized" within (integrated into) a larger system.

The author, van Gigch, here suggests that as the scope of a system is enlarged to include other subsystems, the influence of externalities of the subsystems would be diminished. The more such subsystems are integrated into the whole system itself, the less costly suboptimization becomes.

Principle 4:   The scope of the system should be enlarged to the point where the advantages, stemming from the internalization of spillover effects

---

[4] Russell L. Ackoff, "The Future Is Now," *Systems Practice,* March 1988, pp. 1–6.

and externalities, outweigh the disadvantages of dealing with a system whose complexity may "outrun analytical competence."

The warning here concerns the level of systems analysis chosen for study. Too much complexity in a problem leads to generalities and ultimately to less success in problem solving because of the available tools. The more finite the problem the more powerful the tools that can be applied. However, a naively simplistic level of analysis may well ignore important systems components and their interrelationships, thus disallowing a fruitful analysis.

Principle 5:   The scope and ramifications of certain problems may prevent their being dealt with at any other than the highest-system level. The lower the level at which a system is considered, the greater will be the likelihood that important interactions with other systems will be missed or omitted.

This principle reflects our earlier discussion of the systems approach versus the analytical approach. In the analytical process of reducing complex problems in order to acquire a better understanding of the various parts, important properties may be lost. These properties exist by virtue of the holistic nature of the system. When the system is broken down, it loses these properties. An example was previously given of the family. The family has certain properties or relationships by virtue of its being a family. If one were to study family members individually, one would miss the rich interactions that exist in the family as a whole. Or, to study the earth while ignoring its interrelationships with the rest of the universe would definitely not illustrate good problem solving.

Principle 6:   In scoring suboptimizations, those that satisfy the requirements of the systems approach should be preferred to those that result from "disjointed incrementalism" or "sector improvement."

Disjointed incrementalism refers to those improvements in systems that result from small incremental changes and that are made with or without adequate information and understanding. This is at variance with the systems approach which suggests a study of total systems for uncovering alternatives rather than the adoption of marginal improvements. Disjointed incrementalism leads to suboptimization by necessity rather than by choice. Sector improvement relies on the same logic as that of incrementalism. To make improvements in the entire system one must understand the whole. Otherwise improvements can only be made in specific sectors of the system. While this may be optimal for that particular systems sector, it is suboptimal for the entire system itself. The systems approach must improve more than just the subsystems; it must improve the holistic system.

Principle 7:   The Principle of Bounded Rationality is interpreted by Miller and Starr as suggesting that we should "not assume an irrational extreme of rationality." It may be more fruitful to consider "immediate optima" in our quest for the "ultimate optimum" and revise our goals and our values many times along the way.

According to the principle of bounded rationality, a term ascribed to Herbert Simon,[5] when making a decision a person will seldom make the effort to find the best solution, but will rather choose a strategy that is good enough. What is suggested here is a revisionist approach to goal setting which calls for periodic reassessment of the system and its environment. The concept of bounded rationality is given more attention later in this chapter.

Principle 8:   The degree of suboptimization can be improved over time if decisions are made that permit a broad range of eventualities. Our present knowledge of the decision process does not allow us to state at which point the decision tree branching should be considered in our quest for the path leading to the global optimum.

This principle suggests that suboptimization can be improved if we consider more scenarios or alternatives. This is obviously dependent on the alternatives chosen to be considered.

Principle 9:   Seeking the global optimum can be likened to a means-ends model. We do not know whether we should work forward or backward in the means-end chain; however, we should always strive to establish means (intermediate ends) which are in accordance with ultimate ends.

The author implies here that means and goals are not absolutes and that goals in one instance could be, in another instance, means to other goals. For example, marketing and production can both be goals as well as means for profitability. The linking pin concept can theoretically tie in all the means-goals of all the subgroups in an organization.

BOUNDED RATIONALITY

The conventional economic theory of the firm assumes perfect rationality. For decision makers to act rationally in a perfect manner, they must be aware of both the entire range of choices available to them and the consequences of such choices. In addition, they must have a clear objective or goal that they can pursue. The decision makers will then objectively and rationally choose the goal that will maximize profit. This highly rational approach tends to reduce decision makers to automatons devoid of human variability and indecisiveness. Because of the limited

---

[5] Herbert Simon, *Administrative Behavior* (New York: Macmillan Co., 1961).

usefulness of the perfect rationality assumption, scholars have turned to a more realistic modification of the rationality concept.

Herbert Simon is credited with developing the term *bounded rationality* which pertains to organizational decision makers. Bounded rationality is behavior that is *"intendedly* rational, but only *limitedly* so."[6] On the one hand, individuals are physically limited in their receiving, storing, retrieving, and processing powers. Simon notes, "It is only because individual human beings are limited in knowledge, foresight, skill, and time that organizations are useful instruments for the achievement of human purpose."[7]

On the other hand, human beings are not always able to articulate their knowledge in ways that allow them to be understood by others. This limitation is a language-bound impediment. Some knowledge may defy articulation while in other instances individuals may not possess the vocabulary to impart the knowledge. When language difficulties exist, as they often do when people travel in foreign countries, for example, demonstration, gesticulation, and other imitative techniques can help. When dealing with complexity and uncertainty one may find it nearly impossible to generate all the consequences of possible alternatives. There simply aren't any rules for the development of alternatives. Another concern is that there is no easy way to judge the consequences of the various alternatives. As pointed out in the literature, decision making under certainty falls mainly within the realm of simplistic problems. Decision making under uncertainty or complexity seldom allows for estimating all the alternatives. Thus, approximation must replace exactitude in decision making. Bounded rationality makes possible an efficient decision rather than an optimal one. The decision will be expected to yield an outcome that is "good enough" for the situation. The individual chooses not to consider all the possible parameters of the decision-making situation. To do so would require an exceptionally in-depth understanding of the many complex relationships involved in the decision, as well as an enormous amount of information that may be too costly to acquire or unavailable at any cost. A decision embracing all of the elements would require that a person know all the possible states of nature. This requirement may prohibit even a formulation of the problem, let alone its analysis. In everyday life we often follow the same pattern. When buying a new car, for instance, we do not seek the best possible deal, but one that is "good enough" under the circumstances. When choosing a spouse, we do not wait till the best possible life companion has been found; we generally choose a mate that is "good enough" under the circumstances.

Limiting the boundaries of a problem is akin to determining the focal system discussed in an earlier chapter. When choosing to leave certain factors out of a

---

[6]*Administrative Behavior,* p. xxiv.

[7]Herbert Simon, *Models of Man* (New York: John Wiley & Sons, 1957), p. 199.

problem, the decision maker is content to *satisfice.* He will choose a strategy that is expected to result in outcomes that are "good enough" given the circumstances. The same goes for an organization's mix of objectives. The blend of objectives sought may not be optimal but is "good enough."

Complexity in the above discussion may be deterministic complexity or uncertainty complexity. In the first instance, a problem may be deterministic, such as a game of chess. The chess problem, however, according to Simon, is prohibitively complex. The second situation, that of uncertainty complexity, does not allow for the consideration of all the alternatives.

Aside from the computational difficulty, language may limit the generation of alternatives. Indeed it is this variation among individual talents that gives rise to hierarchical structuring in organizations.

The concept of satisficing naturally flows from the assumption of bounded rationality and consists in taking a course of action that is "good enough." Simon puts it this way: "Administrative theory is peculiarly the theory of intended and bounded rationality — of the behavior of human beings who *satisfice* because they have not the wits to *maximize.*"[8]

In real life, not only do people not perform the computational payoffs for various outcomes, but they also often choose to simplify their choices. One way of doing this is for sellers to determine a minimum price that they are willing to accept for an item, which in effect says that anything above this level is satisfactory.

Another problem experienced in determining payoffs, according to Simon, is that it may be impossible to compare the payoffs.[9] Three instances of this exist: (1) where decisions are made by a group and the payoff preferences of the individual group members differ; (2) where the payoffs are of a heterogeneous nature, such as salary and climate, and the individual is unable to compare the incomparable items; and (3) where each alternative has an indeterminate number of possible payoffs.

Predicting the behavior of decision makers who satisfice is a complicated task since the individual attributes of the decision maker may vary significantly. The decision maker's perception, aspiration levels, persistence in striving to reach those levels, and other attributes moderate the notion of satisficing.

Organizations, like individuals, can learn. They learn to avoid uncertainty, they learn to simplify problems, they learn to use feedback, and they learn to use alternative goal criteria. In short, they learn to satisfice.

Problems are closely related to goals and usually result from a deviation from goals. In cybernetic terms, it is the control mechanism that communicates with us, telling us that a deviation exists now or will soon exist. In those organizations where a constant goal exists, problem identification is ultimately simple. In those with multiple goals, problem identification may not be as simple.

---

[8] Simon, *Administrative Behavior,* p. 198.
[9] Simon, *Models of Man,* pp. 250–52.

PROBLEM IDENTIFICATION

Though textbooks are replete with problem-solving techniques, seldom do we find any detailed discussion on how to formulate a problem. Churchman was one of the first to recognize the importance of this issue.[10]

Before a problem can be formulated, one must have some idea of the various dimensions of the problem. The first component in the problem-formulating process is the decision maker in charge of the system under consideration. Obviously, this person must be dissatisfied enough with some aspect of the current situation to consider it problematic in the first place. But this dissatisfaction with some element of the situation is not all. The decision maker must have available a course of action for ameliorating the situation. The first without the second does not constitute a problem; it is but a source of perplexity, vexation, or frustration.

The second component concerns the decision maker's objectives. This must be viewed in the context of the other objectives that are to be maintained. Thirdly the problem must be viewed as embedded in an actual environment, which may include people, financial or material resources, or machines. Lastly, there must be alternative choices of action — at least two — if any choice is to be freely made. With only one alternative, one cannot be considered free to choose.

The Decision Maker

According to Churchman, it is first necessary to identify the person or persons who have the authority to make the decision. Although organizational charts may be of some help, they may not provide the needed information. In some instances, a group may have to give its final approval, or someone higher up in the organization may have the right to overturn or veto the decision.

The Decision Maker's Objectives

One way to uncover the real objectives is to formulate a list of all possible outcomes of the problem. If the decision maker is unable to live with certain outcomes, then restraints on certain objectives may be identified, or else other objectives may still be looked into. Churchman makes the point that the decision maker must also consider current objectives that are to be maintained. For example, if the objective to be maintained is to keep the present employment level and the new objective is to lower production costs, then an obvious conflict exists. This is why an objective must always be considered in relation to other existing objectives.

---

[10] C. West Churchman, Russell L. Ackoff, and E. Leonard Arnoff, *Introduction to Operations Research* (New York: John Wiley & Sons, 1957), pp. 128–29.

The System and Its Environment

The organization as a system embedded in a specific environment comprises managers, employees, machines, materials, consumers, competitors, government, and the public. Earlier it was stated that objects by themselves do not make a system; rather they must be organized for a purpose. It is this organization that makes these components a system. Which of the above components are to be included in the system will depend on the system being studied. For example, the production system of an organization would probably not include competitors as an integral component; however, the marketing system would.

At this point Churchman introduces objectives of other participants in the system. If a solution to a problem is unacceptable to other parties in the organization, then their ideas should also be considered. Understanding these other interests may extend the range of possible solutions.

Alternative Courses of Action

To evaluate the desirability of one alternative over others, one must determine the outcomes of the other feasible solutions. Ideally, the outcomes ought to be expressed in a quantitative form with probabilities assigned to each. When it is not possible to derive objective outcomes, then subjective ones will suffice. In these instances one tries to develop surrogates which provide measures of effectiveness for the outcomes. For example, the extent of repeat business could be a realistic measure of customer service. Turnover could well be a measure of employee satisfaction. In any event, measures of effectiveness must be chosen by the decision maker.

If the outcomes are quantitative, one might possibly develop a model that would incorporate the major components of the problem. As will be noted later in the text, there are innumerable types of models and solution methods for solving problems. One common approach is to construct a matrix that lists the objectives (O), the alternative courses of action (A) and the calculated outcomes. For example, Table 10–1 shows two objectives, O1 and O2 and two alternative strategies A1 and A2, with the outcomes listed in the body of the table. If both objectives have the same relative importance for the decision maker, it is clear that O2 should be the objective chosen and A2 the selected course of action. However, if O1 were of greater importance, and this importance were to be assessed on a

TABLE 10–1   Payoff Matrix

	Alternate Strategies	
Objectives	A1	A2
O1	5	3
O2	6	7

scale of 0 to 1, then different options might emerge for the decision maker. Table 10–2A shows the relative importance of objective 1 as 0.6 and the relative importance of objective 2 as 0.4. Of course, the importance of the various objectives will somehow be determined from the decision maker's experience or the superior's list of priorities.

TABLE 10–2A   Payoff Matrix

Objectives	Importance Probability	Alternate Strategies	
		A1	A2
O1	(.6)	5	3
O2	(.4)	6	7

By weighting these courses of action we can arrive at the efficiencies of the outcomes. The sum of the weighted efficiencies (efficiency times its relative importance) is a measure of effectiveness.

TABLE 10–2B   Weighted Payoff Matrix

Objectives	Importance Probability	Alternate Strategies	
		A1	A2
O1	(.6)	.6 × 5 = 3.0	.6 × 3 = 1.8
O2	(.4)	.4 × 6 = 2.4	.4 × 7 = 2.8
	Expected Total Efficiencies	5.4	4.6

Instead of choosing O2 as the objective and A2 as the alternative strategy as before, the decision maker would now choose O1 as the objective and A1 as the planned course of action. We see that by restricting our consideration to only one objective, we can easily arrive at an incorrect solution. While the above primitive example is offered only as an illustration, still it can serve as a starting point for more sophisticated decision models.

Measures of Effectiveness

While elsewhere in the text we treat effectiveness of organizations, here we will discuss effectiveness as a measurement problem. The particular choice of a measure of effectiveness rests on two factors—the relative importance of the objective and the outcomes of the alternative courses of action. It is decidedly more difficult to develop measures of effectiveness when qualitative objectives are chosen than when quantitative objectives are selected. If all the objectives are qualitative, then they can be measured in terms of the probability of attaining the objectives. Effectiveness would be represented by the sum of the weighteddefficiencies which is always a single value. The course of action would be based on a comparison of these values.

The efficiency of each course of action for each objective is shown as

$$\sum_{j=1}^{k} E_{ij}v_{j}$$

where (v_j) is the standardized value of each objective and (E_{ij}) is the efficiency of each of the courses of action for each objective.

It is also possible to determine the effectiveness of a mixture of quantitative and qualitative objectives but this is beyond the scope of this coverage.

PROBLEM SOLVING

Different problems call for different solution techniques. In Chapter 3 on cybernetics, systems were classified according to their complexity. Unfortunately complexity alone is not the only characteristic to be considered in problem solving. More important than complexity in problem solving is the problem's structure or lack thereof. A well-structured problem meets the following three criteria:

1. The various variables can be quantified in some manner.
2. There is a clearly specified objective function.
3. Algorithms exist that allow for numerical solutions.

In light of the above criteria one can readily see why the systems approach is difficult to opertionalize even in simplistic situations. Not too surprisingly the problems that are most susceptible to modeling are those whose structure is well understood. In loosely or ill-structured problems the variables may be hard to identify and impossible to quantify. This is more obvious when the system or problem being modeled includes people. The second criterion limits problem solving to certain types. Few organizations have profit as a single objective function. Multiple objectives are the norm, and seldom have these authors seen multiple objectives prioritized in an organization. Suffice it to say that organizational survival rests upon profitability, and thus is ranked implicitly.

Algorithms provide for computational solutions of well-structured problems. An algorithm, more specifically, is a precise formulation of procedural instructions so organized as to result in the optimum solution of a specific problem. Once algorithms are developed, they can be used over and over again to solve similar problems. It is no mere coincidence that the well-structured problems are the ones most often cited in the literature. Reason: their algorithms exist. This is not to imply that the development of such algorithms is an effortless task. Far from it. All that is meant is that when the first two criteria of well-structured problems have been met, the last is much easier. In some instances researchers have applied an available algorithm to problems without knowing the behavior of the variables. For instance, the linear programming algorithm may unwittingly be applied to a problem where the variables are not linear. Of course, ignoring a limiting constraint of an algorithm will invalidate the results of its use.

Just because an algorithm exists for solving a problem, one should not assume that no other suitable algorithms are possible. In the theory of algorithms it is axiomatic that given a solvable problem, there are many algorithms to solve it, but not all of equal quality. The job of finding these others, however, may not be worth the time and effort.

What makes a good algorithm? System scientists, especially general systems theorists, are interested in this question just as they are interested in what makes a good analogy. The quality of algorithms, however, is usually judged not by theoretical considerations but by practical criteria. Among them are: time requirements, computer memory requirements, the accuracy of the solutions, and their generality.

The time requirement rules out many possible algorithms. To examine, for instance, all possible moves in a perfect game of chess, even though the number of moves is finite, would require millions of years, even at today's computer speeds. The memory requirements too would be very taxing. But as the speed of computers and their memory capacities keep improving with the years, the time and memory requirements will play a lesser role in algorithm generation.

The accuracy potential of algorithms is also often related to time. With further refinements of an algorithm and with double-precision accuracy, the accuracy of a solution can be enhanced but again at the expense of more time and memory. The generality criterion can turn a good algorithm into a better one. As with concepts so too with algorithms: the wider the range of application, the more valuable the algorithm — the greater its scientific import.

Problems that are not well structured and that cannot be solved by algorithms can, however, be solved by heuristics. A heuristic is a computerized rule of thumb that searches for a satisfactory solution rather than an optimal one. A heuristic comprises a computational procedure that limits the number of alternative solutions. These procedures reduce the amount of time spent searching for a solution. The computer closely parallels the human mind when processing a heuristic program. In much the same way it learns to improve its performance by adjusting its parameters. Because it does not provide the optimal solution, it does not induce as much confidence as an algorithm. For this reason the results of this method are often compared to the solutions derived from alternative methods of analysis. While algorithms demand quantitative inputs in a well-structured problem within a restrictive framework, heuristics can be applied to more complex, ill-structured problems. Solutions derived heuristically need not be optimal to be of value; they need only represent some improvement on the already available solutions.

The decision to use either an optimizing algorithm or a heuristic program which will provide a good solution is a trade-off between the error introduced by the restrictive algorithmic constraints in modeling, and the suboptimality achieved in the heuristic program. Heuristic approaches to problem solving have been used for many years in dozens of areas from designing plant layouts and balancing assembly lines to managing inventory and scheduling airline arrivals and departures.

In closing, we should perhaps mention that some scholars hold that general systems theory is a precise mathematical theory, yet completely general, and that the properties and behavior of systems can be investigated in a precise manner. This suggests that one can have specificity via mathematics without losing the generality of its derived mental constructs. When writing of general systems theory, Mesarovic, an advocate of this approach, states: "What is clearly needed is a *simple, elegant, general,* and *mathematically precise* concept of a (general) system which provides a point of departure for more detailed and more complex notions and problems."[11] While Mesarovic admits that the formal approach is but the starting point in the development of general systems theory, he believes that it will eventually evolve into one with a mathematical structure. As he conceives the developmental process, in the formalization approach one first defines the concepts verbally as specialists in their particular fields intuitively understand them. Then the concepts are incorporated into propositions with a minimum of mathematical structure and then stated axiomatically (as being self-evident). By adding more axioms and studying the corollaries of these new assumptions, one can, by this method, study large-scale systems with their rich interrelationships.

SUMMARY

In the economic man theory, one assumes perfect rationality, but in the real world consisting eminently of people, one cannot firmly hold to this view. Goals must definitely enter into the picture.

Perrow identifies social, output, systems, product, and derived goals. Social goals are those that pertain to society's needs and well-being; output goals relate to certain types or classes of citizens, clients, or consumers; system goals pertain to the system's structure and processes; production goals are those that an organization would like to see realized in its goods and services; and finally derived goals relate to the organization's impact on its environment.

The optimization process was then contrasted with that of suboptimization. Optimization is the process whereby one finds the best balance of factors in a situation. Suboptimization is the process of settling for less than the best. Various mathematical models for optimizing were recounted—value analysis, linear and goal programming, transportation methods and others. Four features common to all optimizing models were: the definition of the problem, the formulation of a model, the stating of the objective, and the finding of an algorithm for solving the problem. While optimization is the ideal, in the real world one must generally be content with suboptimizing because of the constraints imposed by costs. The drug

---

[11] Mihajlo D. Mesarovic, "A Mathematical Theory of General Systems," in ed. George J. Klir *Trends in General Systems Theory,* (New York: John Wiley & Sons, 1972), pp. 152–53.

example was cited again as illustrative of what can happen when suboptimization enters into the discussion.

Optimization applies to models and not to the real world; it applies to models of system components and not to total systems. It is often difficult not only to identify all the feasible alternatives needed for optimization but also to evaluate what makes a solution optimal. In optimization parameters are givens; in the real world, the problem lies in finding out what the actual problem is. The nine principles for optimization by van Gigch were then explained. These merit serious study.

Herbert Simon's concept of bounded rationality was first advanced to take into consideration human limitations and to counter the economic man concept that has proven untenable. Bounded rationality is purposely rational but only in a limited way. It dictates an efficient decision rather than an optimal one, by limiting the parameters that can enter into the decision-making situation. This is what transpires in everyday life—in choosing a spouse, buying a car, shopping in the supermarket, deciding on a college in which to enroll, etc. This is what is meant by satisficing—settling for the "good enough" under the circumstances. This concept follows naturally from that of bounded rationality. To predict the behavior of a decision maker who satisfices is extremely hazardous because individual values enter into the picture.

Problems generally occur when a course of action deviates from its goal. Organizations with multiple goals may have special difficulty with problem identification.

Very few guidelines exist for problem identification, but Churchman's four pointers may help. The first is the dissatisfaction of the decision maker with some aspect in the environment. The second concerns the objectives of the decision maker. The third relates to the environment in which the problem is embedded. The fourth element is the availability of more than one choice if the choice is to be free. One must be aware of these alternatives and one must have some way to evaluate the alternatives and their consequences. A weighted payoff matrix can help when the problem can be quantified.

Effectiveness as a measurement problem rests on two factors—the relative importance of the objective and the outcomes of the alternative courses of action.

As for problem solving, one ought to look at the problem's structure or lack of it, not just at its complexity. Well-structured problems have three characteristics: their variables can in some way be quantified; their objective function can be clearly specified; and an algorithm exists for their solution. Problems with well understood structures lend themselves to modeling. Few organizations have a single objective function that can be clearly specified, since many have multiple objectives which are generally not prioritized. Algorithms make possible the solution of well-structured problems. Where these exist, the danger is that they may be applied to problems for which they were never intended. Where algorithms do not as yet exist, their development is a major undertaking. The tool for solving poorly structured problems is heuristics—the discovery or problem solving by experimental methods, especially by trial and error. Heuristically derived solutions need not aim at optimality as do those algorithmically derived; they need only represent some improvement in the already available solutions. Heuristic approaches to problem solving have been used for many years in many different areas of business.

In closing the first three parts of this text, we noted Mesarovic's view that general systems theory, while a precise mathematical theory, is still completely general unlike those of the physical sciences each with its specific content (concepts and propositions) and that the properties and behavior of systems must be investigated in a precise manner. In effect, this view holds that one can impose specificity via mathematics without losing any of the generality of its mental constructs. While he details the stages whereby this mathematization of general systems theory will come about, he has to admit that GST is still in its early stages of development.

KEY WORDS

Economic Man	Satisficing
Optimization	Algorithm
Suboptimization	Outcomes
Bounded Rationality	Payoff Matrix
Goal Impermanence	Heuristics
Objective Function	Disjointed Incrementalism
Problem Identification	Well-Structured Problem
Problem Solution	

REVIEW QUESTIONS

1. Categorize Perrow's various types of goals as to internal-organizational or external-organizational. What benefits might be achieved by doing this?

2. Organizations have many goals. If only one goal can be optimized at any point in time, how can an organization develop a cybernetic system to encompass many goals?

3. Each new president promises to reduce the budget deficit. Show how suboptimization takes place in the presidential budget.

4. Show how the dangers of suboptimization are greater in a narrow perspective of a system than in a wider perspective.

5. How does the principle of Bounded Rationality enter into the decision of taking a job?

6. Why does Churchman state that an objective must always be considered in relation to other existing objectives?

7. Define the criteria of a well-structured problem.

8. Why are system problems generally not well structured?

9. How does Churchman suggest that one go about measuring the effectiveness of qualitative goals?

10. How does one explain the paradox that suboptimization leads to optimization?

Additional References

Harrison, Frank. *The Managerial Decision-Making Process.* Boston: Houghton-Mifflin, 1981.

Huber, George. *Managerial Decision Making.* Glenview, Ill.: Scott, Foresman, 1980.

Lyles, Richard L. *Practical Management Problem Solving and Decision Making.* New York: Van Nostrand Reinhold, 1982.

Margerison, Charles. *Managerial Problem Solving.* London: McGraw-Hill, 1974.

Moody, Paul E. *Decision Making: Proven Methods for Better Decisions.* New York: McGraw-Hill, 1973.

Plunkett, Lorne C. V. and Guy A. Hale. *The Proactive Manager: The Complete Book of Problem Solving and Decision Making.* New York: John Wiley & Sons, 1982.

Pounds, William F. "The Process of Problem Finding." *Industrial Management Review* 11 (Fall 1969), pp. 1–19.

Radford, K. J. *Modern Managerial Decision Making.* Reston, Va.: Reston Publishing, 1981.

Roach, John. "Simon Says: Decision Making Is a Satisficing Experience." *Management Review* 68 (January 1979), pp. 8–17.

PART FOUR

Application of Management Systems

*Man is a prisoner of his own way of thinking
and of his own stereotypes of himself.*

*His machine for thinking
the brain
has been programmed to deal with a vanished world.*

*This old world was characterized by the need
to manage* things —
stone, wood, iron.

*The new world is characterized by the need
to manage
complexity.*

Complexity is the very stuff of today's world.

*The tool for handling complexity is
ORGANIZATION.*

*But our concepts of organization belong
to the much less complex old world
not to the much more complex today's world*

*still less are they adequate to deal with
the next epoch of complexification —
in a world of explosive change.*

Stafford Beer

PART OUTLINE

CONNECTIVE SUMMARY

The preceding three parts of this book have laid the foundation for this last part concerning the applications of the systems approach. Part One provided us with an understanding of the interdisciplinary nature of the systems approach as well as with a vocabulary of terms and concepts. Part Two gave us an in-depth view of the characteristics of cybernetic systems along with the tools for understanding and managing these exceedingly complex, probabilistic, and self-regulating wholes. In Part Three we considered the organization's environment and the tools and methods the organization uses for learning about its environment, thus enhancing its effectiveness.

The stage has thus been set for the implementation of the systems approach to managing organizations as large, complex and probabilistic systems. One of the cardinal rules of management is that management decisions deal with the future of an organization rather than with its past. Most modern management scholars and practitioners would, in fact, define their jobs as to anticipate the future consequences of today's decisions and actions. For this reason, in Part Four: Application of Management Systems, we adopt a futuristic viewpoint.

There are two dimensions of this approach. The first is that of *time*. How far into the future must managers think about the consequences of their decisions? The second dimension is the degree of *uncertainty* about these consequences. Would there be significant differences between the intended or planned consequences, which are estimated at the beginning of the planning process (ex ante), and the actual consequences which are measures at the end of the planning period (ex post)? What factors will determine the type and size of these differences? Will they originate inside the organization or will they be the result of independent changes in the organization's external environment?

Systems methodology — as seen in the previous chapters — attributes most, if not all, the causes of uncertainty in the systems behavior to changes in the external environment. That is precisely why so much time and effort was spent in dealing with the organization-environment interaction system in general, and the external environment in particular. The source of the manager's inability to predict accurately the final state of the system from the knowledge of its initial state is that the future of the external environment will not be what managers assumed it to be.

Given the uncontrollable nature of the external environment, management must realize that the only way it can estimate the future impact of environmental changes on its strategic plan is to anticipate or estimate its most likely changes and to incorporate their potential impact into the strategic plan. The key to management's ability to do this is knowledge. The more management knows about the external environment the more control it has over it.

The two dimensions of futuristic thinking are, of course, interrelated. The longer the time dimension (that is, the farther into the future a manager wishes to look), the higher the probability that the environment will change. By the same token, the manager's attempt to increase certainty by looking only to the immediate future disqualifies the plan as a strategic one.

Thus contemporary managers are caught in a double bind: If they wish to provide a certain plan for the organization by avoiding long-range thinking they will be accused of being myopic. If, on the other hand, they do take a long-term view they stand a chance of having their plans being upset by the inevitable changes in the external environment.

Today's managers must take the long-term view. Indeed, having a strategic plan is no longer a luxury affordable only by large organizations. Every organization, from the largest private enterprise like General Motors Corporation to the smallest public college, develops a strategic plan. While the time dimension of these plans vary, the good ones will have annual strategic plans within a five-year strategic plan.

Because management appraisals by organizational stakeholders are based not on the quality of the plans but on how well they materialize, management hedges against the damaging effects of external environmental changes. It does so by (a) increasing the degree of knowledge about the external environment and (b) by shortening the response time required to correct a given deviation attributable to an unanticipated turn in the environment. Thus, the long-term perspective is counterbalanced by the quick response to external environmental changes. Organizations with these two qualities in their management style are known as "fast-cycle" organizations.

The key to minimizing uncertainty about the future consequences of today's decisions and actions is information. Since the source of this uncertainty is the external environment, the remedy is accurate, timely, and reliable *external* information. Thus the task of the systems-oriented manager is to continually scan the external environment and incorporate its potential impacts into the organization's strategic plan.

The systems approach provides an excellent framework for developing systems to enhance executives' ability to minimize the uncertainty inherent in the strategic plans. Accurate and current knowledge of the external environment allows the executive to extend the planning horizon to a logical medium-to-long range period. Well-designed and accurate internal systems allow organizations to minimize the time required to evaluate deviations from plans and to take corrective actions to bring them back to predetermined acceptable limits.

Part Four, Application of Management Systems   provides a step-by-step framework for the application of the systems methodology dealt with previously.

Chapter 11.   In the Systems Approach to Management we synthesize the material covered in the previous three parts and organize it into a three-step framework that constitutes our version of the systems approach.

Chapter 12.   Future Forecasting briefly exposes the reader to some of the most commonly used forecasting techniques for dealing with the future state of systems whose environments change frequently. Strategic planning is an exercise in forecasting ill-defined situations in which the environment changes independently of the management's intentions and actions.

Chapter 13.   Executive Support Systems surveys the state of the art in the field of decision support systems (DSS) aimed at providing on-line support for executives and their staff. Recent developments in information technology enable executives to access and "drill down" into conventional Management Information Systems dealing with transactional processing of organizational data.

Chapter 14.   Artificial Intelligence (AI) presents a brief survey of some of the most promising attempts to develop a science of cognition. It has always been one of the greatest aspirations that systems would be developed which would mimic human reasoning and decision making. Expert systems represent hopeful steps toward that goal.

Chapter 15.   Environmental Scanning for Strategic Management and Planning describes an executive support system whose aim is to provide the executive with the opportunity to *(a)* gain knowledge about the external environment, and *(b)* to incorporate the potential impacts of environmental changes into the organization's strategic plan.

CHAPTER 11

The Systems Approach to Management

The gull sees furthest who flies highest!

Richard Bach, *Jonathan Livingston Seagull*

CHAPTER OUTLINE

A RECAPITULATION

A basic postulate that underlies this text is that the second half of the 20th century is characterized by systems thinking—a trend that began with science and which has spread into other spheres of human activity. The study of human organizations has also been noticeably affected by this trend. Most contemporary writers in organization or management theory either implicitly or explicitly advocate a systems approach to the management of today's complex organizations. Therefore, it is not only desirable but also necessary to give students, and through them a wider sector of the society (i.e., practicing managers), at least some inkling of the profound conceptual changes that have been set in motion. Familiarity with certain systems concepts is indeed fundamental for an understanding of modern managerial thinking.

Systems thinking denotes both a technological revolution and a conceptual revolution. The latter cannot be understood without clearly tracing its origin and development. To this end Chapter 1 followed the origin and development of systems thinking as it evolved from the speculative ideas of the early biologists to the presently developed disciplines of general systems theory and cybernetics.

While a completely detailed historical account of systems thinking could not possibly be attempted, nevertheless, the earlier chapters of this text included enough information to enable the inquisitive student to adequately sample this new and exciting area of intellectual activity. A brief summary of the main ideas set forth in these chapters should prove beneficial in applying systems thinking to the study and management of organizations.

There are basically two main things associated with the age of systems: (1) the systems approach, or conceptual systems, and (2) management information systems, or applied systems. The systems approach is a philosophy or a viewpoint that conceives of an enterprise as a system—i.e., a set of *objects* with a given set of *relationships* between the objects and their *attributes,* connected or related to each other and to their *environment* in such a way as to form a *whole.*

The tremendous increase in the size and complexity of 20th-century organizations has forced students and managers into adopting a point of view that skirts traditional analytic thinking employed by the researchers of the so-called hard or physical sciences—primarily physics and chemistry. The analytic thinker, when confronted with a complex phenomenon, attempts to understand it by breaking it into smaller and less complex parts; by studying the parts separately; and subsequently by putting the findings together to gain an understanding of the whole. This is the "age of analysis," landmarked by the works of some of the greatest 20th-century philosophers, such as Bergson, James, Russell, Dewey, Santayana, and Whitehead.

Systems thinking, or the systems approach, represents the "age of synthesis." Here, the researcher's approach to understanding complex phenomena is one of synthesizing the findings of various disciplines, with the ultimate aim of developing a method or technique applicable to several seemingly different phenomena. All phenomena, whether physical or social, are treated by the systems thinker and researcher as systems. The age of systems is landmarked by the works of the late

Ludwig von Bertalanffy, a biologist; the late Norbert Wiener, a mathematician; the biomathematician, Anatole Rapoport; Kenneth Boulding, a noted economist; Herbert Simon, a Nobel Prize laureate in economics; and numerous others.

THE MANAGER-SCIENTIST SYMBIOSIS

How do manager's manage? A not improbable answer seems to be that managers manage by experience. Another would suggest knowledge. But is not experience to be equated with knowledge? Still other writers would advocate intuition as a factor governing the manager's decision making and controlling actions. Kenneth Boulding even proposes that managers, like all other human beings, base their managerial activities upon *images:* their subjective knowledge of what they believe to be true. The images develop as a result of all the past experiences of their possessors.[1]

Whatever the experts' consensus or lack of it, one can safely argue that managers manage by experience and knowledge. While there may be considerable philosophical controversy about the importance of knowledge as compared to experience, the truth remains that managers use both for decision making and control.[2]

Experience alone (i.e., the process of personally observing, encountering, or undergoing something) will prove inadequate. So too will the mere acquaintance with facts, truths, or principles (i.e., knowledge). A combination of the two, however, will provide the synergy needed for managers of today's complex organizations. For such complex systems one desires to know what the system is; what the logical internal relationships and the external relationships are—those with the rest of the world; and how the system is quantified. Here personal experience is supplemented and augmented by knowledge of facts, principles, and laws applicable to similar systems.

The task of science is considered to be the systematization of knowledge about the world. This systematizing involves the codification of personal experiences and knowledge of mankind as well as the organization of knowledge and experience into a form transmittable to others. This culturally transmittable organized body of knowledge and experience serves as a prototype against which new ideas may be compared and into which they may be incorporated.

As managerial problems become more complex, the need for a systematized body of knowledge becomes more imperious. Thus, the scientist can be of invaluable services to the manager. There is plenty of evidence to support this assertion. Early applications of the science of mechanics to production management proved very successful as far as the technological aspects of the processing

---

[1] K. Boulding, *The Image: Knowledge in Life and Society* (Ann Arbor: Ann Arbor Paperbacks, University of Michigan Press, 1969).

[2] S. Beer, *Management Science: The Business Use of Operations Research* (New York: Doubleday Science Series, Doubleday, 1968).

of raw material and their conversion into marketable products were concerned. Equally successful has been the application of economics to the monetary (e.g., pricing, costing, and so forth) aspects of the production and distribution of economic goods. Considerably less successful has been the attempt to utilize social science principles and postulates to deal with the so-called human side of the enterprise.

Recently attempts have been made to apply higher mathematics and sophisticated statistical methods to the management of business enterprises and most recently to the management of social nonbusiness types of institutions. The extreme enthusiasm of the so-called *quant man* has been matched with an equally strong skepticism of the practitioner. The general consensus seems to be that problems of this sort will become amenable to solution if and only if (to use the profession's lingo) managers become mathematicians or mathematicians are turned into managers. Of course, it is quite unlikely that either event will occur.

The systems approach begins with the assumption that the manager and the scientist have something in common: their viewpoint of an organization as a system. The only difference is that the manager's knowledge of that system is based upon his or her own experience with the system, whereas the scientist's knowledge is based upon experience with *other similar* (i.e., analogous) systems. Thus the two have different conceptual frameworks that govern their study of the system. This difference in conception is primarily the result of different educational backgrounds and training.

Consider, for a moment, Mary Dobson, a manager who is confronted with an inventory problem.[3] She knows a lot about it; she has been with this particular job for some time, and before that she had experience with similar systems generating similar problems. If she were asked to describe the inventory problem to someone else, her description of it would be, by necessity, through use of a conceptual model. This model would represent an accurate account of the situation but it would nonetheless be incomplete and somewhat "nonscientific" to the extent that it is too person-bound. In other words, although the manager usually knows what she is dealing with, still her being so close to the real situation may result in a distorted view of it. The manager's conceptual model of the real phenomenon might be unnecessarily detailed in some respects while lacking sufficient detail in other respects. In any event, the manager's too close view may interfere with her grasp of the overall problem.

Now, let us introduce James Bradley, a scientist. As already noted, his conceptual model of the situation will be somewhat different. He most likely has had no previous experience with the particular inventory setting that the manager is concerned with. Nevertheless, he develops a conceptual model of the situation. His modeling approach will draw heavily upon a storehouse of knowledge and scientific experience. Most likely, he will begin to quantify his crude conceptual model right away. Most verbal statements that make up the manager's conceptual

---

[3] This discussion is based primarily upon Beer's treatment of the subject. Beer, *Management Science.*

model will be replaced with some kind of number system. The scientist's model will have to be tested or experimented with. Upon the satisfactory performance of the scientist's model, it is then converted into a system for dealing with the real situation.

The modeling process described in the two previous paragraphs is diagrammed in the following schematic adapted from Beer's *Management Science* (Figure 11–1).

The management scientist's modeling process actually constitutes a hierarchy of models that begins with the manager's conceptual model (CM), goes through a state in which the model has virtually no resemblance to the original (it represents the scientist's way of conceptualizing), and finally ends up as a fairly realistic model that *can* be interpreted by the manager. It is this process of creating a rigorous scientific model that can be understood and appreciated by the practicing manager that makes the manager's transformation into a scientist and the scientist's transformation into a manager a *conditio sine qua non* of successful application of the systems approach to managerial problems and opportunities.

FIGURE 11–1   The Management Scientist Modeling Process

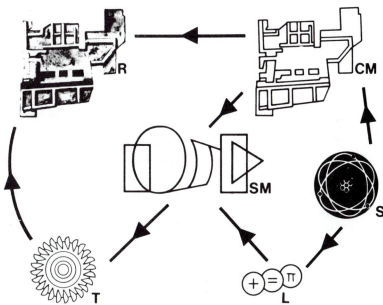

Science (S) contributes to the formation of the conceptual model (CM) and furnishes languages (L) that, together with the conceptual model, permit a scientific model (SM) of the real system to be formed. The scientific model furnishes techniques (T) that permit the real situation (R) as well as the scientific model, to be manipulated.

SOURCE: Adapted from Stafford Beer, *Management Science: The Business Use of Operations Research* (New York: Doubleday Science Series, 1968), p. 69.

APPLYING THE SYSTEMS APPROACH TO MANAGEMENT

What does this new way of thinking mean to the contemporary student of organizations? To put it differently, why should one be concerned with the systems approach? Or, assuming that one sees the need and relevance of a systems-oriented study of organizations, how does one begin to apply systems thinking to the study and management of today's exceedingly complex organizations? This present chapter is devoted to the development of a skeletal framework for the "how-to" portion of the systems concept.

The application of the systems approach to management can be conceived as consisting of the following two phases:

Phase A.   Awareness and conviction.

Phase B.   Implementation.

Phase A: Awareness and Conviction

In this phase, the manager becomes aware of the existence of a new way of thinking and approaching organizational problems and opportunities. Subsequently, the manager becomes convinced that the systems approach does indeed represent an appropriate method for the kinds of problems that the manager is confronted with.

This awareness-conviction process consists of the following three steps:

Step 1. Viewing the organization as a system.

Step 2. Building a model.

Step 3. Using information technology as a tool both for model building and for experimentation with the model; i.e., simulation.

Step 1: Viewing the Organization as a System.   Developing a systems viewpoint of an organization is primarily a matter of the manager's adopting a new philosophy of the world, of one's organization and its role within this world, as well, as a new viewpoint of oneself and one's role within this organization and this world. The manager's philosophy here advocated is, of course, systems thinking. There can be no doubt that this is a new philosophy for the practicing manager. The basic postulate of systems thinking (i.e., securing adequate knowledge of the whole relevant system before pursuing an accurate knowledge of the working of the "parts") is definitely against everything that the manager has been taught or has learned through personal experience.

Traditionally, organizations are departmentalized along functional lines. In business enterprises, for instance, one finds such departments as production, sales, finance, and accounting. Nonbusiness organizations follow a similar pattern. The organization or agency is divided into subagencies denoted by such names as districts, divisions, and sectors. In all these cases, individual managers or administrators perceive their own niches as the whole and consequently strive for their improvement and optimization. In reality, however, the scope of a particular manager's territory is determined by the behavior of the whole organization of which it is a part.

There are numerous ways of looking at an organization. Depending on the manager's educational background, position, the role he or she plays within the organization, and the issues at hand, he or she can adopt several viewpoints and focus on several key concepts. In this section, a brief description of an organization from the systems viewpoint is presented.

As was pointed out earlier in this book, the systems viewpoint emphasizes the relationship between the organization and its external environment. In managing this relationship, the manager must identify the main inputs, processes, outputs, and feedbacks. Figure 11–2 depicts an organization as an open system. In this system, inputs are drawn from the external environment, they are processed, and the resulting outputs are sent to the environment.

There are three main inputs to any organization: people, materials and equipment, and money. These three inputs are processed in accordance with certain economic principles (such as least-cost combination). The results of this processing are three outputs: products, waste, and pollution. All three of these functions (input, process, output) are coordinated by the management through such functions as goal setting, decision making, and controlling. Management, in carrying out this combination of the input, process, and output functions, receives inputs from the external environment regarding scientific and technological development and governmental and public attitudes and policies.

The dotted line around the system represents its boundary. As was already stated, the boundary delineates the system from its external environment. The environment represents the totality of factors that have a marked impact on the organization's function but are beyond its immediate control. For example, as can be seen from Figure 11–2, the organization has considerably more control over the input variables labor and money than it has over material and equipment. On the output side, the same situation can be observed with respect to products (goods and services) and waste (reprocessable) on the one hand and pollution on the other.

Step 2: Building a Model.   The terms *system* and *model* are sometimes used interchangeably with perhaps little harm done. However, it is convenient and more useful from the educational viewpoint to distinguish between the two.

Models and systems are human ways of dealing with the real world. The real world is considerably more complex than the mind can comprehend. Therefore, a person first constructs a picture that contains a number of the real-world characteristics great enough to provide an understanding of what is involved. *This picture, which is an abstraction of reality and which contains the most important elements of reality, is called a model.* The purpose of constructing a model is to *understand reality* by organizing it and simplifying it. The model *represents* reality but it *is not* reality.

FIGURE 11–2   The Organization as a System

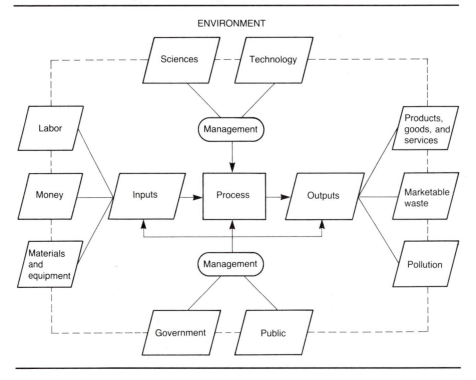

If the model is an organized abstraction of reality, the system is organized reality. Real-world phenomena, such as organizations, are exceedingly complex. As a rule they are not organized in a manner that will conform with a person's pursued goals. For this reason, people create real-world systems. Because it is an abstraction of reality, the model is always less complex than the real phenomena. Because the system is a tool for dealing with reality, it must be as complex as the real phenomena.

Thus, the relationship between the model and the system is of the following nature. Because human cognitive capacities are limited, real-world phenomena are first reduced to models. From these models people design real systems to deal with them. Figure 11–3 explains this relationship in more detail.

Systems-oriented managers who look at phenomena from the holistic viewpoint perceive them as an orderly summary of those features of the physical and/or social world that may affect their own behavior. Thus, the box labeled "the Manager's Conceptual Model" (MCM) represents the manager's interpretation of "what is really out there." The Real-World Phenomenon is the manager's own "reality," of course. At a later time the picture may be modified to accommodate fresh evidence or new data. Another observer, given the same or similar background, may entertain a different "picture" of the real phenomenon. In short,

FIGURE 11–3   The Real World, the Model, and the System

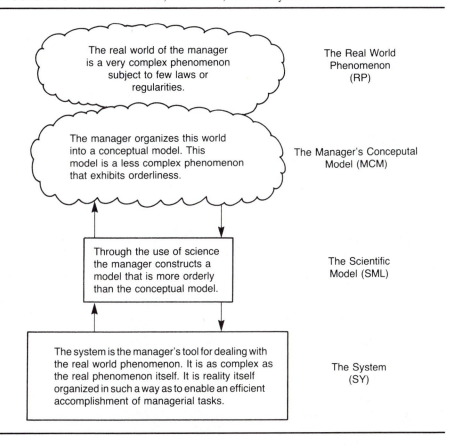

The real world of the manager is a very complex phenomenon subject to few laws or regularities.

The Real World Phenomenon (RP)

The manager organizes this world into a conceptual model. This model is a less complex phenomenon that exhibits orderliness.

The Manager's Conceptual Model (MCM)

Through the use of science the manager constructs a model that is more orderly than the conceptual model.

The Scientific Model (SML)

The system is the manager's tool for dealing with the real world phenomenon. It is as complex as the real phenomenon itself. It is reality itself organized in such a way as to enable an efficient accomplishment of managerial tasks.

The System (SY)

the conceptual model represents the ultimate outcome of the manager's sequence of mental activities (observation plus conceptualization) and not the outcome of the sensor system alone.

As previously stated, the systems-oriented manager of real phenomena will treat them as systems. To entirely comprehend the real phenomena (RP) is impossible because of their sheer complexity. Hence, the manager actually models the RP: first, conceptually; then, scientifically. The scientific model (SML) is a representation of the RP but with much less detail than the RP itself. Again, it should be recalled that the systems thinker is most of all interested in acquiring an adequate knowledge of the RP. Hence, the models, both conceptual and scientific, include only those factors or elements that are absolutely necessary for a rough description of the RP. The modeling process is not a once-and-for-all exercise. Rather, it should be conceived of as consisting of several provisionary models that adequately but roughly describe the manager's conception of the RP.

To sum up the four main parts of Figure 11–3, the real world perceived as the real phenomena (RP) is studied as a system by first being converted into a manager's conceptual model (MCM). By working between the RP and MCM, the systems-oriented manager constructs a scientific model (SML) and from this base eventually arrives at a system (SY) that will be as complex as the real phenomena (RP) themselves. This last point cannot be overstated. Systems thinking does not advocate conceptual simplicity. The apparent simplicity involved in the modeling process is only of a temporary nature. It is used as a means of comprehending the complexity inherent in the RP. The ultimate system that will be used to deal with the real-world situation must be as complex as the real phenomena (SY = RP). That is, of course, dictated by the universal law of requisite variety; one deals with complexity through complexity.

Let us consider an example to illustrate this relationship between a model and a system. Suppose an engineer desires to build a pipeline to carry oil from the northern part of a geographical area to the southern part. The first step is to construct a model by drawing a map of the area; then a line is drawn across this terrain (a conceptual model). This modeling process is continued until a physical model of the pipeline is constructed (a scientific model). Upon completion of this model, equipment, men, material, money, and so on are assembled and pipeline construction begins. The end result will be an oil-transportation system (pipeline), which will be as complex as the terrain itself. It will not only follow the exact characteristics of the real world but it also will be able to adapt to changes in environmental conditions.

Step 3: Using Information Technology.   The end result of the previous step is a model. Surely, the manager would like to test the model before converting it into the system. In testing the model, the manager discovers that the task is rather difficult, cumbersome, and time consuming. For these reasons the manager utilizes the computer to experiment with the model. Experimentation with a model is called simulation.

In simulating a particular model, the investigator deliberately changes certain parameters of the model, certain key variables or relationships, in the hope of gaining some knowledge of the degree of sensitivity of the model to such changes. The numerous books written on the subject of simulation indicate that the process is useful, albeit not simple. However, the basic concepts of simulation which are of interest to the manager are simple. Given that every model is based upon certain assumptions regarding an uncertain future which the model is supposed to organize and eventually predict, how would the model's organizing, heuristic, and predictive power change in the event of changes in some of the assumed conditions?

In general there are three kinds of simulation: (1) human simulation, (2) computer simulation, and (3) man-machine simulation. The first kind of simulation can range all the way from the practicing session of a sports team, animated war battles, or managerial meetings to sophisticated "sensitivity analysis." In managerial meetings the process of simulation begins by asking "what if" questions to proposed plans of action. The team proposing the plan will recompute the model's most likely performance under the different conditions imposed upon it, and so on.

Computerized simulation involves essentially the same process as human simulation, the only, but big, difference being that changes in certain parameters are initiated by the computer, which in turn recomputes the most likely results of these changes. This kind of simulation, although it can be very interesting as well as very informative, is generally of scant interest and small utility to the practicing manager because of the mysticism attached to the internal workings of the machine. As a result of this romanticized attitude, management's reliance on computer simulation results is still very limited. This attitude is to some extent reinforced by the simulation expert's unwavering preoccupation with simulating more and more abstract problems which have an intrinsic interest for him but are of little practical consequence for the manager.

The third kind of simulation, man-machine simulation or business gaming, is of paramount importance to the practicing manager. The logic underlying business gaming is essentially the same as in the two previous kinds of simulation: explication and understanding of the process of problem solving via experimentation with a model of this process, as well as testing the impact of possible variations in certain assumptions upon model outcome.

In gaming, the investigator takes the managerial decision function (the decision to change certain parameters) out of the computer program and restores it to the manager, while the computation of the possible results of the decision is left to the computer. The time between the change in a parameter (managerial decision) and the outcome can vary from several hours in the more traditional games to instant replay in the most advanced simulation games (on-line computer-management interaction). Figure 11–4 represents a typical business game situation involving four teams, each with four members. The fact that the manager initiates changes rather than being forced to accept certain arbitrary and random variations in certain market or firm conditions takes a lot of the mysticism out of the computer simulation, thereby making it more realistic and believable.

In summary, one necessarily begins with conceptualization and ends with conceptualization. As in all phenomena, the manager's intellectual tasks of policy setting, decision making, and control are, naturally enough, cyclical. In cyclical phenomena, as Heraclitus averred centuries ago, the beginnings and the ends are the same. Again, one must not begin delving into complex managerial situations involving thousands of relationships by quantifying first, simulating second, and applying third; rather one should begin with rigorous thinking about the logical relationships among the elements of the whole, then quantify them, and so on.

Before discussing in the last section of this chapter a paradigm of systems science in action, one ought to point out again the danger of excessive and premature analysis and quantification antecedent to logical conceptualization of the problem. Managerial problems by nature involve human experience and when

FIGURE 11–4   Business Gaming

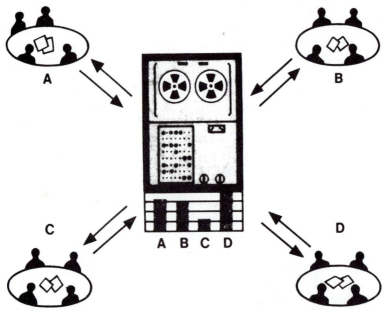

A competitive simulation situation. Four teams of "managers" operate competitive compa-nies, trying market and production strategies, etc. The computer, furnished with a model of the complete industry and market, feeds back information to the "managers" and also keeps score.

SOURCE: Adapted from Stafford Beer, *Management Science: The Business Use of Operations Research* (New York: Doubleday and Company, 1968), p. 86.

one deals with human experience, what L. Mumford once said about the so-called scientific method (analytic thinking) and its ability to deal with total human experience is still relevant:

> Admittedly the sciences so created were masterly symbolic fabrications: unfortu-nately those who utilized these symbols implicitly believed that they represented a high order of reality, when in fact they expressed only a higher order of abstraction. Human experience itself remained, necessarily, multidimensional: one axis extends horizontally through the world open to external observation, the so-called objective world, and the other axis at right angles, passes vertically through the depths and heights of the subjective world; while reality itself can only be represented by a figure composed of an indefinite number of lines drawn through both planes and intersecting at the center, the mind of a living person.[4]

---

[4] Lewis Mumford, *The Myth of the Machine: The Pentagon of Power* (New York: Harcourt Brace Jovanovich, 1970), p. 74.

Phase B: Implementation

The Conceptual Framework: The Systems Science Paradigm. The task of systems science, like any other science, is to develop and maintain some kind of a consensus among its practitioners regarding (1) the nature of legitimate scientific problems and (2) the methods employed for dealing with these problems. Kuhn employed the term *paradigm,* long familiar to students of classical languages, to connote "universally recognized scientific achievements that for a time provide model problems and solutions to a community of practitioners."[5] These paradigms, then, represent basic milestones in the development of a discipline.

Just as the invention of the telescope when combined with Newton's and Leibniz's calculus was significant for the development of classical physics, so too are the inception and development of systems science for the study of organizations. And just as F. W. Taylor's work at the beginning of this century provided a working paradigm for managers, so too does systems science in the 1980s provide a paradigm for sound management of complex organizations. However, the paradigm of systems science is not a blueprint for application of systems to organizations; a rather substantial amount of exciting mop-up work must first be done in the form of matching the facts to the paradigm and in further articulation of the paradigm itself.

There are several foci for factual systemic investigation of organizations, and these are not always nor need be distinct. First, there is the question of the philosophical predisposition of systems enthusiast/practitioners: *theirs is a world of organic-open systems.* Two main processes are of paramount importance in studying organic-open systems: growth and control. Growth is a necessary condition for the survival of any system; at the same time, control (the ability of a system to sustain a rate of growth in keeping with its capacity and the environment's tolerances) is a necessary condition for balanced growth.

Second, the organic system is investigated from the *holistic* viewpoint. However, operationally speaking, holism does not necessarily imply that the systems scientist must investigate everything about everything. What holism implies is that enough thought will be given to determining the critical variables influencing the growth and control patterns of an organization as well as to establishing ways of monitoring the critical parameters in the organization-environment interface. Holism means, to paraphrase Fuller, that one should begin with the universe. Again, the holistic approach does not imply that the manager should be concerned with everything that goes on within the department or division or the whole company; rather it demands that one always go one step beyond what up until now has been thought of as being satisfactory—i.e., step out of the circle that the job description has drawn.

Every manager can be a systems manager as long as one's approach is governed by the two following principles formulated by B. Fuller:

---

[5] Thomas S. Kuhn, *The Structure of Scientific Revolutions,* 2d ed. (Chicago: University of Chicago Press, 1970).

1. I always start with the universe: An organization of regenerative principles frequently manifest as energy (and/or information) systems of which all our experiences and possible experiences, *are only local instances.*
2. Whenever I draw a circle, I immediately want to step out of it.[6]

Managers whose styles are directed by these two principles begin their investigation of the world about them, not by gathering and analyzing the facts pertaining to happenings within "their" department, but rather by identifying their universe—i.e., their department as it affects and is affected by its environment. This definition of the manager's department along with its environment will provisionally determine the boundary (the circle, in B. Fuller's terms) of the system. About this system the manager will want to know its inputs, processes, outputs, feedbacks, relationships, as well as their attributes. The search for these system determinants begins with the construction of a conceptual model.

Third, the apparently unsurmountable task of holistically investigating an organization as an open-organic system under constantly changing conditions is facilitated through modeling processes. The modeling process begins with a considerably gross conceptualization of the system and ends up with a more or less precise model of an econometric nature.

Finally, the last focus of the systemic investigation has something to do with the most likely outcome of this kind of ambitious endeavor. The most likely outcome will involve an *understanding* of the focal situation as it relates to the rest of the organization and its environment. Once this understanding is achieved, certain quantification techniques can be utilized to calculate possible outcomes of proposed courses of action or inaction.

The Golden Rules

Rule 1: Understand First, Diagnose Second, Prescribe Third.   Managerial problems are to a large extent futuristic: they call for solutions whose implementation will affect future events. To the extent that the future is unpredictable, the manager must infer from incomplete information. Inferences from incomplete information will be the more realistic the more the manager understands the complete problem.

In discussing the differences between analytic and systems thinking, it was emphasized that the systems viewpoint advocated that the systems-oriented investigator should strive for an *adequate* knowledge of the whole relevant phenomenon rather than for an *accurate* knowledge of it. Now that we have reason to believe that managerial problems are by nature games of incomplete information, one can see the relevance of the realistic quest for adequate knowledge and the futile nature of the analytic thinker's drive for accurate knowledge.

---

[6] B. Fuller, *I Seem to Be a Verb.* (New York: Bantam Books, 1970).

Understanding managerial problems presupposes the realization that (1) life in an organic system such as a business enterprise is an ongoing process, (2) that one gains knowledge about the whole not by observing the parts but by observing the process of interaction among the parts and between the parts and the whole, and (3) that what is observed is not reality itself but the observer's conception of what is there.

Once this understanding of the whole relevant system is secured, then an understanding of a specific situation (problem and/or opportunity) is relatively easy. The diagnostic process should at least point to an array of alternative prescriptions, of which the systems scientist must choose one.

It cannot be overemphasized that the use of systems science to solve managerial problems or to create managerial opportunities proceeds from understanding to prescribing and not the other way around. The most common practice of the management scientist and operations research expert usually follows the opposite direction. It is also well known that this attitude, in addition to being illogical, is also very unpredictable. Beer's aphorism seems as appropriate today as it was when he first wrote it:

> This warning about confusing particular solutions to stereotyped problems with a proper understanding of management science seems very necessary today. No one would confuse the pharmaceutical chemist's dispensing of a prescription with the practice of medicine. Yet there is today a widespread attempt in many industrial companies, and to some extent in government, to make use of the powerful tools of O.R. trade without undertaking the empirical science on which their application should alone be based. This is like copying out the prescription that did Mrs. Smith so much good, and hopefully applying it to oneself.[7]

Rule 2: Conceptualize, Quantify, Simulate, Reconceptualize, Apply.   In modeling and systematizing managerial phenomena, the manager must go through a series of modeling attempts, all of which are arranged in a thoroughness-abstraction hierarchy. The apex of this hierarchy is occupied by the most abstract thinking, while the basis of it houses the most detailed models of the managerial phenomena. Beer calls this hierarchical arrangement of models "cones of resolution." Figure 11–5 shows how each level of resolution contains more and more detail. The tourist seeking to visit certain points of interest in the world may end the modeling process with a mere visit to the Eiffel Tower in Paris. An architect, however, might go further in modeling. Thus, the same objects or phenomena will occupy different levels within the same cone of resolution depending on the individuals' interests.

Conceptualization of a managerial problem or opportunity begins at the top of the cone of resolution. Clearly, at the top the level of abstraction is highest and the degree of thoroughness is at a minimum. The primary concern of the systems scientist here is to comprehend the logic of the basic elements as well as the

---

[7]S. Beer, *Management Science,* p. 26.

FIGURE 11–5   Cones of Resolution

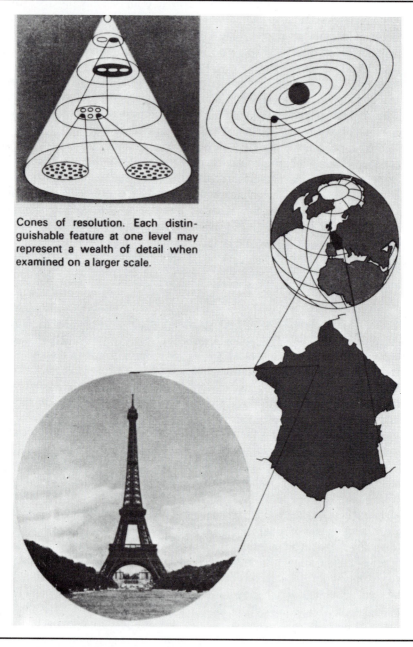

Cones of resolution. Each distinguishable feature at one level may represent a wealth of detail when examined on a larger scale.

SOURCE: Adapted from Stafford Beer, *Management Science: The Business Use of Operations Research* Copyright © Aldus Books Limited, London, 1967.

relationship among the elements—i.e., the logic of the system. Usually the investigator would be very satisfied to discover a common yardstick by which to measure the impact of one element's interaction with the other. In the business world we employ money as the common denominator of all relevant activities of the firm and its market, as is shown in Figure 11–6. The inadequacy of this top view along with its definitely monetary flavor becomes clearer the more one descends the cone of resolution from the balance sheet toward the isomorphic relationships between the factory and the market.

In summary, this is what we mean by *conceptualization* — understanding and organizing the interactions among the elements making up the phenomenon under scrutiny into a logical network of relationships in such a way as to reveal the direction of the underlying structure.

This framework derived from general systems theory is then converted into a quantitative network whereby the logical relationships are assigned economic values (i.e., costs and/or benefits). In this way the original abstract arrangement of relationships becomes an econometric model (i.e., a mathematical structure of economic relationships). The systems scientist is now ready to experiment with these highly particularized econometric models. Experimentation with a model over time is referred to—it will be recalled—as simulation.

The Main Stages.   Operationally, the application of the systems approach to management is a sequential process that involves the following three interrelated stages or phases:

Stage 1. Conceptualization.

Stage 2. Measurement/quantification.

Stage 3. Computerization.

Stage 1: Conceptualization.   All too often the statement is made that "the systems approach [or the systems concept or systems in general] has not developed enough to lend itself to employment in rigorous study of organization." This statement is, of course, true only if one perceives systems as a grab bag of unrelated cliches thrown together in a list of items under the heading of systems characteristics or systems attributes or simple buzz words. From the moment, however, that one begins to look at the systems approach as a theory or as a grown discipline with its philosophical premises and concepts (e.g., information, positive and negative feedback), hypothesized or propositional relationships among the concepts and its approach (e.g., holism, synergy), then the systems approach is quite more mature and operational than most theorists and practitioners tend to think.

Let us illustrate the point of systems operationally by examining once again the problem of assessing the relationship between the firm and its environment, using the concept of the cones of resolution discussed earlier in this chapter. To begin with, let us reiterate again the steps in the systems scientist's thought

FIGURE 11–6   Cones of Resolution for the Firm-Market Interface

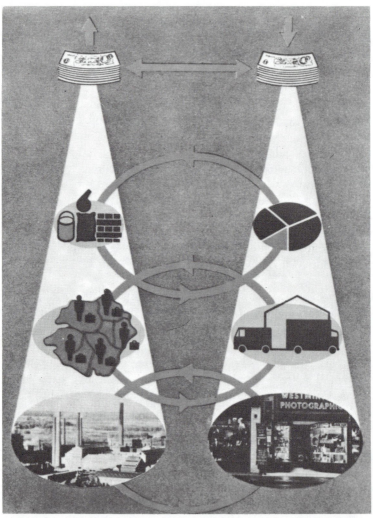

For some purposes, comparison of cash income with expense (top level) adequately describes the interaction of a company with its market. For other purposes the proportion of income derived from each product is relevant, for others the number of trade representatives, etc., is required, and so on down the cone of resolution until we come to the actual company and market.

SOURCE: Adapted from Stafford Beer, *Management Science: The Business Use of Operations Research* (New York: Doubleday and Company, 1968), p. 112.

process. This person begins by looking at the organization as an open-organic system which is in constant interaction with its environment. This holistic approach to the study of the organization dictates that one should focus on both the organization and its environment as they interact with each other. To deal with this complexity one is forced to model this interaction. Finally, the researcher tries to understand as much of the interaction process as possible without regimenting the phenomenon to a meaningless two-member relationship.

Let us begin at the top of the cone of resolution (Figure 11–7). There the organization-environment interface is pictured as just two boxes interacting with each other via two feedback loops (Feedback 1 and Feedback 2). This, of course, is the easiest and the most economical way of gaining an understanding of what is involved. However, the informativeness of this model is exceedingly limited. Both the organization and its environment are represented by T-accounts ($) indicating the financial positions of both subsystems vis-a-vis each other.

The second level in the cone of resolution focuses on the environment in greater detail. Thus, the box "environment" is dissected into some of its most important (from the firm's viewpoint) sectors. The firm produces certain products which are sold through the market. Production is realized by combining certain factors of production (e.g., technology); sales are accomplished by competing with certain rivals (e.g., world competition) and by complying with certain regulatory agencies (e.g., the Environmental Protection Agency).

In all these interactions the firm will choose a specific relationship within a certain environmental sector (will make an offer, so to speak). The environmental sector will then indicate whether the proposed state satisfies its needs (which are, of course, organizational constraints). In cases of incongruences – situations where the proposed state does not fully satisfy the market's desires – the firm must propose another state, and so on. The point made here is that stated relationships between the two subsystems are commonly determined rather than arbitrarily chosen by either subsystem. Dominant relationships are only in the short run viable; in the long run, dominance must give way to cooperation. The firm strives for a dynamic equilibrium (an equilibrium under constant change of the rates and levels determining it) between itself and the sectors of the external environment.

The third level in the cone of resolution involves a further elaboration of the sector of the external environment labeled market. Here the systems manager can see quite a bit more about the market and its activities. Of course, many of the boxes are still pretty much "black boxes" to the extent that the manager does not know everything there is to know about, let us say, the competitor's activities so as to design an effective strategy. However, the manager knows enough about them to be able to identify them as well as conjecture their possible impact. Looking at Figure 11–7 on the third level, one can perhaps identify possible decision/control points, those critical effective parameters which require extensive monitoring.

Stage 2: Measurement/Quantification.   So much for conceptualization. If that were all there is to systems, then, of course, no scientific status could possibly be claimed for it. While the modeling process is necessary, it is by no means sufficient. The stage has now been set for still another modeling process to begin, although of a slightly different nature.

Quantification begins when the need for measurement of changes in the state of systems elements has arisen. Mensuration has been a human preoccupation

FIGURE 11–7 Modeling of the Environment-Organization Interaction System through the Cones of Resolution Technique

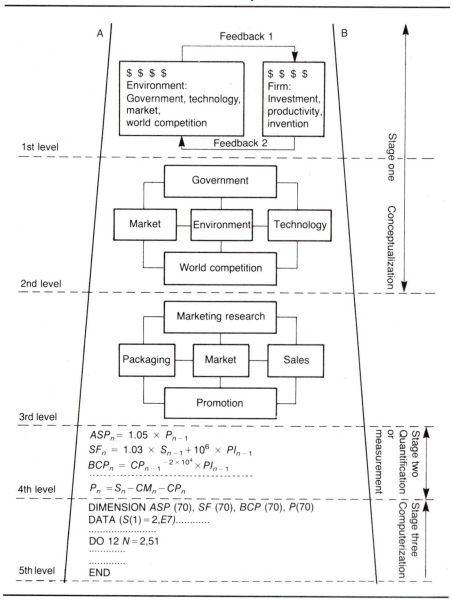

from the beginning until now. In organizations measuring inputs (e.g., cost) and outputs (e.g., revenues) has been one of the earliest applications of science to the management of organizations; the process is known as accounting and/or finance. Beer seems to think that "the origins of a scientific approach to management were connected with the measurement process." One cannot but agree that measuring changes in the firm's internal and external states is as old as organizations themselves.

At the beginning of this century another quantification attempt got off the ground. This time the logic of measurement concentrated on quantifying the workers' contributions to the firm's goals. The rationale behind this was that if one knew how to measure the potential output of a worker, then, at least theoretically, one would be able to utilize the factor more effectively and efficiently. Time and motion studies were once very popular and, of course, still are important. Despite the innumerable criticisms, measuring workers' contributions did show that a better combination of person and tool can indeed increase productivity. Although only the physiological side of the human being was measured, the outcome of this measuring process did result in better tool or machine design, thereby making the exertion of human muscle energy less and less necessary. The science of biomechanics epitomizes the giant strides made since the early time and motion studies first made history.

At the beginning of the second quarter of this century the measuring efforts in organizations were concentrated on the nonphysiological aspects of the human factor of the enterprise. Configurations of monetary and nonmonetary incentives were designed to motivate both workers and, to some extent, lower supervisory personnel so as to utilize a greater portion of their potential in achieving organizational objectives. Just as scientific management in the early 1900s assumed that healthy workers operating a better designed tool would be more productive, so did human relations assume that happy employees working within a better and happier environment would utilize their potentials more productively.

The second half of the 20th century ushered in another measurement process. This process aims at assessing the contributions of (1) managerial personnel and (2) information toward accomplishing organizational goals. The measuring process is, of course, of a slightly different nature. It aims at measuring the measurer. For this reason the process is considerably more difficult and delicate than the previous processes. Its domain of measurement more or less encompasses the entire organization as it relates to its environment. Since the expected payoffs of this measurement process are much greater, one ought to take a closer look at this measurment process. The implications of measuring the measurer are so great and the cost of doing so so high that it behooves every modern manager to reexamine the entire measuring process.

Thus far we have been using the terms *measurement* and *quantification* more or less interchangeably. The reason for this is that measurement is frequently defined as the "assignment of numerals to elements or objects according to [certain] rules."[8] From such a definition one can easily get the impression that

---

[8] For a more detailed treatment of the subject of measurement see: *Measurement: Definitions and Theories,* ed. C. West Churchman and P. Ratoosh (New York: John Wiley & Sons, 1959); also C. W. Churchman, "Why Measure?" in *Management Systems* 2d ed., ed. P. P. Schoderbek (New York: John Wiley & Sons, 1971); also R. W. Shephard, "An Appraisal of Some of the Problems of Measurements in Operation Research," in *Management Systems,* ed. Schoderbek.

measuring means quantifying, as most of the literature in what is emphatically called quant methods seems to indicate. From this definition one might conclude that whatever cannot be quantified cannot be measured; or to carry the logic a step further, what cannot be thusly measured cannot be of any consequence for management. Practicing managers and administrators do, however, know better. They know that quantification is only one way of measuring. Another way of measuring, known as qualitative measurement, does exist and is as meaningful, and under certain conditions as useful, if not more so, than quantitative measurement.

With this in mind, it may perhaps be better to refer to the next (fourth) level of modeling in the cone of resolution depicted in Figure 11–7 as the measurement process rather than the quantification process. Operating at that particular level, the systems scientist becomes a measurer. Acting in that capacity, she or he must decide:

1. In what language to express results (language).
2. To what objects and in what environments results will apply (specification).
3. How results can be used (standardization).
4. How one can assess the "truth" of the results and evaluate their use (accuracy and control).[9]

In Figure 11–7, fourth level, a small portion of Bonini's Model is illustrated.[10] The model indicates that the aspired profit for the period or year n (ASP_n) is equal to the actual profit of the previous period (P_{n-1}) multiplied by a factor of 1.05. Estimated sales for the same period n (SF_n) are equal to actual sales of the previous periods (S_{n-1}) multiplied by a factor of 1.03 plus a pressure index which reflects the growth of the industry (PI_{n-1}) times 10^6. Budgeted production administrative expenses for period n (BCP_n) equal actual production administrative expenses for the previous period (CP_{n-1}) minus 2 multiplied by the pressure index times 10^4. Finally, actual profit for the same period n (P_n) equals actual sales (S_n) minus actual manufacturing cost (CM_n) minus actual sales administrative expenses (CS_n) minus actual production administrative expenses (CP_n).

A further elaboration on measurement and the measuring process would carry us beyond the intended scope of this work. For our purposes, it will suffice to restate two conclusions: (1) The function of measurement is to develop a method for generating a class of information that will be useful in a wide variety of problems and situations. This method may involve either a qualitative assignment of objects to classes or the assignment of numbers to events and

---

[9] Churchman, "Why Measure?" p. 123.

[10] The Bonini Model is described in C. McMillan and R. F. Gonzalez, *Systems Analysis*, rev. ed. (Homewood, Ill.: Richard D. Irwin, 1968), pp, 424–28. ©1968 by Richard D. Irwin, Inc.

objects; in most instances, both quantitative and qualitative measuring processes will be employed; (2) The process of measurement is facilitated by rigorous conceptualization along the three levels within the cone of resolution, and these will definitely help or hinder the descent to the next level of computerization.

Stage 3: Computerization.   Ideally the ultimate product of Stage 2 will be a mathematical model that will be translated into a computer-consumable project. It should be recalled that mathematical notation is the language understood by the machine. Therefore, qualitative considerations must be inferred from the functioning and the output of the model by the human being who compares the outcome of this phase of the process with the aspirations and expectations formulated during Stage 1 (conceptualization). Computer simulation provides the least expensive way of performing this comparison. The outcome of this comparison, expressed in terms of simulation results, will lead to either a reconceptualization (back to Stage 1) or to remeasurement and quantification (back to Stage 2) or to both. In any event, the ultimate outcome of Stage 3 will be a computer program along the lines of the portion depicted in level 5 of Figure 11–7.

To recapitulate, it is imperative that the systems scientist proceed from conceptualization to computerization and not vice versa, as is indicated by the upside-down pyramid in Figure 11–7. A better grasp of the problem along the lines of Stage 1 will enable the researcher to develop a better measurement method and a better computer program rather than force the problem into a preconceived computer program.

As can be seen from Figure 11–7, the systems scientist initially loses quite a bit of the real phenomenon of the environment-organization interaction system because of the abstract and aggregative nature of the manager's conceptual model. These losses, however, are rather moderate when compared to those that would be incurred were the manager to follow the "model-up" method rather than the "model-down" approach which is advocated here. The conventional modeler will begin at the base of the cone of resolution and then will proceed upward.

The typical analytic thinker's cone of resolution can be imagined as being the mirror image of the systems scientist's cone of resolution. A quant person's model will most likely begin at the fifth level of the cone of resolution depicted in Figure 11–7. In working up the cone, certain losses will occur because of the narrow and precise quantification of certain relationships of the organization-environment interaction system. However, unlike the outcome for the systems scientists, these losses tend to increase with the height on the cone of resolution until eventually the quant person crosses the boundary of reality (as perceived by the manager). From there on this person is out of touch with reality, eventually reaching a point of maximum irrelevance by pursing certain solutions to problems which he or she alone can understand and interpret.

It is unfortunate that some, parading under the banner of operations research and management science, use their techniques as procrustean beds upon which managerial problems are amputated and distorted so that they exactly fit the sacred box of the quantifier. The discontent of members of The Institute of Management Science (TIMS) with the institute's orientation attested to this phenomenon some two decades ago.[11]

SUMMARY

Figure 11–8 summarizes the logic of the application of the systems approach to the study of real-world phenomena. In general, the logic is the same as in Figure 11–1 at the very beginning of this chapter: a real phenomenon (RP) must necessarily be studied via a model (ML) which is used in the design of a system (SY) which in turn represents organized and systematic reality. Two novelties are added in Figure 11–8: (1) the several subapproaches or subdisciplines under the name of the "systems approach" arranged in a philosophy-science or qualitative-quantitative continuum; and (2) the three subsystems of the grand system — viz., the scanning subsystem, organizing subsystem, and the decision subsystem.

A systems-oriented manager or student can design a system for studying an organization as it interacts with its environment by drawing from any of the four subdisciplines beginning with the general-qualitative considerations at the philosophy end of the continuum (GST) all the way to specific quantitative considerations (OR or SE) at the right end of the same continuum. Again, it must be emphasized that one begins with the general and proceeds to the specific — i.e., from left (GST) to right (SE) and not vice versa. It is imperative that the systems approach be understood and utilized as an integrative "linkage" discipline and not as a grab bag of specific techniques of some "quick and dirty" steps for troubleshooting. As Laszlo put it, "the system thus created [containing both quantitative and qualitative considerations] feeds on information."[12] Its inputs will be information (primarily data) of a relatively crude nature and of relatively small value; its outputs will also be information, although of a much higher value and usefulness. This information processing and transformation system functions as an integrated whole consisting of the three subsystems that interlock through feedforward and feedback mechanisms. Raw data about the the external environment as well as about the internal working of the firm are gathered by the scanning subsystem, analyzed and evaluated by the intelligence-organizing subsystem, which more or less separates data into those having immediate and high information content, and those with future utility. Finally, the information thus generated is transmitted via the regular channels of

---

[11] See, for instance D. F. Heany, "Is TIMS Talking to Himself?" *Management Science* 12, no. 4 (December 1965), B146–55.

[12] E. Laszlo, *The Relevance of General Systems Theory,* Papers Presented to Ludwig von Bertalanffy on His Seventieth Birthday, G. Braziler's Series, *The International Library of Systems Theory and Philosophy* (New York: George Braziler, 1972).

FIGURE 11–8   The Application of the Systems Approach to the Study of Real-World Phenomena

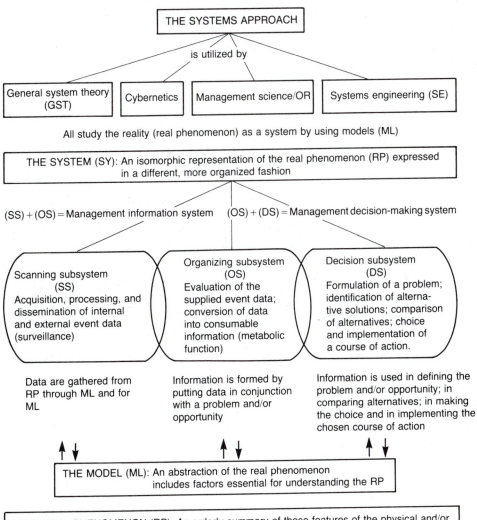

THE SYSTEMS APPROACH

is utilized by

| General system theory (GST) | Cybernetics | Management science/OR | Systems engineering (SE) |

All study the reality (real phenomenon) as a system by using models (ML)

THE SYSTEM (SY): An isomorphic representation of the real phenomenon (RP) expressed in a different, more organized fashion

(SS) + (OS) = Management information system   (OS) + (DS) = Management decision-making system

Scanning subsystem (SS)
Acquisition, processing, and dissemination of internal and external event data (surveillance)

Organizing subsystem (OS)
Evaluation of the supplied event data; conversion of data into consumable information (metabolic function)

Decision subsystem (DS)
Formulation of a problem; identification of alternative solutions; comparison of alternatives; choice and implementation of a course of action.

Data are gathered from RP through ML and for ML

Information is formed by putting data in conjunction with a problem and/or opportunity

Information is used in defining the problem and/or opportunity; in comparing alternatives; in making the choice and in implementing the chosen course of action

THE MODEL (ML): An abstraction of the real phenomenon includes factors essential for understanding the RP

THE REAL PHENOMENON (RP): An orderly summary of those features of the physical and/or social world that affect behavior.

the decision subsystem. It is this "metabolic power"[13] of a well-organized firm that guarantees its long-range survival.

---

[13] Stafford Beer, "Managing Complexity," in *The Management of Information and Knowledge* (Washington, D.C.: U.S. Government Printing Office, 1970), pp. 41–61.

KEY WORDS

Systems Thinking

Age of Synthesis

Manager-Scientist Symbiosis

Manager's Conceptual Model

Simulation

Conceptualization

Measurement

Computerization

Cones of Resolution

Quantification

Models

REVIEW QUESTIONS

1. Briefly outline the three basic stages involved in applying the systems approach to management.

2. What is the role of the "model" in the application of the systems approach to the study of the real world?

3. It is postulated that in managing real-life organizations the manager or administrator is assisted by the management scientist. Scientists have little knowledge of managerial situations as they unfold in real life. The manager, on the other hand, usually has little knowledge of scientific techniques. How do the two manage to solve managerial problems?

4. Discuss the two "rules" which are to be followed when using science to deal with real-life managerial questions.

5. Explain the concept of "cones of resolution," and show how it can be used in studying or examining the acquisition of a small retail shop by a big chain-store operation.

6. What is the systems-science paradigm?

7. Briefly outline the three basic phases of the systems paradigm.

8. An insurance company headquartered in the eastern part of the United States is contemplating "branching out" into a new venture. You are part of the team which is assigned to take the systems approach to this managerial problem. Briefly outline the basic stages of this approach to your team members.

9. XYZ Electric company, faced with an increased demand for electricity, is planning the construction of a new power plant. The management desires to take the systems approach to this investment. You have been hired as a consultant for this project and asked to outline the conceptualization, analysis, and measurement phases of this problem.

10. Continue the above problem with the final phase of computerization. Name some basic management-science techniques which you find can be useful in the above situation.

The Electra Company Case[14]

Electra Company was founded in the mid-1950s by a group of young engineers and accountants. Over the last two decades the company has been very successful. During the last five years alone Electra grew at an average annual rate of some 10%. Over the same period a number of new departments and divisions were created.

Last year Electra hired Systemics, Inc., a large consulting firm, to take a look into the company's present operation and future plans. Systemics, Inc., spent about six months studying Electra's documents and interviewing top managers. The general impressions and specific recommendations of Systemics were put forth in a report that was sent to Electra's president.

The *Executive Digest,* a one-page summary highlighting Systemic's main recommendations concluded: "There appears to be a lack of understanding by the company's personnel in general and by most of the top managers in particular regarding the main goals and objectives pursued by Electra. Specifically, most executives do not see how their particular job affects the duties of the others and vice versa. In addition, nobody seems to know what the external environment of the company is and how it is affected by what the company does and how the environment affects the operations of the company. We, therefore, highly recommend that Electra Company adopt the *systems approach.*"

After reading this last paragraph, the president of Electra Company called a meeting for the next day with all top executives. The president opened the meeting with the following statement: "I trust that all of you have a copy of Systemic's report and that you have had the opportunity to take a look at it. Well, I don't know what you got out of it, but it seems to me it all boils down to our using the systems approach. Now, I was an engineer for quite a long time and I think I know something about engineering systems. In any case, what's wrong with our engineering department?"

"Well, sir," said Bill Billings, the young vice president for corporate affairs, "I wish it were that simple. With all due respect to you, sir, I don't think you've interpreted Systemic's recommendations correctly. Because of the nature of my job, I spent a lot of time with these young gentlemen from Systemics. I think I am correct in saying, sir, that the systems approach that is recommended by Systemics has little to do with engineering systems. It rather has . . ."

Billings was interrupted by Joe Smith, the vice president for marketing, who said, "Bill has got a point, sir, if I may interrupt for just a second. I just returned from an executive development program where we had a two-day workshop on the

[14] Reprinted from H. R. Smith, Archie B. Carroll, Asterios G. Kefalas, and Hugh J. Watson. *Management: Making Organizations Perform* (New York: Macmillan, 1980) by permission of Macmillan Publishing Co.

systems approach. Well, I don't want to put you through the entire ordeal I went through, but what I think Systemics wants us to do is to set up a new department and then buy a large computer, and that will take care of us."

The president noticed that Billings was a bit skeptical. He turned to him and said, "Well, Bill, you don't seem to be in complete agreement with Joe. What's on your mind?" Billings thought for a while and then said, "You're right, sir. See, Joe is partially right, but only partially. Computers and computer systems are a part of the systems approach, but only a part. And to be honest with you, sir, that's the last thing we have to consider at this point. I tell you what we ought to do. I think we ought to do some homework. In other words, we ought to do some reading on the systems approach, talk to our friends who might have had some experience with it, and so on, and then have another meeting some time next week. I'll volunteer to present an introductory paper to get us started."

"That sounds like an excellent idea," said the president. "Thank you for volunteering, Bill. You're on a week from today. Good luck to all of you. Do some homework. I myself am going to skip my usual weekend fishing and dig out some of my old books on engineering and systems. Good day, gentlemen."

[A week later]. The president entered the meeting room smiling and full of excitement. "Good morning, gentlemen," he said. "This is going to be a lot of fun. This systems thing is really very exciting. It makes a lot of sense to me. I don't understand why we couldn't have come up with something like Systemics did. It's mostly common sense . . . but let Bill tell us all about it."

Billings got up, turned on the overhead projector and then turned it off. "I just wanted to test this thing," he said. "Good morning, gentlemen. Today I'll give you, as I promised, a brief talk on what the systems approach is all about. In a later meeting I'll show you how we, as a manufacturing company, can begin implementing it."

"Let me first point out that the systems approach is a way of looking at an organization. In other words, it is a viewpoint . . ."

Continue Mr. Billing's presentation.

How would you conceptualize Electra as a system? What are Electra's main inputs? Explain to Electra's management how every manager must develop a way of thinking of Electra as an open system.

Additional References

Ackoff, Russel L. *Redesigning the Future: A Systems Approach to Societal Problems.* New York: John Wiley & Sons, 1974.

Baker, Frank. *Organizational Systems: General Systems Approaches to Complex Organizations.* Homewood, Ill.: Richard D. Irwin, 1973.

Buckley, Walter, ed. *Modern Systems Research for the Behavioral Scientist.* Chicago: Aldine Publishing, 1968.

Checkland, Peter. *Systems Thinking, Systems Practice.* New York: John Wiley & Sons, 1981.

Chen, G. *The General Theory of Systems Applied to Management and Organization.* Seaside, Calif.: Intersystems Publications, 1980.

Chin, Robert. "The Utility of Systems Models and Developmental Models for Practitioners." In Warren Bennis; Kenneth Benne; and Robert Chin. *The Planning of Change.* New York: Holt, Rinehart & Winston, 1961.

Cummings, Thomas G., ed. *Systems Theory for Organizational Development.* New York: John Wiley & Sons, 1980.

DeGreene, Kenyon B. *The Adaptive Organization: Anticipation of Management Crisis.* New York: John Wiley & Sons, 1982.

Kast, F. E., and J. E. Rosenzweig. *Organization and Management: A Systems Approach.* New York: McGraw-Hill, 1970.

Kenney, Ralph L. "Measurement Scales for Quantifying Attributes." *Behavioral Science* 26, no. 1 (January 1981), pp. 19–36.

Lanford, H. W. *System Management Planning and Control.* Port Washington, N.Y.: Kennikat Press, 1981.

Lockett, Martin, and Roger Spear, eds. *Organizations as Systems.* Milton Keynes, England: Open University Press, 1980.

Miller James G. *Living Systems.* New York: McGraw-Hill, 1978.

Negoitsu, Constantin V. *Management Application of Systems Theory.* Basel: Birkhauser Verlag, 1979.

Phillips, Denis C. *Holistic Thought in Social Science.* Stanford, Calif.: Stanford University Press, 1976.

Schlenger, William E. "A Systems Approach to Drug User Services." *Behavioral Science* 18, no. 2 (March 1973), pp. 137–47.

Schoderbek, Peter, and Charles Schoderbek, eds. *Management Systems.* New York: John Wiley & Sons, 1971.

Schrode, William A. *Organization and Management: Basic Systems Concepts.* Homewood, Ill.: Richard D. Irwin, 1974.

Scott, W. Richard. *Organizations: Rational, Natural and Open Systems.* Englewood Cliffs, N.J.: Prentice-Hall, 1981.

Smith, August W. *Management Systems: Analysis and Application.* New York: CBS College Publishing, 1982.

Stephaou, Stephen E., ed. *The Systems Approach to Societal Problems.* Malibu, Calif.: Daniel Spencer Publishers, 1982.

Sutherland, J. W. *Societal Systems: Methodology, Modeling and Management.* New York: North-Holland Publishing, 1978.

Thompson, Mark. "A Systems Approach to Environmental Engineering." *Behavioral Science* 20, no. 5 (September 1975), pp. 306–24.

Toronto, Robert S. "A General Systems Model for the Analysis of Organizational Change." *Behavioral Science* 20 (May 1975), pp. 145–56.

Vemuri, V. *Modeling of Complex Systems.* New York: Academic Press, 1978.

Zelany, Milan, ed. *Autopoiesis: A Theory of Living Organizations.* New York: North-Holland Publishing, 1981.

CHAPTER 12

Future Forecasting

The trouble with our times is that the Future is not what it used to be.

Valery

CHAPTER OUTLINE

INTRODUCTION

It is not uncommon nowadays to hear it said or to read in a business magazine that a particular organization "went under" because management failed to judge the environment correctly and to take the appropriate actions to alleviate the steering of the corporate ship into some very shallow waters. The usual explanation offered by management is that "nobody can really tell what the future is going to bring" . . . "you just cannot prepare or plan for what is going to come" . . . "how could we have known that the public would change its attitude or taste for our product" . . . and so on.

All of this may sound like the poor excuses of people who, for some reason or another, failed to do their homework and who are trying to justify this failure by blaming the so-called external business environment. Be what it may, the fact is that statements of this kind point to a very real and legitimate problem of today's management and have little to do with intellectual alertness and willingness to do the homework.

Without going into great detail to prove it, one can very easily argue that the external business environment changes with a rate which is far beyond the reach of the intellectual and perceptual capability of the average manager or, for that matter, a group of managers. Alvin Toffler defined this gap between human perceptual capacity and the speed of change in the world as "Future Shock" almost 20 years ago.[1]

Management education, both formal and experiential, both at school and at in-house management-development programs, has paid little attention to this important phenomenon. The motto seems to be "getting today's job done today and not tomorrow . . . for tomorrow's job cannot be done today," or something similar to that.[2]

How can a manager develop a sense of the future? In other words, how can a manager relate present actions taken as a result of managerial decision making *to their consequences?* This is the focus of this chapter.

FUTUROLOGY: ORIGIN AND MEANING

Until very recently, futurology—*the study of possible future worlds and the societal adjustments to convert these possible worlds into preferred situations*—was thought of as the domain or preoccupation of eccentric individuals who could not hold a job because of their constant obsession with tomorrow's dreams rather than with today's realities. Since most organizations regarded these individuals as misfits, they broke off, so to speak, and formed their own organizations, some of which have been exceedingly successful. For example, the World Future Society

---

[1] Alvin Toffler, *Future Shock* (New York: Bantam Books, 1971). See also Toffler's *The Third Wave* (New York: William Morrow, 1980).

[2] See, for example, Thomas H. Naylor, "Management is Drawing in Numbers," *Business Week,* April 6, 1981, pp. 14, 16.

numbers some 50,000 members in 80 countries all over the world. Other groups, such as The Committee for the Future, Resources for the Future, Inc., The Hudson Institute, The Club of Rome, and so on, are less popular, albeit more powerful in terms of influencing public and corporate policy setting and decision making.[3]

The need for developing a sense of the future, of relating today's actions to their future consequences, has gone through a two-stage process:

Stage One. Awareness.

Stage Two. Action taking.

The Stage of Awareness

In general, the process of gaining a sense of the future begins with an organization's recognition that the future consequences of given managerial or administrative actions might be as important as, if not more important than, the present results that the organization has achieved. In retrospect we all, of course, know now that the above statement has been true with respect to the once controversial Alaska pipeline project and the Alaska oil spill, the National Environmental Policy Act with its Environmental Protection Agency, the OSHA and ERISA legislation, and numerous other governmental actions.

An organization expresses its "future awareness" through such actions as sponsoring or supporting futures conferences and workshops, running future-minded student contests, or sending executives to the meetings of the World Future Society, the Club of Rome, or the Committee for the Future. Or, it can become involved in one of the "State 2000" groups; invite a futurist to talk to a group of high-ranking executives; and so on.

A few examples will illustrate and highlight the pervasiveness of the contemporary state of awareness of futurology. On the world level, interest in the future has been great. There have been at least a dozen world conferences on the future. Television specials, artistic exhibitions, and other manifestations of human interest literally have grown exponentially.

In the United States, Congress has shown a tremendous interest in the future. In 1976 it established the Congressional Clearinghouse of the Future, whose purpose is to help Congress carry out its "foresight responsibility" and to assist individual congressional committees to do the "future impact" analysis required by the House's so-called "foresight provision." As a result of this future awareness, some 20 items of future-related legislation were enacted since 1970.[4] For essentially the same reasons, citizens in the United States have formed and helped

---

[3] A. G. Kefalas, "On Human Organizations: An Overview," *Human Systems Management* 1 (1980), pp. 79–84.

[4] Ren Renfro, "How Congress Is Exploring the Future," *The Futurist,* April 1978, pp. 105–12. Also, Anne W. Cheatham, "Helping Congress to Cope with Tomorrow," *The Futurist,* April 1978, pp. 113–15.

recognize some two dozen "State 2000" groups, including Iowa 2000, Georgia 2000, and California Tomorrow.

In the private sector, individual businesses as well as business associations have organized conferences and have published the results along with strong messages regarding the necessity of future orientation by all industry. For example, in 1975 the Conference Board, a business research group, organized a two-day conference on the future of organizations; top industrial leaders were invited to listen to futurists and also to present their own views on the future of the organization.[5] In 1976, the American Management Association ran a survey "The Future of the Corporation" and presented the elicited views of some 644 business leaders.[6] In 1974 the Savings and Loan Associations Institute in Chicago published a pamphlet entitled "Future Alternatives Report."[7] (The title is a slight misnomer.)

The best example, however, comes from the food-service industry. In 1977 *Institutions/Volume Feeding,* the trade magazine for the industry (a $100 billion business that includes the Big Mac as well as your corner deli or health-food shop) invited and paid 13 well-known personalities ranging from Walter Heller, chairman of the Council of Economic Advisors under Presidents Kennedy and Johnson, to Paolo Soleri, the eccentric futuristic architect, to present their views to some 100 presidents and vice presidents of the food-service industry about the future and its impact upon their business. The meeting took place in Big Sky, Montana, which was chosen, in the words of Jane Young Wallace – the magazine's editor-in-chief and the chairperson of the two-day conference – "precisely because it is almost impossible to reach. We wanted nothing to interfere with the business at hand."[8] To report the conference, the magazine put out a special issue. In introducing it, publisher David Wexler attempted to convince the readers who did not attend that they had really missed something, for, as he put it: "While some of this may seem rather highflying, it's probably more important to you, and probably more basic than the state of your warehouse."[9]

The Stage of Action Taking

Once the stage of awareness has been reached and the organization has indicated a desire to develop a sense of the future, it begins to realize that something must be done about attempting to (1) assess the external business environment which most likely will determine the state of the future, and (2) estimate or forecast the possible impact of changes in the environment upon the organization's structure, function, and evolution. In the academic jargon, this is

---

[5] Lilley W. Kay, ed., *The Future of Business in Society* (New York: Conference Board, no. 710, 1977).

[6] John Palutsek, *Business and Society 1976–2000: An AMA Survey* (New York: American Management Association, 1976).

[7] "Future Alternatives Report" (Chicago: United States League of Saving Association, 1974).

[8] "Reconnaissance II," *Institutions/Volume Feeding,* 40th anniversary issue, November 3, 1977, p. 8.

[9] "Reconnaissance II," p. 45.

called futures research. No matter what the name is, the purpose is to identify the interrelated issues that may occur in the future and may have an impact upon the organization.[10] The first step in identifying the correct issues is to have an accurate perception of the future. This can only be obtained by increasing the portion of reality that is perceived and decreasing the portion that is unknown or perceived incorrectly. Therefore we can say that the primary focus of futures research should be on closing the gap between reality and that which is perceived.

WHAT IS THE FUTURE?

We all have some notion of what the future is. We remember from our school days when the teacher explained the future tense to us by saying that it is something which is not happening now and did not happen yesterday, but which will happen some time from now. We were also taught the future is always the "unknown" in the equation, making the entire situation under consideration very uncertain. Depending on cultural upbringing, most of us were taught to think of the future either as something which will happen no matter what we as individuals or groups of common mortals may do or something that we can make happen. The truth, of course, is that the future is neither of these two extremes. The ancient Greeks were advised by Epicurus to

> remember that the future is neither ours nor wholly not ours, so that we may neither count on it as sure to come, nor abandon hope of it as certain not to be.

For the manager, considerably more important than this philosophical conviction is the pragmatic consideration of the time frame of the future. This is what is known in technical nomenclature as "the planning horizon." Most schools of business and public administration teach that there are three futures: short range (not more than 1 year), medium range (not more than 5 years), and long range (from 5 to 10 years). The manager then is advised to plan only for the short and medium ranges because dealing with any planning horizon over 5 years is extremely difficult, if not impossible, while any planning for more than 10 years is exceedingly hazardous to one's life.

We have found the following classification of possible future time horizons very useful. The future can be divided into five basic periods, suggests computer scientist and futurist Earl C. Joseph.[11] He notes that both the immediate future and the very distant futures are largely uncontrollable from today; the portion of the future over which a person today has the most control is the period 5 to 20 years from now.

The future starts now, this moment, and extends forever. The future as viewed from today, any today, is made up of a multiplicity of alternative futures

---

[10] For some examples of some successful environmental scanning systems or issues management systems, see chapter 15 of this book.

[11] Earl C. Joseph, "What Is Future Time?" *The Futurist,* August 1976, p. 178.

TABLE 12–1 The Five Futures a la E. C. Joseph

Futures	Range	Prime Characteristics
1. Now — the immediate future	Now to the next year	Frozen/dictated by the past. Uncontrollable. Unlockable except by a major event — usually a catastrophe. Present decisions or actions have little or no effect over this time frame. Minor choices available (food, clothing, vacations, etc.).
2. Near term–short term futures	Next 1–5 years	Past-programmed future. Crisis-programmed change; incremental reactive change possible. Evolutionary advances can be implemented during this time frame. Partially controllable from today. Decisions made today can cause major shifts in this time frame; major efforts are required to bring about revolutionary change. Policy choices available; new programs, systems, institutions, and leaders can cause impact in this time frame.
3. Middle-range futures	5–20 years from now	Choice over futures is available among alternative opportunities the future offers, if awareness of alternatives exists in the present. Almost completely controllable and determinable today. Revolutionary change implementable in this time frame from directed evolutionary (small) changes initiated today. Almost anything imaginable may be brought about in this time frame. The future available in this time frame is inventable and malleable today. Today's decisions can solidify this future time frame.
4. Long-range futures	20–50 years from now	Opportunities and/or crises triggerable/seedable today. Many alternative opportunities can be made visible. Largely uncontrollable (from today). Open futures.
5. Far futures	50 years from now and beyond	Largely invisible (to today). Uncontrollable future (from today). Utopian and dystopian speculation possible.

toward which we can move with or without our control. These alternative futures can be considered as planning horizons available to us for the purpose of expanding our control over the futures we bring to fruition. The future can be broken down into five basic periods (as an extension of today), as Table 12–1 shows.

Whatever time frame one chooses to employ, one should not forget the following "principles," which might prove very useful in thinking about the future. First, one should remember that much of the organization's future (and one's own place within it) will depend upon the future of society. So the future of a given organization is "outer" determined more than "inner" designed. Second, there is the "pipeline" principle. According to this principle, certain things once set in motion will continue to produce an effect even if the forces behind that motion have somehow ceased to operate, just as some water flows out of the shower head *after* it is turned off. Finally, there is the principle of accelerated, or self-feeding, change. That is, the more things change, the more they will change. Most of your "vice presidents in charge of resistance to change" will, of course, instruct the change agent to think about the old adage "The more things change the more they remain the same." This is one of those principles which may be true of natural systems but is definitely untrue of social systems.

METHODS USED IN FUTURES RESEARCH

In futures research there exists no generally acceptable study method or approach. The nature and complexity of organizations and their environments are so different that a generally accepted method is impossible. However, all methods of futures research do have a fundamental premise on which the research is based. That premise is that changes in organizations and their environment *occur very gradually.* It is only because of limited perception that these changes seem to occur suddenly.

Table 12–2 summarizes some of the most common methods of futures research. Not all the methods are equally accepted or equally successful.

As disciplines of human inquiry mature, they develop sophisticated methods for contemplating and explaining the world around us with ever-greater accuracy and believability. Of the multitude of methods employed by futurists, the following seem to be the most commonly used and most useful to futures managers.[12]

---

[12]The literature on the types or kinds of methods for studying the future, or future forecasting, is rather confusing. Both the nomenclature and the use of each technique or a combination of several techniques have been reported to be in a rather unclear state. For example, extrapolation is by some used synonomously with trend analysis, whereas for others the two terms are distinctly different. For an adequate explanation of six future research or studying methods, see Stuart Sandow, "The Pedagogy of Planning: Defining Sufficient Futures," *Futures,* December 1971, pp. 328–34. For an empirical investigation of the frequency of use of forecasting techniques with an emphasis on social forecasting, see Kenneth Newgren, "Social Forecasting: An Overview of Current Business Practices," in Archie B. Carroll, ed., *Managing Corporate Social Responsibility* (Boston: Little, Brown, 1977). The description of method presented here draws heavily from H. R. Smith et al., *Management: Making Organizations Perform* (New York: Macmillan, 1980) Ch. 17.

TABLE 12–2   Method of Futures Research

Methods	Description
1. Trend extrapolation	Assumes that what has been happening in the past will continue to happen and that the direction of change and the rate of change can be extrapolated into the future.
2. Analogical forecasting	Determines the relationships of the system under study by drawing inferences from a different system which is similar.
3. Genius forecasting	Results when a well-informed and bright individual examines the present and the past and makes predictions about the future.
4. Delphi technique	In this improved version of genius forecasting, experts forming a panel are asked individually their opinions. The results are tabulated and presented to the panel. The purpose of this additional information is to possibly alter their opinions.
5. Cross-impact matrix method	In this approach, the probabilities of an item in a forecasted set can be adjusted in view of judgments related to potential interactions of the forecasted items.
6. Simulation	This method defines the initial stages of a process of interaction and then has individuals play out roles according to certain rules, interacting with the play of others.
7. Scenario analysis	A sequence of possible events is selected which determines the inputs that are allowed to interact in order to assess the total consequences if such events do take place.

1. Extrapolation.
2. Scenario writing.
3. Delphi method.
4. Cross-impact analysis.
5. Cross-trend analysis.
6. Multivariable utility measurement.

Extrapolation

Extrapolation (or trend analysis) is perhaps the oldest, simplest, and, under certain circumstances, the most useful technique or method for thinking about and forecasting the future. In its simplest form, according to the *Random House Dictionary, extrapolation* means "inferring the unknown from something which is known." In other words, one considers past experience to be a good guide for future expectations. Obviously, extrapolation requires that both the past and present have been perceived, studied, and understood accurately and that the movement from the present to the future point of interest will follow a certain path in accordance with a well-defined, well-documented, and tested "law."

Extrapolation is very often used to project or estimate the population of a given species, such as humans; or the quantity of total goods and services that will be produced by a given society by the year 2000; or, for that matter, the amount of money accruing to an initial investment. All of these examples refer to a well-known phenomenon of growth of a given quantity, be it people, money, resources, or fish. Let us assume that a person has the possibility of putting $10 a year under a mattress for future use. This person wants to know how much money there will be by the end of the 10th year. Obviously, the sum will amount to $100. Now, let us assume that the same person takes the $100 and, instead of placing the money under the mattress, deposits it in a bank as an initial one-time investment getting 7 percent annual interest. Now the question is, how much money will there be by the end of the 10th year? Figure 12–1 depicts both forms of growth.

Curve A in Figure 12–1 shows the case in which a miser puts $10 under the mattress every year. At the end of the 10th year the miser has accumulated $100, at the end of the 20th year, $200, and so on, until the end of the 50th year when the wealth of the miser is $500. A look at Curve B of Figure 12–1 reveals that if the person invests an original sum of $100 at 7 percent interest and lets the interest accumulate, the principal investment will have doubled after 10 years and will continue to double every 10 years; at the end of the 50th year, the original $100 will have grown to $1,600. The kind of growth exhibited by B is called *exponential growth,* and it is defined as that kind of growth in which a given quantity increases by a constant percentage of the whole in a constant time period. In other words, each increment is kept and allowed to increase virtually forever at the same percentage as the initial quantity.

The mathematics of exponential growth is simple and is a convenient shorthand for a rather complicated process that can be found in many aspects of everyday life. Let us take a well-known problem that is at the forefront of societal concerns. That is the subject of population growth. For some people, population is growing too fast. For others, it is not growing fast enough. In any case, the manager who is interested in estimating the size of the population at some point in the future could do so by using the following formula and acquiring the appropriate data.

$$N_t = N_o e^{rt}$$

in which

N_o = Population now
N_t = Population in the future time (t)
e = Constant = 2.718
r = Rate of growth
t = Time

By putting the appropriate figures in this formula and then showing the results graphically, one will produce the curve exhibited in Figure 12–2.

FIGURE 12–1   Linear and Exponential Growth

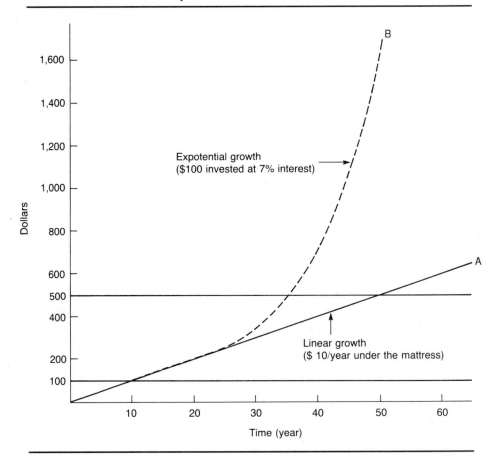

The easiest way to think about future quantities of certain things is to think in terms of the number of years required for a given quantity to double, that is, the doubling time. See Table 12–3 for some examples of doubling times for different rates of growth.[13]

To determine the doubling time, one must set

$$\frac{N_t}{N_o} = 2$$

$$\frac{N_t}{N_o} = e^{rt}$$

$$2 = e^{rt}$$

[13] See, for example, Paul R. Ehrlich and Ann H. Ehrlich, *Population, Resources, Environment: Issues in Human Ecology* (San Francisco: W. H. Freeman, 1976), p. 10.

FIGURE 12–2   World Population Estimates

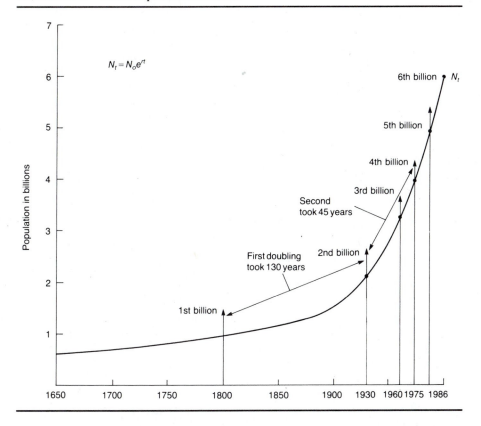

Taking the natural logarithm for both sides gives

$$ln\ 2 = rt$$

$$\frac{ln\ 2}{r} = t$$

$$t = \frac{.6931}{r}$$

or the doubling time

$$t = \frac{70}{r}$$

Let us call this the *Rule of 70:* to find the doubling time of an exponentially increasing or decreasing quantity, one divides 70 (or .6931) by the rate of growth.

A look into Figures 12–2 and 12–3 reveals that the doubling time for both population and Gross National Product (GNP) has been cut to one third of the 1900 rates. This is indeed one of the most insidious characteristics

TABLE 12–3   Doubling Time: The Rule of Seventy

Growth rate (r) (% per year)	Doubling time (t) year $t = \dfrac{70}{r}$
0.1	700
0.5	140
1.0	70
2.0	35
3.0	23
4.0	18
5.0	14
6.0	11.6
7.0	10
8.0	8.75
9.0	7.78
10.0	7

of exponential growth or decay. In later sections, we will return to some of the tremendous managerial implications of both exponential growth and decay processes.

Scenario Writing

Unlike extrapolation, which is almost as old as recorded civilization, scenario writing is the product of post–World War II futuristic thinking. Although no one is sure of the exact origins of the technique, Herman Kahn is perhaps the man most people credit with its discovery and popularization, during the 1950s while he was with RAND Corporation. The initial rationale for the development of scenario writing was the need to incorporate into the formal, rational, or analytic forecasting techniques a more intuitive set of assumptions and probable events to supplement and augment the validity of the analytic models. It must be clearly pointed out that scenario writing is a supplementary forecasting and futuristic thinking method, the introduction of which is justified on the grounds that the futures thinker desires to maximize the ability to grasp as much as of the *whole* as possible. Because analytic or explicit futuristic methods are constrained by the conditions of the formal symbolic system used (mathematics), scenarios that can be written in any natural language such as English, Japanese, or Greek provide for flexibility and for comprehensibility by the largest populace. Table 12–4 provides an arrangement of the most common types of scenarios.

Scenarios are hypothetical sequences of events constructed for the purpose of focusing attention on causal processes and decision points; or, alternatively,

FIGURE 12–3 GNP – Total Outputs of Goods and Services in the United States (1972 dollars)

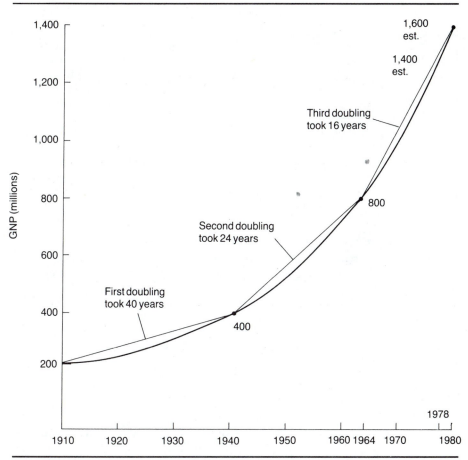

a scenario is a sequence of possible (but not necessarily highly probable) events and sociocultural choices. Robert Ayres explains two main advantages of scenario writing.[14]

1. Scenarios are an effective tool to counteract "carry-over" thinking, and to force the analyst (or policy maker) to look at cases other than the straight-forward "surprise-free projections."

2. Scenarios are an antidote for concentrating exclusively on the forest and ignoring the trees: analysts who limit themselves to abstract generalizations may easily

---

[14] Mihaljo Mesarovic and Edward Pestel, *Mankind at the Turning Point* (New York: Signet, 1974), p. 37.

TABLE 12–4   Four Types of Scenarios

	Type of Scenario	Aim(s) of Scenario	Premises of Scenario	Procedures Used
Exploratory scenarios	1. Tendential scenario	Seek to determine a possible future.	Suppose that "heavy trends" are permanent and predominant	Examine the continuation in the future of those trends and the mechanisms that explain them.
	2. Framework scenario	Try to delimit the space of range of possible futures	Suppose that "heavy trends" are permanent and predominant.	Make very varied (extreme) hypotheses concerning the evolution of these trends.
Anticipatory scenarios	3. Normative scenario	Seek to produce an image of a future that is possible and desirable.	Suppose that we can determine at the outset a range of possible objectives to achieve. Establish a procedure relating the future to the present.	Make a synthesis of these objectives to achieve and relate this image of the present.
	4. Contrasted scenario	Outline a 'desirable' future located at the frontier of possibilities.	Suppose that we can determine at the outset a range of possible objectives to achieve relating to the objective desired.	Make a synthesis of objectives to achieve and relate this image of the future to the present.

SOURCE: Peter Hall, *Europe 2000* (New York: Columbia University Press, 1977), p. 7.

overlook crucial details and dynamics (because no single set seems especially worthy of attention) even though looking at some random specific cases can be quite hopeful.[15]

---

[15] Quoted in Robert V. Ayres, *Technological Forecasting* (New York: McGraw-Hill, 1969), p. 143.

In order to illustrate how a scenario-writing technique works, let us take a look at a study commissioned by the Ford Foundation. The project, entitled *Exploring Energy Choices*,[16] was designed to study the future of energy availability for the United States. The research team constructed three plausible but very different energy futures for the period through the year 2000. The alternative futures, or scenarios, are based on different assumptions made about the energy growth patterns our society might adopt for the years ahead and the policies and consequences that each would entail. Of course, an infinite number of futures is possible, and it is unlikely that the real energy future of the United States will conform closely to any of the three scenarios. The scenarios are not predictions but tools for rigorous thinking.

The first scenario, which is called *historical growth,* assumes that the use of energy will continue to grow much as it has in the past. It assumes that the nation will not deliberately impose any policies that might affect our ingrained habits of energy use but will make a strong effort to develop supplies at a rapid pace to match rising demand.

The *technical fix* scenario shares with *historical growth* a similar level and mix of goods and services. But it reflects a determined, conscious national effort to reduce demand for energy through the application of energy-saving technologies. Research so far has revealed that the slower rate of energy growth in *technical fix* — about half as high as *historical growth's* — permits more flexibility of energy supply but still provides for an adequate quality of life. Only one of the major domestic sources of energy — Rocky Mountain coal or shale, or nuclear power, or oil and gas — would have to be pushed hard to meet the energy growth rates of this scenario.

Zero energy growth is different. It represents a real break with our accustomed ways of doing things. Yet, it does not represent austerity. "It would give everyone in the United States more energy benefits in the year 2000 than he enjoys today, even enough to allow the less privileged to catch up to the comforts of the American Way of Life. It does not preclude economic growth."[17] This obviously requires the use of less energy-intensive equipment and methods of production and operations. Zero energy growth would emphasize durability, not disposability of goods. It would substitute for the idea that "more is better" the ethic that "enough is best."

All three scenarios share certain characteristics. They all assume household comforts and conveniences greater than today's; no one must live, because of energy scarcity, in a lightless shack or a sweltering tenement. Every American would have a warm home in winter, air conditioning in hot climates, and a kitchen complete with appliances. People would still drive cars and have jobs, although

---

[16] *Exploring Energy Choices: A Preliminary Report of the Ford Foundation Energy Policy Project* (Washington, D.C.: Ford Foundation, 1974).

[17] *Exploring Energy Choices,* p. 40.

they might drive less or have different jobs, depending on the scenario the nation followed.[18]

Delphi Method

Another intuitive technique of thinking about and forecasting future events is the so-called Delphi method, named after the Greek oracle of Delphi, situated some 70 miles north of Athens. The Delphi method is usually referred to as a structured interaction among members of a group known as the "experts," who are either verbally, electronically (through a computer terminal), or by questionnaire interrogated with regard to their expectations for a series of hypothetical futures.[19] The ultimate outcome of the Delphi method is a consensus of expert opinion concerning the possibility and timing of certain technological events. For example, the following question may be asked of a panel of experts: In what year (if ever) do you expect nuclear power to supply more than 10 percent of the electricity needs of the United States?

Delphi was developed by the RAND Corporation during the early 1960s. In its most common form, Delphi consists of rounds of a series of questionnaires, distributed by the investigator to the experts. The first round is a series of questions, the responses to which are then statistically compiled. The results of this first round are then fed back to the original panelists, who are asked to submit revised answers together with reasons for agreeing or disagreeing with the initial consensus. In the third (and most often the last) round, the procedure is repeated, often with additional discussion and debate. The ultimate outcome desired is always an expert consensus, which is arrived at after considerable arguing, clarifying and, sometimes, compromising.

An example of a Delphi study will illustrate the general procedures employed.

Figure 12–4 shows a world population projection based on "naive" trend extrapolation and, beneath it, an envelope of projections obtained from a panel of experts interrogated separately by T. J. Gordon and Olaf Helmer.[20] The experts in this case clearly took account of the existing trend but almost unanimously agreed that the future rate of population growth would slow down for various reasons, notably the continuing invention and wider acceptance of birth-control measures. The panel was also cognizant of contrary trends, such as a continued decline in the death rate as a result of medical progress, and the possibility (but not probability) of greater improvement in the technology of food production and distribution. The result is a balanced forecast in which the best information available has been utilized in a way that no simple model or statistical extrapolation could hope to duplicate.[21]

---

[18] *Exploring Energy Choices,* p. 40.

[19] *Technological Forecasting,* p. 149.

[20] "Report on a Long-range Forecasting Study," RAND Corp., September 1964, p. 2982.

[21] Ayres, *Technological Forecasting,* p. 150.

FIGURE 12–4   Delphi Forecasts of Population (1960–2000)

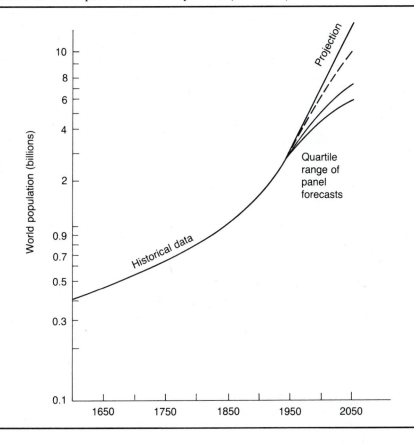

Cross-Impact Analysis (CIA)

Cross-impact analysis was developed by Theodore Gordon, The Futures Group, in the late 1960s. It was designed to assess the interaction between trends and future events. Until this time, forecasting techniques (for example Delphi) dealt mainly with single events. It became evident that any single event resulted from the interaction of numerous factors. For example, political, scientific, technological and economic factors influenced the first nuclear reactor. With CIA, the goal became systematic evaluation of these relationships providing new insights and accuracy in forecasting. The name comes from the *interrelation between events and developments,* i.e., the "cross-impact," and the actual structure is a matrix—thus the term cross-impact matrix method or cross-impact analysis.

In general, the most basic generic CIA model involves the following steps. First, predicted developments/events (i.e., those under consideration) are listed across the top and down the side of a matrix. For each block in the matrix a probability is estimated (indicating the strength of the interaction), as well as a coefficient (plus shows enhancement and minus inhibition). Initially the set of events and probabilistic data represent expert opinion. The time delays between events affecting one another is also considered.

Next, one development is randomly selected. Based on its assigned probability, a decision is made as to whether this could occur. As a result of this, the probabilities of the remaining events may be adjusted, still keeping the original proportion between interactions as a guideline. Now another event is selected and a decision made on whether it could occur. This is continued until a "decision" has been made on all events. In this way, the matrix is "played" over and over again so that the probabilities can be computed based on the percentage of times that a yes decision is made (i.e., the event would occur).[22] Usually one or more computer programs handle the processing.

Helmer[23] and DeGreene[24] describe a more complex version. In this case there are four types of impact: (1) event on event (E on E), (2) event on trend (E on T), (3) trend on event (T on E), and (4) trend on trend (T on T). In each case, different information is entered in the box. Events are defined as "singular occurrences," for example, a scientific discovery. In comparison, trends are gradual developments such as GNP or population growth.

The *event-on-event* (*E*-on-*E*) interaction looks at the way the occurrence of event 1 will affect the probability that event 2 will occur in the next cycle. This is simplified to two considerations: the time delay (i.e., how many cycles will occur before the effect is felt), and the number which represents the probability that event 2 will occur—the multiplying factor.

The *event-on-trend* (*E*-on-*T*) interaction considers the effect on a trend if an event occurs. This is expressed as the time delay, as well as the amount that will be added or subtracted to the trend values.

The *trend-on-event* (*T*-on-*E*) situation occurs when a trend does not follow the expected values, and results in the probability that certain events will change. As with event on event, it is expressed as lag time and multiplying factor.

Finally, the *trend-on-trend* (*T*-on-*T*) interaction occurs when a trend does not follow its expected values and causes other trends to shift. It is described in terms of the lag time and the amount to be added or subtracted to the value of trend 2.

An Example.   Helmer[25] provides the following example to demonstrate CIA. Note that it is an extremely simple one that does not claim to be realistic in terms of results. The subject is the world food situation; five 5 year scenes were selected

---

[22] Alvin Toffler, ed., *The Futurists* (New York: Random House, 1972), p. 182.

[23] Olaf Helmer, "Cross-Impact Gaming," *Futures,* June 1972, pp. 150–53.

[24] Kenyon B. DeGreene, *The Adaptive Organization—Anticipation and Management of Crisis* (New York: John Wiley & Sons, 1982), pp. 127–29.

[25] Olaf Helmer, "Problems in Futures Research," *Futures,* February 1977, pp. 21–26.

for the time period (1977–2002). There are two events (E₁E₂) and three trends (T₁T₂T₃), making a total of five developments:

E₁ is the introduction of a major new source of energy.

E₂ is the beginning of large-scale manufacture of artificial protein.

T₁ is the world population (in billions).

T₂ is the food supply per capita (1977 value = 100).

T₃ is birth-control acceptance (percentage of world population).

A scale from −3 to +3 was chosen for the cross-impact coefficients:

+ ½ = small

+ 1 = medium

+ 2 = large

+ 3 = very large

Table 12–5 shows the forecasts that an expert panel might give for the five developments:

TABLE 12–5 Expert Forecasts

Development \ Scene	1	2	3	4	5	6
E₁	0.10	0.30	0.50	0.60	0.70	
E₂	0.50	0.10	0.15	0.20	0.25	
T₁	4.0	4.5	5.0	5.4	5.7	6.0
T₂	100.0	95.0	90.0	95.0	100.0	110.0
T₃	25.0	30.0	35.0	40.0	45.0	50.0

The number given for each event (E₁ and E₂) represents the probability P_{ij} (i.e., the probability that the event i will occur during scence j if it has not already occurred), while the number given for each trend (T₁, T₂, and T₃) represents the value v_{ij}, which the trend i is expected to assume at the beginning of scene j. Since there are only five scenes, the end of scene five is equated with what would have been scene six. For each trend, 0 is the lower boundary. T₃'s natural upper boundary is 100. Bounds for T₁ and T₂ were set at 10 and 200 respectively. Central values must also be defined. Since medians are convenient, the central value for trend one (T₁) will be 5, T₂ will be 100 and T₃ 40. The natural lower and upper boundaries for the event probabilities will be 0 and 1 and the central value .5.

Certain parameters or problems must be dealt with before the analysis can begin. First, a scale transformation must be introduced to keep values within their boundaries. A cross impact causes the event probability or trend value to be raised or lowered. Given several additions, for example, the boundary could be exceeded.

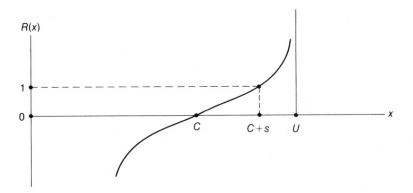

The scale transformation, $y = R(x)$ translates the lower bound 0 into $-$, the upper bound (U) into $+$, and the central value into 0. The function to do this is:

$$R(x) = \frac{(x - c)}{x(u - x)} \quad (0 < x < U)$$

The constant is included to deal with the exogenous effects. That is, the value of T_1 in scene three $(V_1,3)$ is simply an estimate. The "real" value of T_1 will be affected by cross impacts but could also be affected by external influences outside the realm of the model. To account for this, the value $(V_1,3)$ is replaced with a number drawn from the probability distribution around $(V_1,3)$ which is assumed to be normal and symmetric. The dispersion is fixed by deciding that quartiles be set at $R(v) + 1$.

To quote Helmer,

At the central value $v = C$, where $R(C) = 0$, let s be the quantity such that $R(C + s) = 1$. This value of s corresponds to the upper quartile of the exogenous-uncertainty distribution just described, and it can, for each trend, be intuitively estimated by answering the following question: If the estimated trend value is C, what is the value s such that the true value of the trend has a .25 probability of lying between C and $C + s$?

Thus, taking into account the near symmetry about C, s is chosen so that, for a trend value v near the central value, the interval from $v - s$ to $v + s$ is the approximate interquartile range (i.e., there is about the same chance for the true value to lie within or outside this interval).

The parameter s will be referred to as the "surprise threshold," meaning that, intuitively speaking we will or will not be surprised depending on whether the true value turns out to lie outside or inside the interquartile interval.

Since a cross-impact upon a trend, resulting either from the occurrence of an event or from the unexpected fluctuation of another trend, also represents some form of "surprise," it is not unnatural to use s (the surprise threshold associated with the trend) as the unit in which to measure the cross-impact effect on it. Thus, for example, a cross-impact of 2 represents an effect comparable to twice the median deviation caused by unexpected exogenous influences.

We note that, in the case of trends, because of the stipulation that $R(C + s) = 1$, we have

$$R(C + s) = k \; \frac{s}{(C + s) \; (U - C - s)} = 1, \text{ hence}$$

$$k = \frac{(C + s) \; (U - C - s)}{s}$$

In the case of events, the notion of a variable surprise threshold could also be introduced but would have less of an intuitively obvious interpretation. For simplicity, we shall assume instead that s_E will uniformly be equal to .25. Thus, if a probability of .5 is changed to a value between .25 and .75 we will not be "surprised"; otherwise we will. Since, for events, $C_E = .5$, $U = 1$, and $s = .25$, it follows that $k = .75$.[26]

Table 12–6 shows some estimates of carryovers *(d)* and surprise thresholds *(s)* with the corresponding R-transforms.

Finally the cross-impact matrix itself must be inputted into the model. This is shown in Table 12–7.

The model now has the needed inputs and parameters and is ready to run on the computer. The runs follow the steps outlined above. As Helmer[27] puts it:

In scene 1, "decide" which of the events occur (by a standard Monte Carlo drawing of random numbers); adjust the event probabilities and trend values for scene 2 according to the cross-impact matrix; then proceed to scene 2, having adjusted the trend values further by adding random deviates to them that were drawn from the appropriate exogenous-uncertainty distributions.

Again decide which further events are now occurring, and adjust event probabilities and trend values for scene 3 according to the prescribed event cross-impacts; observe the deviations of trend values in scene 2 from their predicted values, and adjust event probabilities and trend values for scene 3 further in accordance with the prescribed trend cross-impacts; apply carryovers where appropriate. Repeat the procedure for scenes 3, 4, and so on. The result will be a "scenario," that is, a sequence of event occurrences, by scenes, and of adjustments in trend values.

Table 12–8 shows event-occurrence frequencies and average trend values after a computer run of 100 replications. The completed model can be used for several purposes to be described below.

In general, CIA aids long-range planning. It does this by making probability and contingency forecasts given a multidisciplinary context. An appreciation of the variety of possible futures is gained as well as an understanding of how the occurrence of some events affect the probability of occurrence of others. More specifically, however, a CIA model can be used for sensitivity analysis, generating scenarios, improving forecasts, and providing comparative policy analysis.

---

[26] Helmer, "Problems in Futures Research," pp. 23, 24.
[27] Helmer, "Problems in Futures Research," pp. 25–26.

TABLE 12–6 Carryover, Thresholds and Transforms

	d	s	$R(x)$
E_1	0.8	0.25	$.75 \dfrac{x - .5}{x(1 - x)}$
E_2	0.7	0.25	$.75 \dfrac{x - .5}{x(1 - x)}$
T_1	0.95	0.25	$499.95 \dfrac{x - 5}{x(10 - x)}$
T_2	0.2	3	$\dfrac{9991}{3} \dfrac{x - 100}{x(200 - x)}$
T_3	0.6	2	$1{,}218 \dfrac{x - 40}{x(100 - x)}$

TABLE 12–7 Cross-Impact Matrix

	E_1	E_2	T_1	T_2	T_3
E_1		0.5		2	
E_2				4	
T_1	0.5	0.4	0.95	$-0.$	0.3
T_2		-0.1	2	0.2	-0.1
T_3			-1		0.6

Sensitivity analysis examines how sensitively the outcome depends upon the trend and event inputs. Each computer run generates a different scenario. By generating numerous scenarios, a planner can appreciate the potential contingencies. Improving forecasts is self-explanatory. Comparative policy analysis refers to utilizing CIA to test policies and implementation strategies.

Trend Impact Analysis

Trend impact analysis was also developed by Theodore Gordon and the Futures Group. Specifically, it is a form of events-on-trends and events-on-events CIA and involves three main steps. First, a computer is used to extrapolate the past history of a trend. Various methods used to obtain this baseline projection are "simple regression equations, computerized curve fitting, econometric modeling, simulation modeling, or scientific/engineering calculations."[28] Next, experts or users of the method decide potential events that would affect the extrapolation.

---

[28] Helmer, "Problems in Futures Research," pp. 27–29.

TABLE 12–8 Event-Occurrence Frequencies and Average Trend Values

		Scene					
		1	2	3	4	5	6
Frequencies of occurrence	E_1	8	28	32	16	4	
	E_2	4	3	15	22	8	
Average values	T_1	4.00	4.50	5.01	5.38	5.68	6.03
	T_2	100.0	94.8	89.6	95.0	101.1	110.1
	T_3	25.0	29.9	35.0	39.9	44.3	49.1

Probabilities are estimated, and CIA is used to determine the impact of events on one another. Finally, a computer applies the impact decision to modify the baseline extrapolation. Expected impact values are also calculated. Once the users have evaluated the adjusted extrapolation, the input may be changed if it is not realistic.

DeGreene lists the five parameters used to describe the impact:

a. The time in years from the occurrence of the given event until the trend first begins to respond.

b. The time in years from the occurrence of the event until the impact on the trend is largest.

c. The magnitude as a percentage of the baseline curve of the greatest impact.

d. The time in years from the occurrence of the event until its impact reaches a steady-state or final level.

e. The magnitude as a percentage of the baseline curve of the final impact.[29]

The advantage of TIA is that forecasters can consider effects of unusual events on future trends, thereby avoiding the assumption that past trends must continue. Rank-ordering of events is possible so that reordering can easily be accomplished. The median trend line together with upper and lower boundary lines are helpful for planners as well as managers.

Multiattribute Utility Measurement (MAUM)

Multiattribute utility measurement is a planning/evaluative mechanism used in situations involving unilateral (versus authoritarian) decision making. It is based on the fact that outcomes differ in their value dimensions and acknowledges the benefits derived from consensus. It focuses attention on problems, values, and goals. By explicitly stating the values of all participants, it helps to minimize differences. Systematic advocacy and negotiation reduce uncertainty. MAUM can be used to plan future projects, evaluate ongoing ones, or evaluate completed ones.

---

[29] DeGreene, *The Adaptive Organization,* p. 129.

Specifically, it is a rating scale that aggregates individual values. DeGreene outlines the following steps:

1. Identify the decision-making persons and organizations whose utilities are to be maximized. Such might be, say, a government agency. Utility is a function of the evaluator, what is being evaluated, and the purpose of the evaluation.

2. Identify the specific decisions relevant to these utilities.

3. Identify the object of evaluation, namely, the outcomes of possible decisions and actions. An object of evaluation could be a proposed research program, for example.

4. Identify the value dimensions, usually through face-to-face meeting. Often the result is a simple list or lists of goals important to the dissenting parties. Dimensions may be subjective, partly subjective, or objective. Generally 8 dimensions are sufficient and 15 are too many.

5. Rank the dimensions in order of importance to the dissenting parties. This also is often by group process.

6. Rate the dimensions in terms of ratios of importance. Start with the least most important dimension and work upwards. A given dimension might end up, say, four times as important as another.

7. Sum the weights that indicate importance, divide each weight by the sum, and multiply by 100. The result is the normalized importance weight of a given dimension of value.

8. Measure the location of each action outcome on each dimension. A scale of 0 to 100 is simplest to use. Zero is defined as the minimum plausible value and 100 as the maximum plausible value. The scale may later be expanded over the range 0–1,000. Now draw a straight line connecting maximum plausible with minimum plausible values. The result of this step is the rescaled position of a given outcome on a given dimension.

9. Calculate aggregate utilities for the outcomes of possible decisions and actions using a weighted average formula.

10. Decide. If a single outcome is to be chosen, simply maximize the aggregate utility for that outcome.[30]

Multiattribute utility measurement is not without problems. The assumption of value independence may not hold. It depends upon participant goodwill, maturity and openness. Negotiating can be tough, and "values" are not easily changed. Powerful individuals or groups or both may dominate the process.

SUMMARY

This chapter dealt with the subject of futurology as the study of possible future worlds and the societal adjustments to convert these possible worlds into preferred situations. A brief summary of the origins, developments, and extent of

---

[30] DeGreene, *The Adaptive Organization*, pp. 131–2.

future orientations by intellectuals, business organizations, and the government were presented.

The chapter explored briefly extrapolation, scenarios writing, the Delphi method, cross-impact analysis, trend impact analysis, and multivariate utility measurement.

KEY WORDS

Futurology

Futures

Futures Research

Extrapolation

Rule of 70

Scenario

Delphi Method

Cross-Impact Analysis

Trend-Impact Analysis

REVIEW QUESTIONS

1. Briefly describe the two stages of what has been called in this book the "sense of futurity."

2. List and briefly explain the five futures a la E. C. Joseph.

3. Show the difference between linear and exponential growth by using an example from the business world.

4. What is the "rule of 70"? Where does the 70 come from? In other words, what is the mathematical justification for using 70 and not 80 or 90?

5. What is the difference — if a difference does exist — between the scenario and the Delphi method?

6. What is the difference between exploratory and anticipatory scenarios?

7. How many types of impacts are there in a typical cross-impact matrix?

8. Is there any difference between cross-impact analysis and trend impact analysis?

9. Write a short essay on Valery's "The trouble with our times is that the Future is not what it used to be."

10. Convince your superior why the use of conventional economic or econometric and marketing forecasts are not adequate and why they must be either (1) replaced, or (2) supplemented with the type of futures research introduced in this book.

The Tomorrow Car: Everybody Needs It But Nobody Wants It[31]

"I don't know, Ms. Jackson," said Mr. Alton, the president of WMW, the World Motor Works. "You've got it all wrong. You're supposed to think about designing, producing, and marketing a car that everybody wants. What is all this talk about committing 100 million dollars for a car that everybody needs but nobody wants?" "Well, sir," said Ms. Jackson, "it is like this." And Jackson went ahead to explain the merits of the new car. Mr. Alton was listening to Jackson, but he wasn't really hearing anything. He kept thinking, "My God, what do they teach these kids in school nowadays?"

Jackson was about through when Mr. Alton noticed a sheet of paper pushed toward him. Among other things, the paper contained a list of forecasts, a sample of which looked like this:

| | Years | | |
Items	1980	1990	2000
. . . .	. . . .	. . . .	. . . .
Steel	$5/ton	$20/ton	$100/ton
Copper	$10/ton	$50/ton	$150/ton
Gasoline	70¢/gal	$1.50/gal	$5/gal
Lube oil	70¢/qt	$2.20/qt	$6/qt
. . . .	. . . .	. . . .	. . . .
. . . .	. . . .	. . . .	. . . .
. . . .	. . . .	. . . .	. . . .
Electricity	10¢/kwh	15¢/kwh	17¢/kwh
. . . .	. . . .	. . . .	. . . .

(Figures hypothetical.)

"Wait a minute, Jackson. Where did you get those figures?" "You see, sir," said Jackson, "these figures came from the latest report on 'Industrial Outlook: The Next Two Decades.'" Jackson pushed a heavy, thick book to Mr. Alton. Mr. Alton took the book in his hands, turned to Ms. Jackson, and said, "Could you run that by me once again? You want us to build a car that will be different. It will

[31] Reprinted from H. R. Smith et al., *Management: Making Organizations Perform* (New York: Macmillan, 1980), by permission of the publisher.

be small and light, rather ugly in appearance, and will get 60 miles to a gallon? Am I reading you right?" "Well, sir," said Ms. Jackson, "that's one option. We could think of a scenario in which a rather nonconventional car could be built." "Listen," said Mr. Alton. "Tomorrow, I'm taking up the subject with the executive committee. Could you give me a one-page description of at least two scenarios?"

1. Assuming Ms. Jackson's role, write these two scenarios.

Additional References

Ackoff, Russell. *Redesigning the Future: A Systems Approach to Social Problems.* New York: John Wiley & Sons, 1974.

Barron, Michael, and David Targett. *The Manager's Guide to Business Forecasting.* Oxford, England: Basil Blackwell, 1985.

Beer, Stafford. *Platform for Change.* New York: John Wiley & Sons, 1970.

Bell, Daniel. *The Coming of Post-Industrial Society: A Venture in Social Forecasting.* New York: Basic Books, 1973.

————. *Toward the Year 2000: Work in Progress.* Boston: Beacon Press, 1967.

Carson, Rachel. *Silent Spring.* Greenwich, Conn.: Fawcett Publications, 1962.

The Future Agenda: A Workbook for Participatory Democracy. Washington, D.C.: Congressional Clearinghouse for the Future and the Congressional Institute for the Future, 1982.

Gabor, Dennis. *Inventing the Future.* New York: Alfred A. Knopf, 1971.

————. *The Mature Society.* London: Secker and Warberg, 1973.

Heibroner, Robert L. *The Limits of American Capitalism.* New York: Harper & Row, Publishers, 1965.

Jarrett, Jeffrey. *Business Forecasting Methods.* Oxford, England: Basil Blackwell, 1987.

Kahn, Herman. *Thinking About the Unthinkable.* New York: Avon, 1962.

————. *The Next 200 Years.* New York: William Morrow & Co., 1976.

Kefalas A. G. "The Future of the Corporation." *Proceedings,* Society for General Systems Research, 1978.

Laszlo, Ervin, and others. *Goals for Mankind.* New York: E. P. Dutton, 1977.

Laszlo, Ervin. *A Strategy for the Future.* New York: George Braziller, 1974.

Lillian, Kay W., ed. *The Future Role of Business in Society.* New York: The Conference Board, 1977.

Linstone, Harold, and W. H. Clive Simmonds. *Future Research: New Directions.* Reading, Mass.: Addison-Wesley Publishing, 1977.

McHale, John. *The Future of the Future.* New York: Ballantine Books, 1969.

Makridakis, Spyros and Steven C. Wheelight. *Forecasting.* Amsterdam: North-Holland Publishing Company, 1979.

Makridakis, Spyros, and Steven C. Wheelight, eds. *The Handbook of Forecasting: A Manager's Guide.* New York: John Wiley & Sons, 1987.

Paluster, John. *Business and Society: 1976–2000: An AMA Survey.* New York: American Management Associations, 1976.

Snyder, Pearce David, and Gregg Edwards. *Future Forces: An Association·Executive's Guide to a Decade of Changes and Choice.* Washington, D.C.: Foundation of the American Society of Association Executives, 1984.

Toffler, Alvin. *Future Shock.* New York: Bantam Books, 1970.

Wheelright, S.C., and S. Makridakis. *Forecasting Methods for Management.* New York: John Wiley & Sons, 1985.

Willis, Raymond E. *A Guide to Forecasting for Planners and Managers.* Englewood Cliffs, New Jersey: Prentice-Hall, 1987.

Wright, George, and Peter Ayton, eds. *Judgemental Forecasting,* Chichester, England: John Wiley & Sons, 1987.

CHAPTER 13

Executive Support Systems

Most companies see enormous untapped value in the data they are accumulating on paper, but the access to the information is primitive and ineffective.

David Friend, chairman and founder of Boston-based Pilot Executive Software.

CHAPTER OUTLINE

INTRODUCTION

The last two decades of the 20th century will be characterized by the tremendous interest that scholars put on the concept of *information*. A quarter of a century ago when Claude E. Shannon and Warren Weaver published their landmark work, *The Mathematical Theory of Communication,* the concept of information, although central to the theory of communication, was exclusively associated with the freedom of choice when one selects a message. Since the primary concern of the communication theorists then was the maximization of both the accuracy of selecting a message and the number of selected messages that could be transmitted over a given medium, the concept of information remained hidden in the obscure vocabulary of messages, signals, channels, encoders, and decoders.

The practical significance of the concept of information did not become apparent until emphasis shifted from transmitting or "moving" information to "using" it. In other words, while the original intention of the theorists was to provide a vocabulary to talk about information and a measurement unit to handle it, later concerns began to focus on the consequences of having or not having it. Alternatively, theorists began to ask these questions. What difference does information really make? What is information good for? Why do people want information? Who benefits from the information? The sender or the receiver? In general, the question was asked, "What human need does information serve?"

After years of painstaking work scientists finally concluded that information is the raw material in the decision-making process. Taking few liberties, without distorting the original meaning of the concept, scientists popularized the concept to the point that engineers understood its usefulness. The rest is of course history. Virtually overnight—in a mere fraction of the time it took Newton to show the practical uses of gravity—electrical engineers and mathematicians designed machines whose exclusive purpose was to assist humans in their decision-making tasks by providing the information needed. Just like their forefathers, who harnessed energy to enhance muscle power, the creators of computers greatly augmented human capabilities of making decisions.

The remainder of this book zeroes in on the use of these cognition-enhancing machines by organizational decision makers. More specifically, the emphasis is placed on how executives use computer-assisted systems to acquire, process, and use information for strategic purposes. Consequently, this chapter focuses on systems that are designed to provide executives with the information they need to set and carry out the organization's strategic management process. Since strategic management is concerned with charting the long-range path of the organization, the importance of the external environment is paramount. For this reason the last chapter zeroes in on environmental scanning, the module of the strategic plan that assesses the external environment's threats and opportunities.

INFORMATION AND INFORMATION TECHNOLOGY: A RECONCEPTUALIZATION

While information scientists (i.e., an interdisciplinary group of electronic engineers, mathematicians, and computer scientists) focused on the quantitative aspects of information, cognitive scientists (i.e., an interdisciplinary group of psychologists, management scientists, and logicians) studied the qualitative properties of information as an uncertainty remover. Starting with the handling of data in the internal workings of the organization, concern for information is currently reaching a climax and information is becoming a strategic weapon.

Figure 13–1 provides a conceptual framework that relates the concept of information to the purpose for using it and to the related technologies thus far developed.[1]

As Figure 13–1 shows, the need for information is dictated by the inexorable law of *uncertainty* that no assertion is ever known with certainty. It is this human, almost complete lack of knowledge, especially about an outcome or result, and the commensurate inability to make a decision, that has made information the contemporary manager's most valuable commodity.

Just like its physical world sibling, energy, whose reason for use is to overcome the human inability to defy gravity, so is information the cure for the human inability to make definite statements about the future state of the system given the knowledge of its current state. In the physical world the amount of energy, which must be supplied to overcome gravity, is easily and accurately measured by calculus developed by classical Newtonian physicists. Whereas, in the world of cognition, the ability to measure the amount of information needed to eliminate the uncertainty is complicated by the difficulty with the concepts of uncertainty and information, as well as the relationship between the two.

The Concept of Uncertainty

The problem with the concept of uncertainty lies in its familiarity. We all know when we are uncertain, but we have no idea of the amount or degree of uncertainty. A typical dictionary definition of uncertainty will allude to the lack of knowledge about the outcome of our actions, the existence of doubt, dubiety, mistrust, skepticism, and suspicion by a given individual about a given situation. In the general management and business vocabulary, uncertainty assumes one of the above definitions. In systems nomenclature, uncertainty is associated with the systems designer's or observer's ability to predict the final state of the system, given the knowledge of its initial or present state. All of these definitions lack

---

[1] This portion draws heavily on the following papers: A. G. Kefalas, "Global Information Technology: A Strategic Weapon for MNCs," presented at the 1988 Annual Meeting of the Academy of International Business, San Diego, October 20–22, 1988; and A. G. Kefalas, "On the Design of a Global Information System (GLOBIS) for an MNC, " proceedings of the 1988 Annual Meeting of the European International Business Association, West Berlin, West Germany, December 10–14, 1988.

FIGURE 13-1 Information: Meaning, Use, and Technology

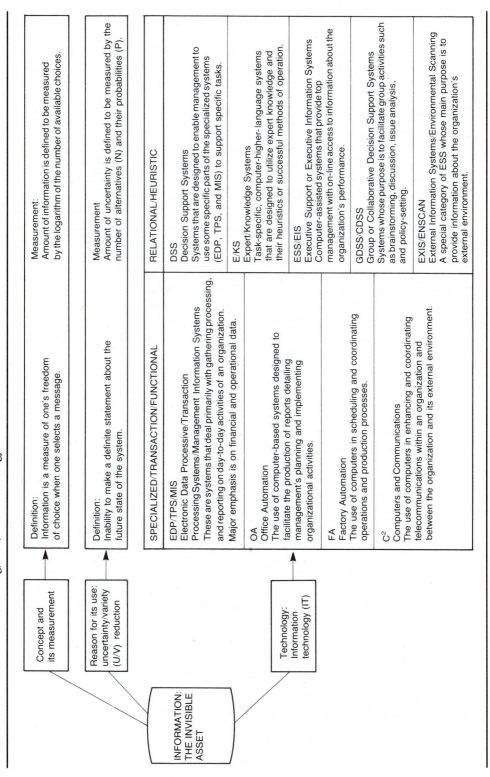

measurement of the concept of uncertainty. Consequently, the systems engineer, i.e., the person who designs systems to reduce and even eliminate uncertainty, is at a loss to find any clues about the "howmuchness" of it. As a result, since engineers do not know how much uncertainty is to be reduced or eliminated, they do not know how much of the medicine (i.e., information) is to be administered to the patient. Consequently, they stumble into the pitfall of "Let them have it all."

In our discussion of cybernetic systems we saw how some management systems scholars relate complexity to probabilistic systems as Beer does. C. West Churchman adds a slight twist of his own and relates complexity to uncertainty: "Another meaning of complexity recognizes that we live in a world of uncertainty." He then goes on to explain uncertainty by tracing its historical roots to Carneades, a Greek skeptic philosopher who lived around 213–129 BC. In keeping with his philosophical views on skepticism, Carneades denied the possibility of attaining any absolute (objective) certitude regarding knowledge but held that probable knowledge (subjective convictions arising from the senses) was a good enough norm for daily living. Humans, he was convinced, will go on to make assertions in a world of uncertainty. To the Greek skeptic Arcesilaus' criterion of reasonableness Carneades added another measure of confidence in the assertions people make based on his new criterion of persuasiveness (Greek pithanon — something calculated to persuade). Churchman, who may have interpreted this term as meaning *appropriate* or right for the purpose, then goes on to assert (mistakenly) that the word appropriate in English has the same etymological root as probability. "In fact," Churchman avers, "all probability theory is simply an extension of Carneades' idea that we need to be able somehow to measure the appropriateness of the statements we make.[2] Accordingly, our world becomes complex to the extent that it is uncertain for us. The obvious suggestion here is that we need to develop a calculus of uncertainty. We can measure unexpectedness, say, on a scale ranging from zero for the completely unexpected or the impossible, to one for the completely certain. Thus we have developed, in history, a theory of probability and more generally, of one type of uncertainty."[3]

Cyberneticians provided perhaps the purest measure of uncertainty by relating it to the concept of variety in a system. We have already seen that as the components of a system increase in number, their interrelationships typically increase, and the system is said to possess more variety than it did initially.[4] When one asks the question, "How can uncertainty be reduced?", the cybernetician

---

[2] Any student of the history of philosophy would find this unacceptable. Furthermore, to credit Carneades with being "the" or even "a" founder of probability theory would do a great disservice to Pierre-Simon Laplace and especially to Blaise Pascal, Pierre de Fermat, and Christiaan Huygens, whose written contributions gave the field of modern mathematical probability the standing it enjoys today.

[3] C. West Churchman, "A Philosophy for Complexity," in *Futures Research: New Directions,* eds. H. Al Linstone and W. H. Clive Simmonds (Reading, Mass.: Addison-Wesley Publishing, 1977), p. 83.

[4] See example in Stafford Beer, *Platform for Change,* (London: John Wiley & Sons, 1985), pp. 32–35.

responds with "Through information!" Information extinguishes variety and at the same time reduces uncertainty. It appears then that the concept of uncertainty is related to the concept of information, choice, and prediction. As F. C. Frick put it some 30 years ago:

> Our uncertainties are intimately tied up with probability estimates and if we are to fit our intuitive notions regarding choices available we must include the probabilities associated with each.[5]

Since all of these concepts are statistical in the sense that they involve probability considerations, it seems reasonable to suggest that information theory would be the most appropriate field for providing the vocabulary and the calculus for treating uncertainty. (See Chapter 6 on Information.)

The Concept of Information: The Invisible Asset

Most people seem to know that having information is good and having more information is even better. Most managers very early in their careers learn that information is the key to good decision making and that making good decisions is the key to success. In this way the causal link between success and information is definitively forged. What distinguishes data from information, how information is measured, and when the point of diminishing returns is reached for information's acquisition—these have already been extensively treated in the chapter on information.

In that chapter we saw that over two decades ago Russell Ackoff exploded some of the popular myths on information and information systems.[6] He showed that having more information is not necessarily better and that there is a point where the human mind simply goes blank when bombarded with huge amounts of information. While psychologists had previously introduced the concept of "information overload," and Shannon and Weaver[7] provided the calculus for mathematically measuring information and channel capacity, it remained for Alvin Toffler to popularize the concept of information overload and to make his readers cognizant of its consequences.[8]

Figure 13–2 provides a schematic of the information creation process. The process takes place in the mind of the person confronted with an uncertain situation. By bringing into one's mind the *problem* and a set of *data,* a selection

---

[5] F. C. Frick, "The Application of Information Theory in Behavioral Studies," in *Modern Systems Research for the Behavioral Scientist: A Sourcebook,* ed. Walter Buckley (Chicago: Aldine Publishing, 1968), p. 183.

[6] Russell Ackoff, "Management Misinformation System," *Management Science,* December 1967, pp. 147–56.

[7] Claude E. Shannon and Warren Weaver, *The Mathematical Theory of Communication* (Urbana, Ill.: The University of Illinois Press, 1964).

[8] Alvin Toffler, *Future Shock* (New York: Random House, 1970).

FIGURE 13–2   Information Formulation, Knowledge Aquisition and Wisdom

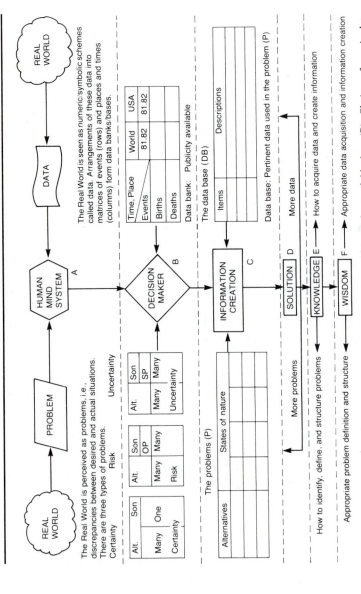

The Real World is perceived as problems, i.e., discrepancies between desired and actual situations. There are three types of problems.
Certainty   Risk   Uncertainty

Alt.	Son		Alt.	Son		Alt.	Son
	One			OP			SP
Many			Many	Many		Many	Many
Certainty			Risk			Uncertainty	

The Real World is seen as numeric/symbolic schemes called data. Arrangements of these data into matrices of events (rows) and places and times (columns) form data banks/bases.

Time, Place	World	USA
Events	81.82	81.82
Births		
Deaths		

Data bank: Publicly available

The data base (DB)

Items	Descriptions

Data base: Pertinent data used in the problem (P)

The problems (P)

Alternatives	States of nature

A — How to identify, define, and structure problems

B — Appropriate problem definition and structure

D — More problems

E — How to acquire data and create information

F — Appropriate data acquisition and information creation

The decision maker confronted with a problem must relate the problem to the available data (B). Choices of certain data (messages) from the database create information (C). Arrival at a solution (D) as a result of using this information creates knowledge (E). Appropriate right for the purpose use of this knowledge of the data leads to wisdom (F).

Steps A and F relate to *effectiveness* (doing the right *thing*). Steps B, C, D, and E relate to *efficiency* (doing the thing *right*).
Legend: Alt = Alternatives, *SON* = States of Nature, *OP* = Objective Probabilities, *SP* = Subjective Probabilities.

SOURCE: A. G. Kefalas, "Towards a Theory of Environmental Forces," Annual meeting of Society for General Systems Research, St. Louis, Mo., May 18–20, 1988.

process (A) begins, which eventually leads to choosing the appropriate data out of a given set of available data (B).

A problem, defined as a discrepancy between desired and actual situations, can be classified as:

1. Problem under certainty, — *many* alternatives but *one* state of nature (e.g., having an indoor or outdoor party on a sunny day).

2. Problem under risk — *many* alternatives and also *many* states of nature which have objective probabilities (e.g., having an indoor or outdoor party under weather forecasted as having a 50 percent probability of precipitation).

3. Problem under uncertainty — *many* alternatives and *many* states of nature for which there are no *objective* probabilities and for which the decision makers must assign *subjective* probabilities (e.g., having an indoor or outdoor party with no official forecast).

The other set of inputs to the decision maker's mind is the publicly available data supplied for example by the Census Bureau. At stage C the decision maker relates the correct choice of the problem to the appropriate data bank and creates a problem matrix on the left and a database on the right. At stage D the decision maker uses this information to find a solution. The decision maker's usage of the relevant data to create the information needed to solve the problem provides the knowledge that he or she will need to better define problems and choose data. Finally, appropriate use of this knowledge constitutes what some authors call "wisdom."

Information Technology: A (R)Evolution

The use of information technology — mechanical devices (hardware) with the appropriate instructions (software) — underwent three main phases. As can be seen from Figure 13–3, most applications of the technology during Epoch A, the data processing era, dealt with the acquisition and processing of data. The entire process was controlled by the EDP professional with very little freedom on the part of the user (manager).

The second phase, Epoch B, began with the increase in the manager's ability to use the technology directly with minimum interaction with the EDP professional. This is the era of management information systems (MIS). The ability of the manager who is confronted with the problem to relate it to a set of data and create information without having to go through the EDP professional greatly enhanced both the use of the technology and its effectiveness.

The future belongs to the era of information and knowledge. Here (Epoch C) the user has almost absolute control over the use of the technology. Direct linkages to large databases thousands of miles away give the user tremendous

FIGURE 13–3   The Information (R)Evolution

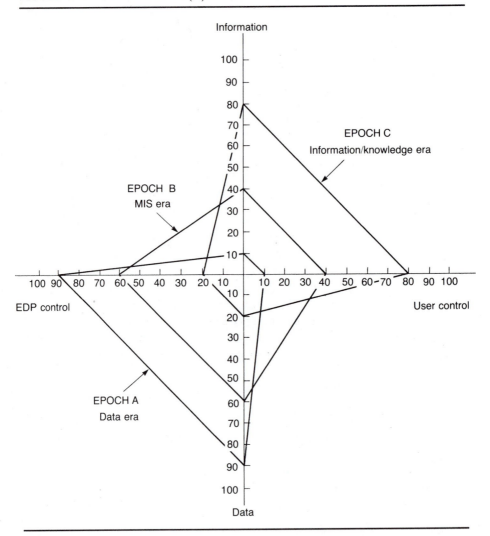

opportunity to create information in a fraction of the time required before. This is the age of the C^2: computers and communications.

In Figure 13–1, in the lower portion under Information Technology (IT) we summarize the tremendous post–World War II developments aimed at commercializing the scientific achievements in information theory. In the literature there are as many classification schemes of these developments as there are textbook writers. The classification scheme used here is designed to facilitate understanding of these developments, rather than to resolve the numerous debates and controversies currently carried out by specialists in the field. The interested reader should consult the available literature in the management information systems area.[9]

Figure 13–4 relates the people of an organization to a computerized reporting system. People are the input providers and the output users of the system. The organization is, of course, a hierarchical structure. The left side of the diagram describes the tasks performed by the three layers of management. These tasks are the uncertainty generators. The right side of the diagram depicts the database (reports) from which data can be drawn into the information system. It should be noticed that as a general rule the people at the top of the pyramid are primarily *synthesizing large information bases* while those at the bottom of the hierarchy *analyze compact information bases.*

The shaded area of the diagram depicts the managerial involvement in the design, use, and management of information systems. There are essentially two main components of the information system: (a) the model and (b) the database. The database contains both (1) historical and (2) judgmental data. The role played by management and staff in the information system is indicated by the amount of the shaded area. For example, top management is mostly involved in the use of judgmental data. Its involvement in historical data and model building is minimal. The lower-level management's role is almost the opposite.

EXECUTIVE SUPPORT SYSTEMS (ESS)

A great deal of debate in the literature is centered on what exactly is and is not an executive information system (EIS). While some writers duel on the distinction between MIS, DSS, and EIS, others are concerned with differences between EIS and an Executive Support System (ESS).

Rockart and Treacy popularized the term *executive information support systems* (EIS) in 1981 in several working papers and a *Harvard Business Review* article.[10] In their first paper, they analyzed the activities of 20 executives and made distinctions between DSS and EIS based on the different management tasks supported by each type of system. Executive activities tended to be less structured, more ad hoc, and wider-ranging than those of middle management DSS users.

---

[9] See for example, D. W. Kroeber and H. J. Watson, *Computer-Based Information Systems* (New York: MacMillan, 1987); D. M. Kroenke and K. A. Dolan, *Business Computer Systems,* (Santa Cruz, Calif: Mitchell Publishing, 1987); E. M. Award, *Management Information Systems,* (Menlo Park, Calif.: Benjamin Cummings Publishing, 1988).

[10] John F. Rockart and Michael E. Treacy, "Executive Information Support Systems," *Working Paper No. 65,* Cambridge, Mass.: Center for Information Systems Research, Sloan School of Management, MIT, November 1989; John F. Rockart and Michael E. Treacy, "The CEO Goes On-Line," *Working Paper No. 67,* Cambridge, Mass.: Center for Information Systems Research, Sloan School of Management, MIT, April 1981; John F. Rockart and Michael E. Treacy, "The CEO Goes On-Line," *Harvard Business Review,* January-February 1982.

FIGURE 13–4   Managers and Information Systems

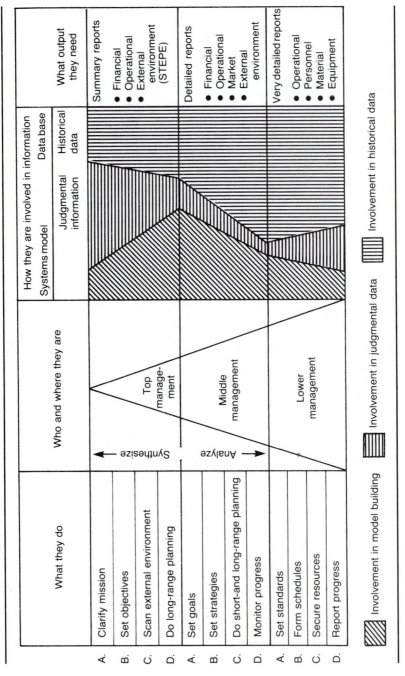

By 1983, the original concept of EIS had been renamed ESS, and Scott Morton[11] made further distinctions between executive and decision support systems. He argued that ESS provide capabilities to meet the "various and variable" information needs of executives, while DSS are generally focused on a single recurring and somewhat structured decision area. In addition, Scott Morton contended that the models so typical of DSS cannot provide the flexibility needed by executive decision makers. ESS, therefore, are data retrieval systems rather than model oriented systems.

Keen continued this focus on hard, numerical data. He saw ESS as a management concept with two basic capabilities – data retrieval and analysis. Contending that these systems reflect the executive's management style, as well as his or her concept of the business, Keen noted that there is no typical ESS. In the end, however, he blurred the distinction between DSS and ESS by recombining them: "One useful way of viewing an ESS, or any similar decision support system, is as a computerized staff assistant.[12]

Later in 1984, the DeLong and Rockart study of 45 randomly selected Fortune 500 companies also broadly defined ESS as:

> the routine use of a computer terminal for *any* business function. The users are either the CEO or a member of the senior management team reporting directly to him or her. Executive support systems can be implemented at the corporate and/or divisional level. (emphasis added)[13]

Today Rockart and DeLong define ESS as:

> the routine use of a computer-based system, most often through direct access to a terminal or personal computer, for any business function. The users are either the CEO or a member of the senior management team reporting directly to him or her. Executive support systems can be implemented at the corporate or divisional level.

Rockart and DeLong believe the three managerial purposes underlying the use of ESS are:

1. The support of particular office functions in an attempt to improve the executive's *effectiveness and efficiency.*

2. Improved support of the organization's *planning and control processes.* The objectives of this type of ESS can range from merely enhancing an existing control

---

[11] See Peter G. W. Keen and Michael S. Scott Morton, *Decision Support Systems: An Organizational Perspective* (Reading, Mass.: Addison-Wesley, 1978); Ralph H. Sprague, Jr., and Eric D. Carlson, *Building Effective Decision Support Systems* (Englewood Cliffs, N.J.: Prentice-Hall, 1982); Michael S. Scott Morton, "The State of the Art of Research in Management Support Systems," in *The Rise of Managerial Computing,* ed. John F. Rockart and Christine V. Bullen (Homewood, Ill.: Dow Jones-Irwin, 1986).

[12] Peter G. W. Keen, "The On-Line CEO: How One Executive Uses MIS," Working paper, Micro Mainframe, 1983.

[13] David W. DeLong and John F. Rockart, "A Survey of Current Trends in the Use of Executive Support Systems," *Working Paper No. 121,* Cambridge, Mass.: Center for Information Systems Research, Sloan School of Management, MIT, November 1984, p. 3.

FIGURE 13–5   Drilling-Down

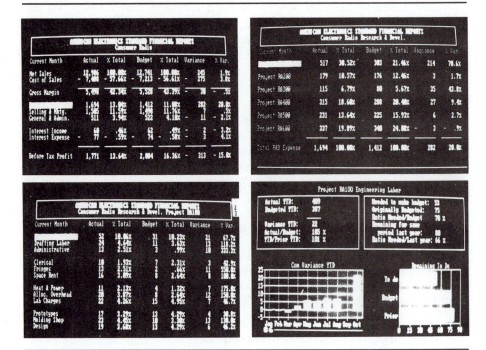

SOURCE: Friend, "The Three Pillars," p. 36.

system, to changing fundamental aspects of the way the organization is managed by redesigning the entire planning and control process.

3. The clarification or *enhancement of the individual manager's mental model* of the firm's business environment.[14]

The Anatomy of an EIS/ESS

No matter what the specific label affixed to systems used by higher management all refer to executive on-line data retrieval and use of existing databases for quick and accurate update of an organization's activities. According to David Friend, chairman and founder of the Boston-based Pilot Executive Software, an EIS's first task in most companies is to leverage the investment that has already been made in collecting and storing a wide range of operating data.[15]

While the specific characteristics of each commercially available EIS/ESS varies all of them share certain characteristics. Friend has identified

---

[14] J. F. Rockart and D. W. DeLong, *Executive Support Systems* (Homewood, Ill.; Dow-Jones, 1988), p. 29.

[15] David Friend, "The Three Pillars of EIS," *Information Center* 3 No. 8 (August 1987), p. 34.

three such characteristics which he calls "The Three Pillars of EIS." The three pillars are:

1. Drill-down
2. Trend analysis and
3. Exception reporting

The reader is already familiar with the conceptual foundations of all three characteristics. The similarity between the cones of resolution explained in Chapter 11 and drilling-down is strikingly clear. Friend explains these three characteristics as follows:

> Since no terms exist to describe these three essential functions of information review, I have created my own. They are drill-down, trend analysis, and exception reporting. How do each of these functions work and why do managers consider them so important? And why can't these functions be accomplished with conventional paper reporting systems?
>
> *Drill-down* (shown in the sidebar on Figure 13-5) is the essence of management reporting in that it provides a logical, high-speed review path. Most print reports are designed to allow for drill-down. Financial statements, for example, start with high-level consolidations. Each line item leads to a supporting schedule with details. The supporting schedule, in turn, leads to another level of detail. A manager can follow a problem from the top sheet down through successive levels of detail, but it is cumbersome to do so in paper format.
>
> *Trend analysis* refers to seeing in what direction the numbers are moving. When a manager finds a variance on a report, the first question is, "Is this a chronic problem or is this simply a one-month anomaly?" If the reporting system does not provide immediate access to historical trends, there is no quick way to answer this crucial management question.
>
> *Exception reporting* is the converse of drill-down. Rather than starting with aggregate data and digging for details, the system monitors the details against user-defined triggers and highlights only those items that are out of line. The computer can track the thousands of detail items a manager is responsible for and highlight only those items that are out of line.
>
> These three pillars of EIS give managers the ability to really *use* the corporation's existing data.[16]

The basic architecture of an EIS/ESS is simply an exploitation of the best features of the personal computer (PC) and the mainframe. In fact, the EIS/ESS mainframe–PC interface is a logical evolution of the conventional dumb terminal hookups to standalone PC-based systems that downloaded data from mainframes to a seamless PC–mainframe integration. Friend describes this architecture as follows:

> An EIS architecture should incorporate the best aspects of both the PC and the mainframe: the PC for its graphics and user friendliness (mouse support, touch

---

[16] David Friend, "The Three Pillars of EIS," p. 34.

TABLE 13–1   A Sample of Commercially Available EIS

Name of the Product	Vendor
CADET EIS	Southern Electric International, Atlanta, GA
Commander EIS	Comshare, Inc., Ann Arbor, MI
Commander Center	Pilot Executive Software, Inc., Boston, MA
CEO	Data General Corporation, Westborough, MA
METAPHOR	Metaphore Computers, Inc., Mountain View, CA
OPN	Lincoln National Information, Fort Wayne, IN
PC/Forum	Forum Systems, Santa Barbara, CA

SOURCE: Efrain Turban and Donna M. Schaeffer, "Executive Information Systems: Functions, Characteristics and a Comparison with DSS and Management Information Systems," University of Southern California, Los Angeles.

screens, and so on) and the mainframe for its database, computational capability, and system-wide integrity. The architecture that ties the PC and the mainframe together, called seamless PC–mainframe integration, offloads most of the processing to the PC but keeps all the source data on the mainframe. Data is served up to individual PCs as needed. The PC formats the tabular or graphics displays.[17]

A sample of commercially available EIS is provided in Table 13–1.

The remainder of the chapter will deal with (a) a brief description of Pilot's Commander Center and a Comshare's Commander EIS by means of examples of EIS, (b) an overview of Lockheed-Georgia's Management Information and Support System as an example of a firm-specific EIS, and finally (c) an exposure to PLEXSYS as an example of a GDSS/CDSS.

At the beginning of this chapter it was mentioned that one of the major problems confronting information systems designers is the difficulty of defining the degree of uncertainty in a given executive task. Consequently, since it is difficult to define the problem, it is equally difficult to determine the data used by executives to solve it. For this reason, information systems specialists shied away from even suggesting that MIS reports provided executive information.

Research in the last decade has demonstrated that even though executives find it difficult to describe their information needs in enough detail to enable a systems designer to restructure an MIS, this condition is "good enough" for the design of an EIS. The justification for this claim rests with EIS, in which the executive is actually the system designer. Just as the driver of a car is the designer of the journey, even though the driver had nothing to do with the design of either the car or the map, so is the executive the designer of a system that allows him or her to conceptualize, check, evaluate, experiment, reevaluate, and reconceptualize an organization's journey into the future. EIS enables the executive to "get

---

[17] David Friend, "The Three Pillars," p. 34.

everybody in on the act and still get some action." In other words, EIS allows the executive to "get it all together" without actually interfering or interrupting anybody's work.[18]

Executive information needs vary with the executive's personality, educational and experiential backgrounds. In addition, the tasks that the organization entrusts an executive with will also influence that executive's information needs. In general, however, an executive is the person whose main task is to assess the organization's position in its journey and the degree of accomplishment of its main purposes. Since executives must be ready to answer, without much deliberation, questions posed by the organization's stakeholders, the system must allow executives to instantly drill into the organization's databases and cut and paste together a realistic picture of the organization's performance that will satisfy the needs of the inquiry.

ESS/EIS allow executives to paint an adequate picture of the organization's performance by providing on-line, real-time retrieval of information in the following:

a. *Key Problem Narratives.* These reports highlight overall performance and key problems and their causes within an organization. Explanatory text often appears with tables, graphs, or tabular information.

b. *Highlight Charts.* These are summary displays that are designed from the user's perspective. These displays quickly highlight areas of concern, and visually signal the state of organizational performance against critical success factors (CSFs).

c. *Top-Level Financials.* These displays provide information on the overall financial health of the organization in the form of absolute numbers and comparative performance ratios.

d. *Key Factors.* These factors which are denoted as Key Performance Indicators (KPIs), provide specific measures of CSFs, at the corporate level. These are flagged as problems on the highlights chart, when they fail to meet some predefined standards, usually according to exception reporting.

e. *Detailed KPI Responsibility Reports.* These displays indicate the detailed performance of individuals or business units in areas critical to the corporation's success.[19]

The Pilot's Commander Center

David Friend, Pilot's founder and major designer of Commander Center, describes the major technical considerations of an EIS as follows:

---

[18] Harlan Cleveland, *The Knowledge Executive* (New York: E. P. Dutton, 1985), p. 12.

[19] E. Turban and D. M. Schaeffer, "Executive Information Systems," p. 7–8.

- To do drill-down, you need access to a lot of data. A typical Pilot EIS database takes up 50 to 100 megabytes of mainframe disk space.
- To use exception reporting, you have to be able to run comparisons of triggers against a lot of detailed items.
- To see trends, you need good color graphics and the ability to treat time series as the natural data form.
- You must ensure accuracy and integrity. The first time the data on the system is wrong or out of date, executives will stop using it.
- Keep it simple. No complex log on precedures, no PF keys, no obscure syntax — in fact no typing at all if possible.[20]

Figure 13–6 presents the main architecture of Commander Center. The system is indeed true to an EIS's main feature: "dig in, cut out, paste and paint."

Comshare's Commander EIS

Comshare defines an EIS as follows: An EIS is a hands-on tool that focuses, filters, and organizes executives' information so they can make more effective use of it. It preserves the benefits of an existing paper reporting system and improves upon it by using electronic information delivery that increases the data's timeliness and usability. An EIS enhances the ease and effectiveness with which an executive can perform information-intensive activities that are inherently part of a company's planning and control processes.[21]

Figure 13–7 presents a summary sketch of Comshare EIS's software architecture. Commander EIS provides electronically implemented applications that mimic the way in which an executive usually handles information. These applications include:

- Status reporting via an electronic Briefing Book.
- Ad-hoc investigation of corporate information models via Execu-View.
- Delivery of up-to-the-minute Dow Jones news and company stock reports via Newswire.

These three applications — Briefing Book, Execu-View, and Newswire — are implemented using a personal computer (PC)-based system that ties into mainframe systems. Briefing Book connects to Comshare's Workstation Manager software on the mainframe computer to obtain up-to-date information that is downloaded to the PC. For Execu-View, the PCs tie into Comshare's System W multidimensional mainframe database. Newswire connects to the Dow Jones News/Retrieval service for current news and stock information.

In addition to these applications, Commander EIS includes a programming language — called Builder's EASEL — designed to create graphic interfaces to

---

[20] Friend, "The Three Pillars," p. 37.
[21] Commander EIS (Ann Arbor, MI.: Comshare, 1988).

FIGURE 13–6   Pilot's Command Center

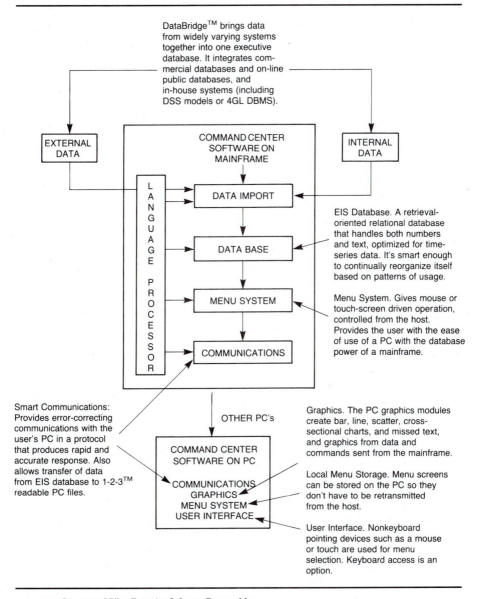

DataBridge™ brings data from widely varying systems together into one executive database. It integrates commercial databases and on-line public databases, and in-house systems (including DSS models or 4GL DBMS).

EXTERNAL DATA

COMMAND CENTER SOFTWARE ON MAINFRAME

INTERNAL DATA

LANGUAGE PROCESSOR

DATA IMPORT

DATA BASE

MENU SYSTEM

COMMUNICATIONS

EIS Database. A retrieval-oriented relational database that handles both numbers and text, optimized for time-series data. It's smart enough to continually reorganize itself based on patterns of usage.

Menu System. Gives mouse or touch-screen driven operation, controlled from the host. Provides the user with the ease of use of a PC with the database power of a mainframe.

Smart Communications: Provides error-correcting communications with the user's PC in a protocol that produces rapid and accurate response. Also allows transfer of data from EIS database to 1-2-3™ readable PC files.

OTHER PC's

COMMAND CENTER SOFTWARE ON PC

COMMUNICATIONS
GRAPHICS
MENU SYSTEM
USER INTERFACE

Graphics. The PC graphics modules create bar, line, scatter, cross-sectional charts, and missed text, and graphics from data and commands sent from the mainframe.

Local Menu Storage. Menu screens can be stored on the PC so they don't have to be retransmitted from the host.

User Interface. Nonkeyboard pointing devices such as a mouse or touch are used for menu selection. Keyboard access is an option.

SOURCE: Courtesy of Pilot *Executive Software.* Boston, Mass.

other software applications. With Builder's Easel, one can develop graphic interfaces to existing systems that contain information users need but that may be difficult to use or that don't present the information as effectively as desired. These systems are typically mainframe-based, but could also be PC-based.

FIGURE 13–7 Comshare EIS Software Architecture

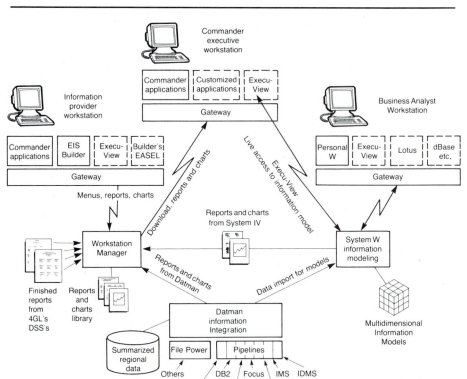

SOURCE: Courtesy Comshare, Inc.: Ann Arbor, Mich.

The MIDS System at Lockheed-Georgia[22]

Senior executives at Lockheed-Georgia are hands-on users of the Management Information and Decision Support (MIDS) system. It clearly illustrates that a carefully designed system can be an important source of information for top management. Consider a few examples of how the system is used.

- The president is concerned about employee morale which for him is a critical success factor. He calls up a display which shows employee contributions to company-sponsored programs such as blood drives, United Way, and savings plans. These are surrogate measures of morale, and because they have declined, he becomes more sensitive to a potential morale problem.

---

[22] This portion draws heavily on H. Watson's work with MIDS, which he so generously discussed with Dr. Kefalas. His contribution is hereby acknowledged.

- The vice president — manufacturing is interested in the production status of a C-5B aircraft being manufactured for the U.S. Air Force. He calls up a display which pictorially presents the location and assembly status of the plane and information about its progress relative to schedule. He concludes that the aircraft is on schedule for delivery.

- The vice president — finance wants to determine whether actual cash flow corresponds with the amount forecasted. He is initially concerned when a $10 million unfavorable variance is indicated, but an explanatory note indicates that the funds are enroute from Saudi Arabia. To verify the status of the payment, he calls the source of the information using the name and telephone number shown on the display and learns that the money should be in a Lockheed account by the end of the day.

- The vice president — human resources returns from an out-of-town trip and wants to review the major developments which took place while he was gone. While paging through the displays for the human resources area, he notices that labor grievances rose substantially. To learn more about the situation so that appropriate action can be taken, he calls the supervisor of the department where most of the grievances occurred.[23]

These are not isolated incidents; other important uses of MIDS occur many times a day. They demonstrate that computerized systems can have a significant impact on the day-to-day functioning of senior executives.

Lockheed-Georgia, a subsidiary of the Lockheed Corporation, is a major producer of cargo aircraft. Over 19,000 employees work at their Marietta, Georgia, plant. Their current major activities are production of the C-5B transport aircraft for the U.S. Air Force, Hercules aircraft for worldwide markets, and numerous modification and research programs.

MIDS is an example of an EIS. It is used directly by top Lockheed-Georgia managers to access on-line information about the current status of the firm. Great care, time, and effort go into providing information that meets the special needs of its users. The system is graphics-oriented and draws upon communications, data storage, and retrieval methods. The initial version of MIDS took six months to develop and allowed Lockheed's CEO to call up 31 displays. Over the past eight years, MIDS has evolved to where it now offers over 700 displays for 30 top executives and 40 operating managers including the president.

The main menu and keyword index are designed to help the executive find needed information quickly. Figure 13–8 shows the main menu. Each subject area listed in the main menu is further broken down in additional menus. Information is available in a variety of subject areas by functional area, organizational level, and project and is organized in a top down fashion. A summary graph is presented at the top of a screen or first in a series of displays, followed by supporting graphs,

---

[23] G. Houdeshel and H. Watson, "The Management Information and Decision Support (MIDS) System at Lockheed-Georgia," MIS Quarterly, March 1987, pp. 127–140.

FIGURE 13–8   The MIDS Main Menu

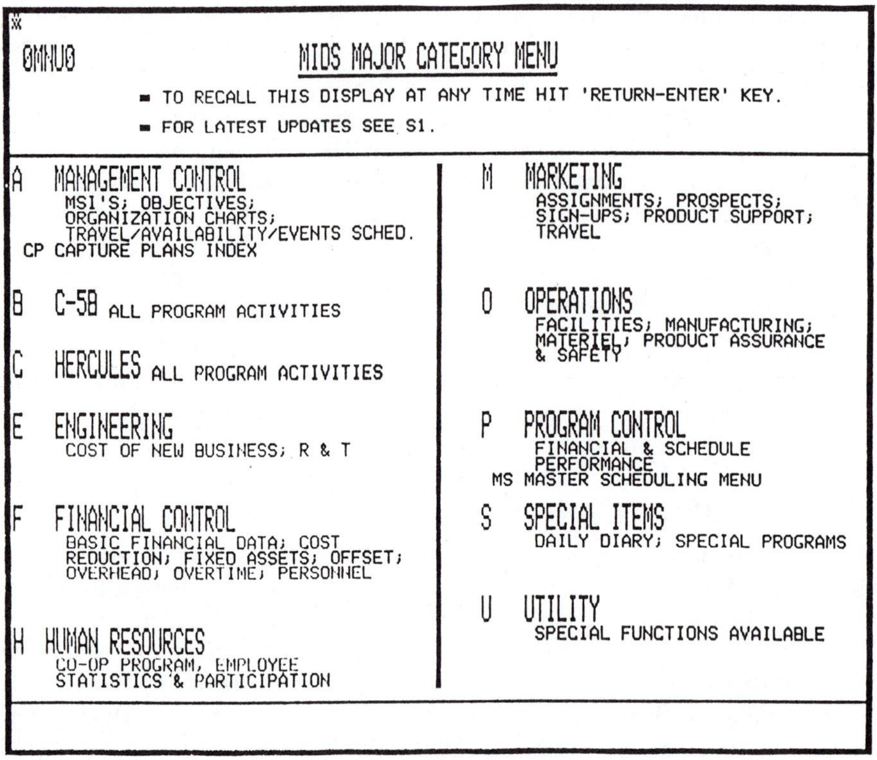

```
X
0MNU0              MIDS MAJOR CATEGORY MENU
         ■ TO RECALL THIS DISPLAY AT ANY TIME HIT 'RETURN-ENTER' KEY.
         ■ FOR LATEST UPDATES SEE S1.

A  MANAGEMENT CONTROL            M  MARKETING
     MSI'S; OBJECTIVES;              ASSIGNMENTS; PROSPECTS;
     ORGANIZATION CHARTS;            SIGN-UPS; PRODUCT SUPPORT;
     TRAVEL/AVAILABILITY/EVENTS SCHED. TRAVEL
CP CAPTURE PLANS INDEX

B  C-5B ALL PROGRAM ACTIVITIES   O  OPERATIONS
                                     FACILITIES; MANUFACTURING;
C  HERCULES ALL PROGRAM ACTIVITIES   MATERIEL; PRODUCT ASSURANCE
                                     & SAFETY

E  ENGINEERING                   P  PROGRAM CONTROL
     COST OF NEW BUSINESS; R & T      FINANCIAL & SCHEDULE
                                      PERFORMANCE
                                 MS MASTER SCHEDULING MENU

F  FINANCIAL CONTROL             S  SPECIAL ITEMS
     BASIC FINANCIAL DATA; COST        DAILY DIARY; SPECIAL PROGRAMS
     REDUCTION; FIXED ASSETS; OFFSET;
     OVERHEAD; OVERTIME; PERSONNEL

                                 U  UTILITY
                                      SPECIAL FUNCTIONS AVAILABLE
H  HUMAN RESOURCES
     CO-OP PROGRAM, EMPLOYEE
     STATISTICS & PARTICIPATION
```

SOURCE: G. Houdeshel and H. Watson, "The Management Information," p. 135.

and then by tables and text. This approach allows executives to quickly gain an overall perspective.

The displays have been created with the executives' critical success factors (CSF) in mind. Some of the CSF measures as profits and aircrafts sold are obvious. Other measures as employee participation in company-sponsored programs are less obvious and reflect the MIDS staff's efforts to fully understand and accommodate the executives' information needs.

To illustrate a typical MIDS display, Figure 13–9 shows Lockheed-Georgia's sales as of November 1986. It was accessed by entering F3. The sources of the information and their Lockheed-Georgia telephone numbers are in the upper right-hand corner. The top graphs provide past sales history, current budgeted and actual sales, and forecasted sales. The wider bars represent actual sales while budget sales are depicted by the narrower bars. Detailed, tabular information is provided under the graphs. An explanatory comment is given at the bottom of the display. The R < and F > in the bottom right-hand corner indicate that related displays can be found by paging in a reverse or forward direction.

FIGURE 13–9   A Sample of MIDS Output: Lockheed-Georgia Sales

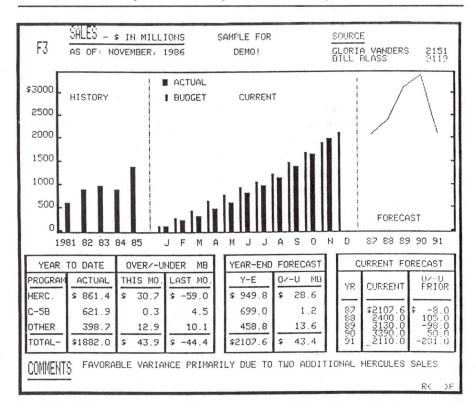

SOURCE: G. Houdeshel and H. Watson, "The Management Information," p. 137.

In order to provide the information needed, a variety of internal and external data sources had to be used. The internal sources include transaction processing systems, financial applications, and human sources. Some of the data can be transferred directly to MIDS from other computerized systems, while others must be rekeyed or entered for the first time. Access to computerized data is provided by in-house software and commercial software such as DATATRIEVE. External sources are very important and include data from external databases, customers, other Lockheed companies, and Lockheed's Washington, D.C. office.

MIDS relies on both "hard" and "soft" data. Hard data come from sources such as transaction processing systems and provide the facts. Soft data often come from human sources and result in information which could not be obtained in any other way; they provide meaning, context, and insight to hard data.

Executives are taught to use MIDS in a 15 minute tutorial. For several reasons, no written instructions for the use of the system have ever been prepared. An objective for MIDS has been to make the system easy enough to use so that written instructions are unnecessary. Features such as menus and the key-word index make this possible. Another reason is that senior executives are seldom willing to take the time to read instructions. And most importantly, if an executive has a problem in using the system, the MIDS staff prefers to learn about the problem and to handle it personally.

A variety of benefits are provided by MIDS: better information; improved communications; an evolving understanding of information requirements; a test bed for system evolution; and cost reductions.

The information provided by MIDS has characteristics which are important to management. It supports decision making by identifying areas which require attention, providing answers to questions, and giving knowledge about related areas. It provides relevant information. Problem areas are highlighted and pertinent comments are included. The information is timely because displays are updated as important events occur. It is accurate because the MIDS staff verifies all information before it is made available.

MIDS has also improved communications by sharing information with vendors, customers, legislators, and others. MIDS users are able to quickly view the same information, in the same format, with the most current update.

MIDS is responsible for cost savings in several areas. Many reports and graphs, which were formerly produced manually, are now printed from MIDS and distributed to non-MIDS users. Some requirements for special reports and presentation materials are obtained at lower cost by modifying standard MIDS displays. Reports that are produced by other systems are summarized in MIDS and are no longer printed and distributed to MIDS users.

Top management's enthusiasm for MIDS is echoed by the current president, Paul French, who recently said:

> I assumed the presidency of the Lockheed-Georgia Company in June 1984, and the MIDS system had been in operation for some time prior to that. The MIDS system enabled me to more quickly evaluate the current conditions of each of our operational areas and, although I had not been an advocate of executive computer systems, the ease and effectiveness of MIDS made it an essential part of my informational sources.

Because French and other senior executives have come to rely on MIDS, middle managers at Lockheed-Georgia and executives at other Lockheed companies want their own versions of MIDS. Within Lockheed-Georgia there is the feeling that "If the boss likes it, I need it." Currently, MIDS personnel are helping middle functional area managers develop subsystems of MIDS and are assisting other Lockheed companies with the development of similar systems.

GROUP OR COLLABORATIVE DECISION SUPPORT SYSTEMS (GDSS/CDSS)

It is well-known that management spends a considerable amount of time in meetings. Their purpose is to bring together a team of several people with different expertises who are from different functional and/or geographical areas.

The team reviews past performance of the firm, develops a plan for the future, or just briefs management on the last happenings in the organization and its environment. Mosvick and Nelson (1987) reviewed a number of studies on meeting productivity and found that 25 percent to 80 percent of a manager's time is spent in meetings. In addition, they estimated that, "Over 50 percent of the productivity of the billions of meeting hours is wasted . . . causing many companies to lose the equivalent of 30 man-days and 240 man-hours a year for every person who participates in business conferences." Their estimate for one Fortune 500 company was that ineffective meetings cost the business a total of 71 million dollars a year.[24]

Systems people play a significant role in these meetings by supplying various reports that are used to illustrate managerial accomplishments numerically and graphically. Over the years systems people have developed elaborate systems that allow meeting participants to not only bring to the meeting "hard copies" of their reports generated "back home," but also to hook up directly to their systems and do on-line interacting with their own databases and support systems.

The 80s ushered in the age of what Robert Johansen has called "groupware — the emerging technology that will expedite meetings, give teamwork a real boost, and cut distances down to size without time-consuming business trips — a generic term for specialized computer aids that are designed for the use of collaborative work groups."[25] These systems are designed to facilitate group activities such as brainstorming, proposal discussion, issue identification and analysis, and policy setting. As Fortune puts it "Like an electronic sinew that binds teams together, the new 'groupware' aims to place the computer squarely in the middle of communications among managers, technicians, and anyone else who interacts in groups, revolutionizing the way they work."[26]

Table 13–2 provides a list of 17 approaches to groupware arranged in order of increasing difficulty.

When a system designed to facilitate teamwork requires that participants *must* all interact simultaneously, it is called a synchronous meeting. An asynchronous meeting allows each participant to communicate according to his or her own schedule in a store-and-forward mode. Table 13–3 arranges the 17 approaches in accordance with the place and time of the meeting requirements.

In the last few years a whole industry has developed with the aim of supplying management with real systems to assist in team-building and teamworking. Table 13–4 provides a list of some of these most well-known systems.

---

[24] R. Mosvick and R. Nelson, *We've Got to Start Meeting Like This! A Guide to Successful Business Meeting Management* (Glenview, Ill: Scott, Foresman 1987).

[25] Robert Johansen, *Groupware: Computer Support for Business Teams* (New York: Free Press, 1988), p. 1.

[26] Louis S. Richman, "Software Catches the Team Spirit," *Fortune,* June 8, 1987, p. 65.

TABLE 13–2 17 User Approaches to Groupware (from least to most difficult)

1. Face-to-face meeting facilitation services . . . Chauffeur
2. Group decision support systems . . . GDSS
3. Computer-based extensions of telephony for use by work groups . . . Telephone Extension
4. Presentation support software . . . Presentation Prep
5. Project management software . . . Team Conscience
6. Calendar management for groups . . . Our Black Book
7. Group-authoring software . . . Group Writing
8. Computer-supported face-to-face meetings . . . Beyond the White Board
9. PC screen-sharing software . . . Screen Sharing
10. Computer-conferencing systems . . . Invisible College
11. Text-filtering software . . . Needle in a Haystack
12. Computer-supported audio or video teleconferences . . . Teleconference Assistant
13. Conversational structuring . . . Say What You Mean
14. Group memory management . . . Picking up the Fishnet
15. Computer-supported spontaneous interaction . . . Electronic Hallway
16. Comprehensive work team support . . . It's All Here
17. Nonhuman participants in team meetings . . . Nonhuman Participants

SOURCE: Johansen, *Groupware*, p. 41.

TABLE 13–3 Display of 17 Scenarios by Geographic and Time Dispersion of Participants

	Time	
	Synchronous	*Asynchronous*
Face-to-face meetings	1. Facilitation services 2. Decision support 8. Beyond white board 17. Nonhuman participants	4. Presentation software 5. Project management 14. Memory management 16. Comprehensive support
Electronic meetings	3. Telephone extension 9. Screen sharing 12. Teleconference aid 15. Spontaneous interaction	6. Calendaring 7. Group writing 10. Computer conferencing 11. Text filtering 13. Conversation structuring

SOURCE: Johansen, *Groupware*, p. 44.

In the remainder of the chapter a brief exposure to the University of Arizona's PLEXSYS is provided. Some examples of applications of PLEXSYS at the University of Georgia will be given in Chapter 15.[27]

---

[27] The discussion on PLEXSYS is based on the generous sharing of the work of Robert Bostrom, who directs the PLEXSYS effort at the University of Georgia. The authors acknowledge Bob's contribution to the exciting area of Collaborative Work Support Systems.

TABLE 13–4   Features of Major GDSS Providers

Provider	System	Functions
Applied Futures, Inc.	CONSENSOR (hardware/software)	Vote tabulation and display
Compshare	SYSTEM W (software) EXPRESS (software)	Data management, modeling statistical analysis graphics, report generation, PC communications
Decisions and Designs, Inc.	Decision conference (complete package)	Interactive decision analysis (six models), conference facility, decision analysis, consultation
EXECUCOM	IFPS (software)	Interactive financial modeling, data management, graphics
Institute for the Future	NA–research	Research/consulting on teleconferencing
MIT Laboratory for Computer Science	MPCAL, CDS, RTCAL, MBlink research only	Support of geographically separated local group work, including calendar management, real-time conferencing, and collaborative document editing
Perceptronics, Inc.	GROUP DECISION AID (hardware/software package)	Interactive decision tree; analysis, tutor, documentation
SRI International	QUICKTREE and APLTYER (software)	Interactive decision tree analysis for individuals
SUNY, Albany Decision Techtronics Group	NA–public service (complete package)	Interactive decision analysis (six models), data management, graphics, decision and process consulting
UCLA Cognitive Systems Laboratory	NA–research (math models/software)	Group decision theory and analysis
University of Arizona Planning Laboratory	PLEXSYS software–research, public service (complete package)	Electronic brainstorming, stakeholder identification and analysis, organization analysis, knowledge management, graphics, report generation, PC network
Xerox PARC	COLAB–research (complete package)	Computer support of face-to-face group work

SOURCE: Kenneth L. Kraemer and John Leslie King, "Computer-Based Systems for Cooperative Work and Group Decision Making," ACM Computing Surveys, 20, No. 2 (June 1988), p. 127.

The University of Arizona PLEXSYS

PLEXSYS was developed at the University of Arizona in 1987. The system includes a number of software tools designed to assist various group activities undertaken in meetings.

A typical PLEXSYS setting is shown in Figure 13–10. The group members interact with each other by typing their views into their microcomputers that are linked together via a local area network (LAN). A public display device is used to show group inputs. A facilitator or server is used to operate the software in accordance with the group's wishes.

PLEXSYS provides a set of tools which allows a group to deal with a task from its inception to the development of specific implementation proposals. Table 13–5 provides a list of the main tools which are included in the PLEXSYS package. In the last chapter we will describe a specific example of using PLEXSYS in conjunction with an environmental scanning class project.

TABLE 13–5   The University of Arizona PLEXSYS: Main Tools

Purpose	Individual Activity	Group Discussion (Facilitator Action)
	Electronic Brainstorming	
Generate initial ideas	Brainstorm ideas	Develop questions
	Issue Identification	
Organize ideas into general issues	Group ideas into issues	Judge issues
	Issue Consolidation	
Sort and consolidate issues	Look for duplications	Consolidate/reorder issues
	Issue Voting	
Evaluate and collectively evaluate issues	Rank issues by voting	
	Policy Formulation	
Generate and refine policy statements	Generate alternative	Develop policy statements Refine policy statements
	Voting	
Collectively make decisions	Vote on policy statements	Develop questions Select voting procedures Discuss aggregate votes
	Opinion Surveys	
Generate opinion	Generate opinion statements	Develop opinion questions Discuss opinion statements

SOURCE: Robert Bostrom, "User Documentation for University of Arizona Collaborative Support Software Tools," PLEXSYS Software, December 1987, pp. 1–2.

FIGURE 13-10   The Electronic Meeting Room Setting

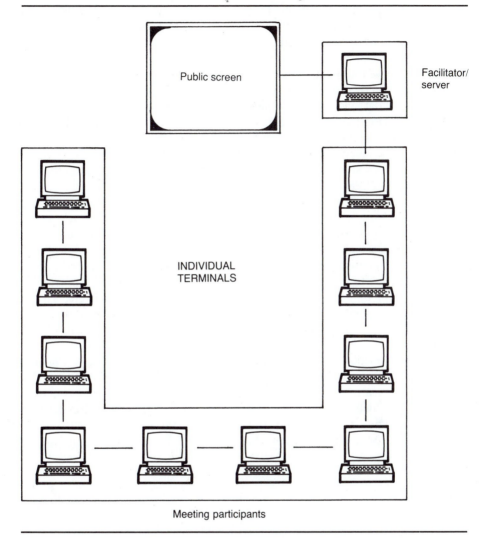

Meeting participants

SOURCE: Robert Bostrom and Robert Anson, "Overview of Collaborative Work Support Systems in an Electronic Meeting Environment," Center for End User Computing, The University of Georgia, May 1988.

SUMMARY

This chapter dealt with the subject of executive support systems. The concern for information has switched from acquiring, processing and disseminating information, to the reasons for and the consequences of its use. Information users are asking themselves What is this information good for? What difference does it make? What problem does it solve? What are the payoffs from its use? The chapter treated information as the cure for uncertainty, a constant concern of executives in organizations. Uncertainty is understood here as the executive's inability to give a precise and definite answer to a given question. Information enhances the executive's ability to provide an answer by (1) enabling the executive to explore a larger number of alternatives (N), and by (2) increasing the executive's ability to better estimate the probabilities (P) for each of these alternatives.

Over the years systems people have developed systems which allow the executive to directly and without the mediation of the EDP and MIS staff to "tap" organizational databases and manipulate data to get the information needed for a specific decision-making problem. These systems have on-line capabilities and involve seamless connections between the executive's PC and the organization's mainframe. The language used to carry through this communication between the executive's PC system and the organization's mainframe is "near natural language."

This chapter dealt mainly with the requirements for an effective EIS or ESS in terms of both the organizational structure needed to accommodate such systems, and their hardware and software architecture. It is becoming increasingly clear that the contribution of well-designed ESS can provide the executive with the ultimate competitive weapon—speed, i.e., a decrease in the time required to turn an idea into a product or service. Competition in time is the new concern in corporate headquarters. Thanks to information technology and ESS, organizations have managed to cut product development and processes implementation to their theoretically possible limits.[28]

The last portion of the chapter examined some commercially available systems currently in use. Comshare's Commander EIS and Lockheed-Georgia's MIDS Systems are the latest developments in executive support systems.[29] Unlike ESS, which are designed to assist individual executives, GDSS/CDSS aim at facilitating team-building and teamwork. Of the many such systems available, PLEXSYS was chosen to represent this category of information technology use.

KEY WORDS

Information	Commander EIS
Uncertainty	Commander Center
Information Technology	Groupware
Executive Support Systems	Collaborative Systems
Drill-Down	Group Decision Support Systems
Trend Analysis	

---

[28] Brian Duchaine, "How Managers Can Succeed Through Speed," *Fortune,* February 13, 1989, pp. 54–57.

[29] Jeremy Main, "At Last, Software CEOs Can Use," *Fortune,* March 13, 1989, pp. 77–80.

REVIEW QUESTIONS

1. What is uncertainty and what is information? How are the two concepts related?

2. Explain Beer's Concept of variety of a system. How do organizations that combine the efforts of thousands of people who generate literally trillions of relationships manage this enormous variety?

3. Explain the concepts of data, problem, information, knowledge and wisdom.

4. Explain the three stages in the information technology evolution. Take a look around your school and interview some of the people in key positions within the university administration to see if they themselves went through these three stages.

5. Visit an organization near you such as a business enterprise, hospital or a government agency and talk to people to see if they seem to fit the arrangement shown in Figure 13–4.

6. What is an executive information system? What purposes do these systems serve? What are the "Three Pillars of an EIS?"

7. Take a look at the commercially available EISs listed in Table 13–1. Are you aware of an organization that uses one of them? Do you know of any new systems that you can add to the list? Or, alternatively, have all these systems survived the test of time or are there any that have disappeared from the market?

8. What are some of the most commonly used types of information that executives use to "paint an adequate picture" of the organization's performance? Interview a firm in your city to determine whether these items are indeed what the executives use to judge their organization's performance.

9. What are Comshare's applications? How do they compare with Lockheed's applications?

10. What are collaborative group support systems? Do a survey of the different departments or colleges in your university to see if there is any type of CGSS. If there is none develop a proposal for linking your school to the University of Arizona's PLEXSYS.

Case: The Knowledge Executive

Peter P. Steller was flying back from a three week tour of the company's European and Far Eastern customers. The young fellow occupying the next seat in the Boeing 747 seemed to be heavily absorbed in the book he was reading. Once in a while he put down the book, closed his eyes, and whispered something to himself in a weird sort of language that sounded like Japanese mixed in with

English. Suddenly, he caught Peter watching him and said "So sorry, I didn't know I was speaking so loud. I didn't mean to disturb you. I know it's a long trip . . . some 14 hours you know . . . that's how long it takes to fly from Tokyo to New York." He extended his hand and said, "Hi, my name is Ken Nishi. I am the CEO for Manumara Co." "Glad to meet you," said Peter, "my name is Pete Steller. I am the CEO for Steller Enterprises, we make and sell pumps."

Ken noticed that Pete was looking at the book, so Ken turned toward him and started explaining what the book was all about. "Actually," said Ken, "it isn't a book. It just looks like one. It was done on our brand new desk-top publishing system. It's a report on a study which we did on our new executive information system, which we acquired and installed last year. This is our first experience with an information system which we purchased that is designed to help our top executives get the information they need to run their companies. We bought the system from a company in California which specializes in executive support systems or what they called ESS. It's a good, reputable company. You might have heard of them. It's called the CEO's Commander, Inc." Pete looked at him for a while pretending that he was trying to recall. "No, I can't say I've heard of them. We don't buy application software from external sources. Our MIS people develop our own in-house software. They say most of the commercially available software is not suitable to our needs. And I must believe them. I don't really get involved. Actually, to tell you a little secret, I don't really like computers. There is so much that you must learn. Heck, COBOL is definitely not for me."

Pete was about to ask a question about the ESS, when Ken turned to him and asked him, "Say Mr. Steller, does your company have an EIS?" Pete looked at him for a while and then replied. "Sure, I get summary reports every week. The MIS boys do a very good job. They managed to cut them down to a manageable size of some 30 to 40 pages . . . good information." "Do you interact directly with the mainframe?" interrupted Ken. "Who me? Mainframe? Oh no, thanks . . . I don't have to do that . . . that's what the MIS guys are for." "Then you don't really have an ESS," said Ken. He continued, "By the time you get the data, they are already 4–5 days old." Pete listened carefully.

In the next few hours, Ken explained to Pete his company's experience with the new ESS, stressing the great benefits that he has seen coming out of this package in the very short period of time they have been using it. Pete observed the enthusiasm with which this young executive was talking about "his" system. . . . "I get up in the morning and while I am having my coffee, I dial in to my PC at the office and get the latest report . . . First the big picture . . . in beautiful living color . . . nice bar charts showing our market shares around the world . . . deviations from yesterday . . . profitability and liquidity ratios . . . changes in employee numbers and their morale . . . customer complaints . . . by the time I get to the office I have already worked out my strategy. . . . Then I call a meeting . . . I am informed . . . nothing escapes me . . . "

Pete thought he was listening to a computer software salesman. It all sounded like a sales pitch. Everything works as planned. Heck, he thought, there must be a catch . . . there has to be . . . no computer package can do all that and not make a mistake . . . .

"Say Ken," asked Pete "what was the name of the package? Who makes it? How can I get a demo or something on it? It really sounds great." Ken smiled gracefully and replied, "Hey, I've got an idea. Why don't you stop by our office in New York and take a look at what we do. It won't take you more than a few hours. . . . I'll give you a real live demonstration. . . . We'll even treat you to a nice dinner in our cafeteria. . . . It's up on the 70th floor . . . from there you can see the entire New York City. . . ." When Pete arrived in the plant the next day he had already put together a list of questions he was going to ask his MIS people to get the information for. He must determine for himself whether he did have an EIS or he must go into the expense Ken's system required.

QUESTIONS

1. Develop a list of questions which will help Mr. Steller determine whether his company has an EIS. The company does have a well-designed conventional MIS department.

2. Assuming that the company's MIS reporting system does not provide the information Mr. Steller needs to have continuous and immediate access to company data, develop a list of (1) the performance criteria needed to evaluate company's performance, (2) the types of information, (3) the sources of information, and (4) the form which information will take to facilitate Mr. Steller's desire to "keep an eye on the business via an ESS."

3. Mr. Steller wishes to expand the use of his EIS to his top managers. Speculate on the kinds of problems he is bound to encounter.

4. Develop a means of detecting "computerphobic" managers and a training program to overcome their fears and enhance their willingness to use the system.

Additional References

Bennett, J. "User-Oriented Graphics, Systems for Decision Support in Unstructured Tasks," in *User-Oriented Design of Interactive Graphics Systems,* ed. S. Treu. New York: Association for Computing Machinery, 1977, pp. 41–54.

El Sawy, O.A. "Personal Information Systems for Strategic Scanning in Turbulent Environments: Can the CEO Go On-Line?" *MIS Quarterly* 9, no. 1, March 1985, pp. 53–60.

Friend, D. "Executive Information Systems: Success, Failure, Insights and Misconceptions," *Transactions from the Sixth Annual Conference on Decision Support Systems,* ed. J. Fedorowicz. Washington, D.C., April 21–24, 1986, pp. 35–40.

Gray, P.; W. R. King; E. R. McLean; and H. J. Watson. *Management of Information Systems.* Chicago: The Dryden Press, 1989.

Hogue, J. T., and H. J. Watson. "An Examination of Decision Makers' Utilization of Decision Support System Output." *Information and Management* 8, no. 4, April 1985, pp. 205–212.

Johansen, Robert. *Groupware: Computer Support for Business Teams.* New York: MacMillan, 1988.

Keen, P.G.W. "Value Analysis: Justifying Decision Support Systems." *MIS Quarterly* 5, no. 1, March 1981, pp. 1–16.

Rockart, J. F. "Chief Executives Define Their Own Data Needs." *Harvard Business Review* 57, no. 2, January-February 1979, pp. 81–93.

Rockart, J. F., and M. E. Treacy. "The CEO Goes on-Line." *Harvard Business Review* 60, no. 1, January-February 1982, pp. 32–88.

Sprague, R. H., Jr. "A Framework for the Development of Decision Support Systems." *MIS Quarterly* 4, no. 4, June 1989, pp. 10–26.

Sundue, D. G. "GenRad's On-Line Executives," *Transactions From the Sixth Annual Conference on Decision Support Systems,* ed. J. Fedorowicz. Washington, D.C., April 21–24, 1986, pp. 14–20.

CHAPTER 14

Artificial Intelligence (AI)

Every aspect of learning and any other feature of intelligence can in principle be so precisely described that a machine can be made to simulate it.

John McCarthy, inventor of LISP

CHAPTER OUTLINE

INTRODUCTION

Some scholars trace the beginnings of artificial intelligence (AI) to the World War II years, when engineers and computer scientists were intrigued with the notion of developing systems that think like human beings, but others place the date in the summer of 1956, when a handful of scientists got together under the leadership of John McCarthy, an assistant professor of mathematics at Dartmouth, to discuss how they could work together in a unified discipline to be called artificial intelligence. Perhaps 1956 can be considered the date of the birth of AI while a prior period may have marked its conception.

What is AI? In an informal sense, it may be understood as that branch of knowledge concerned with making computers behave in a way that for humans would be characterized as intelligent. Though this definition begs the question What is intelligence? – one that even psychologists have difficulty defining – still it is one to which most humans can relate. One could spell out in more detail the intuitive notion of intelligence, as Williamson does, in terms of handling information: taking it in, recalling it, making associations between various sets of information, and finally using this information to make a decision about something.[1]

As to its place in the overall classification of knowledge, AI is generally regarded as a subfield of computer science. As such it is that branch of knowledge concerned with machines that perform such functions as reasoning, learning, and drawing inferences in problem solving at a level usually associated with human intelligence. A more formal definition is given by Feigenbaum and McCorduck:

> a subfield of computer science concerned with the concepts and methods of symbolic inference by a computer and the symbolic representation of the knowledge to be used in making inferences; a field aimed at pursuing the possibility that a computer can be made to behave in ways that humans recognize as intelligent behavior in each other.[2]

Like general systems theory, AI's parentage is interdisciplinary. Like the hero who has grown up in many towns, hometown bragging rights are now claimed by various disciplines. George prefers AI's kinship with cybernetics to all others:

> In this book I have attempted to place Artificial Intelligence in the context of cybernetic thinking and information science.[3]

Jaumard and others treat AI as the chief concern of operations research or of management science. Jaumard's selected bibliography on AI was addressed to

---

[1] Mickey Williamson, *Artificial Intelligence for Microcomputers: The Guide for Business Decision Makers,* (New York: Brady Communications Co., 1986), pp. 3–4.

[2] E. A. Feigenbaum and P. McCorduck, *The Fifth Generation: Artificial Intelligence and Japan's Computer Challenge to the World* (Reading, Mass.: Addison-Wesley, 1983).

[3] F. G. George, *Artificial Intelligence: Its Philosophy and Neural Context* (New York: Gordon and Breach Science Publishers, 1986), p. ix. See also A. M. Andrew, *Artificial Intelligence* (Tunbridge Wells, Kent: Abacus Press, 1983). This is vol. 3 in the Cybernetics and Systems Series edited by J. Rose.

operations research and management science specialists. AI, he thought, could offer them two opposite streams of interest: OR tools could be useful in AI problem solving while AI techniques could be embedded in OR methods.[4]

Even a cursory perusal of copies of the proceedings of the International Joint Conference on Artificial Intelligence, the International Cybernetics Congress, the International Conference on Fifth Generation Computer Systems, the International Conference on Automated Deduction, the International Conference on Computational Linguistics, the International Society of Optical Engineers, and the many less specialized computer magazines ought to convince one that in the brief span of just a few decades interest in this field has been worldwide and interdisciplinary and its advances, truly phenomenal.

AI AND CONVENTIONAL COMPUTING

From the very beginning, computers were designed to be efficient numerical processors—"number crunchers." Their stored instructions were carried out sequentially according to an algorithm. An algorithm, as we saw in a previous chapter, is a step-by-step procedure for solving a problem. A simple algorithm used daily by banks calculates the balance of all accounts at the end of each day. The necessary additions and subtractions are faithfully carried out for each account and the appropriate fees assessed. Although the amounts may differ each day, the same algorithm is used. The software used in AI, however, is not based on algorithmics but rather on symbols which represent words, letters, or numbers. These in turn can stand for (encode) concepts, statements of fact, or events. One can create a knowledge base by linking these facts and concepts with one another. To solve a problem these symbols must be manipulated. Given a particular problem, software in the AI program instructs the machine to search the knowledge base and look for certain patterns (pattern recognition process). When these patterns are identified, i.e., when certain criteria are satisfied, the computer goes on to solve the problem. Actually, the computer simply looks around until it finds the best answer.

Most ordinary everyday programming is done in the well-known languages such as BASIC, FORTRAN, FORTH, C, Pascal, COBOL, and Assembly Language. AI programs typically use either LISP or PROLOG. LISP stands for LIST Processing language and was created by John McCarthy in the late 50s or early 60s. LISP contains only a few basic elements like the atom, list, S(ymbolic)-Expression, function, and argument, but because of its great flexibility it can create higher-level built-in functions. A wide variety of LISP dialects is possible because of their built-in functions. There is a sort of nesting of lists within

---

[4] B. Jaumard, "A Selected Artificial Intelligence Bibliography for Operations Researchers," *Annals of Operations Research: Approaches to Intelligent Decision Support* 12, nos. 1–4 (1988), p. 3. See also J. Pearl, *Heuristics: Intelligent Search Strategies for Computer Problem Solving* (Reading, Mass.: Addison-Wesley, 1984). Also J. R. Slagle, *Artificial Intelligence: The Heuristic Programming Approach* (New York: McGraw-Hill, 1971).

lists in the language which allows for complex problem solving. It uses a procedural language, which describes a sequence of steps to be followed or an algorithm to be implemented.

PROLOG, the other AI language, stands for PROgramming language for LOGic and was developed in France in 1972. Like LISP it is designed for symbolic computation rather than numerical computation, but unlike LISP it is a descriptive language and not a procedural one. A descriptive language accepts a set of facts and rules that describe people, places, and things and their interrelationships, and from these facts and rules derives other facts to automatically solve the problem. When you ask PROLOG a question, it will search its database of facts and relationships for suitable matches. As expected, the offspring of LISP and PROLOG are already on the scene. Since a detailed discussion of LISP and PROLOG is outside the scope of this text, the interested reader can find the particulars of each program in several of the references cited at the end of this chapter.

AI CLASSIFICATION

Different authors include different subjects under the heading of AI. Harmon and King (1985) divide AI into three relatively independent research areas: natural language programming, robotics, and expert systems, while Frenzel (1987) identifies six major areas of AI applications:

- General problem solving.
- Expert systems.
- Natural languages.
- Vision.
- Robotics.
- Education.

Other writers use various other arrangements. Since Frenzel's list seems to be the most inclusive, we will now give some attention at least to all of his categories, but we will concentrate on expert systems, the field of greatest development.

General Problem Solving

Most AI applications are geared to specific problems rather than general problems and although some attempts have been made to develop the latter, not much success is reported in the literature. Of course, the computer in general can handle large mathematical problems that would take man-years to solve by hand. In the development of large complex systems, researchers are frequently required to represent the relationships in mathematical form. Once these are formulated, it is then possible to supply real data and let the computer do what it does well—crunch numbers. These applications, however, only solve one type of problem and really can't be termed general problem solving systems. Mathematics, it must be remembered, is contentless and while it is of use to the widest audience, it can only solve specific types of problems.

Games such as chess and checkers, while thought by some to be general problems, are really quite specific and not of great value for other applications. They do, however, employ a knowledge base and utilize inferencing techniques. In the main nearly all AI software is developed for a specific problem.

Expert Systems

Clearly, the predominant use of AI is in *expert systems.* The foundation of expert systems is the knowledge base of a given field that contains all of the relevant expertise of people from the field. Feigenbaum defines an expert system as follows:

> ... an intelligent computer program that uses knowledge and inference procedure to solve problems that are difficult enough to require significant human expertise for their solution. Knowledge necessary to perform at such a level, plus the inference procedure used, can be thought of as a model of the expertise of the best practitioners of the field. The knowledge of an expert system consists of facts and heuristics. The "facts" constitute a body of information that is widely shared, publicly available, and generally agreed on by experts in the field. The "heuristics" are mostly private, little-discussed rules of good judgment (rules of plausible reasoning, rules of good guessing) that characterize expert-level decision making in the field. The performance level of an expert system is primarily a function of the size and quality of the knowledge it possesses.[5]

It ought to be clear that expert systems are based on knowledge, but not all knowledge-based systems are expert systems. During the early development stage of the concept, an attempt was made to interview experts and to capture their knowledge. In fact, the phrase "expert system" was derived from the use of this approach. However, other expert systems in use today may rely on a large knowledge base, but they do not require any human expertise. "Knowledge engineers" seem to prefer to call these expert systems *knowledge systems.* Expert systems tend to be developed in specific fields such as medicine, geology, imagery, etc. while knowledge systems tend to be more generalized.

Some believe expert systems should be distinguished from decision support systems (DSS); others do not. Both are utilized for decision making. DSS uses three distinctly different approaches to problem solving: decision matrices, linear programming, and Monte Carlo simulation. Scores of DSS packages are already available for use on personal computers.

The Knowledge Base.   An expert system has three major components: a knowledge base, an inference engine, and a user interface. (See Figure 14–1.) The

---

[5] P. Harmon and D. King, *Expert Systems* (New York: John Wiley & Sons, 1985), p. 5.

FIGURE 14–1 Components of Expert Systems

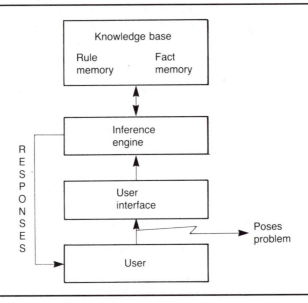

nucleus of the expert system is its knowledge base. The knowledge base comprises the computer's source of intelligence which is used by the inference mechanism to arrive at conclusions. Knowledge can take two forms: 1) facts, concepts, axioms, principles, theories, or laws and 2) heuristics. The first category of knowledge is available from the common fund of knowledge — textbooks, journals, schools, etc. The second form of knowledge is the result of learning from experience. And, of all the people with experience, only a small portion may be considered experts in the field. For example, if 100 scientists were asked to name the top nuclear physicist who could solve a particular problem, the top ten percent named might rightfully be termed experts in the field. In chess, perhaps only a handful would be considered experts. Thus, the knowledge and learning of experts is tapped and stored in the computer. The goal of expert systems is not to replace the experts, but rather to make their knowledge and experience available to others for problem solving. Needless to say, experts come and go, and this is merely one way, and an effective one at that, to capture their knowledge and experience.

In the construction of a knowledge base, the knowledge must first be organized and codified. For example, to develop an expert system for biology, one might begin by organizing the knowledge by plants and animals, then for animals, by vertebrates and invertebrates, etc. The knowledge collection process, termed *knowledge engineering,* is obviously a time-consuming one, even if the knowledge already exists. One must obviously input this knowledge to the computer in some special computer language. However, what is different in AI software from that in conventional systems is that the data must be stored in memory in such a way that it can be used in the reasoning process. It is this characteristic that allows the inference engine to manipulate the knowledge in answering questions or in drawing conclusions. The knowledge represented is in any of the following either deductive or inductive reasoning: trees that show hierarchical knowledge; semantic networks that show the relationships between objects; frames that represent particular blocks of knowledge about an object or situation; scripts that describe a series of events in a situation; and production rules which are two-part statements containing a premise (IF) and a conclusion (THEN). Production rules are one of the most common methods of representing knowledge. Probability and certainty measurements can also be incorporated in knowledge representation.

The Inference Engine.   The inference engine lies between the knowledge base and the user interface. The inference engine is the software that directs the computer to search the knowledge base and match up desired patterns. In doing so it also determines the sequencing of the rules. When one rule is processed, another may be called into use. Rules will reference each other and thus form an inference chain. By adding new facts to the database, the inference engine may modify the triggering of the rules. This inference process continues until the solution to the problem is found.

User Interface.   The last component of the expert system is the user interface. This is the software that allows the user to communicate with the system by posing questions or presenting options, usually from menus, for entering the needed information into the database. It also serves as the medium for communicating the solution to the user. Often the questions asked and the statements made are "canned," though the more modern and sophisticated expert systems are using natural language processing programs that serve as front ends, making the applications almost transparent to the user.

Expert Systems Applications.   Applications of expert systems have sprung up in vast numbers in the past few years. The 1987 Proceedings of the International Society of Optical Engineering list over 70 papers that treat AI applications in a wide variety of disciplines. The single expert system that initiated this massive movement was MYCIN. MYCIN was the first expert system to perform at the level of human intelligence. The system was developed at Stanford University in the mid-1970s for the expressed purpose of assisting physicians in the diagnosis and treatment of meningitis and bacteremia infections. It was the first expert system to combine the features of a knowledge base and an inference engine and apply them to a real problem.

In order to use MYCIN a physician calls up the program at a computer terminal and responds to various questions asked by the computer. Eventually the program will provide a diagnosis and recommend a particular drug therapy. MYCIN reasons about the patient's lab results of body fluids, symptoms, and patient characteristics (age, sex, and race). The particular drug therapy prescribed is tailored to the individual patient. An evaluation of MYCIN, reported in Harmon and King, notes that MYCIN is as good or better than most very skilled medical experts. In one evaluation eight experts compared the choices of MYCIN with the choices of drugs prescribed by nine human diagnosticians for 10 difficult cases of meningitis. MYCIN received a higher rating than any human prescribers of the drugs. MYCIN has proven to be the forerunner of innumerable successful expert systems and has spawned a revolution which will dominate the fifth generation of computers.[6]

The knowledge base of MYCIN, known as EMYCIN (for Essential MYCIN) has been found useful for other expert systems too. It was extracted and used to build systems like PUFF for diagnosing pulmonary diseases and SACON for advising engineers in procedures for structural analysis.

Industry has been very responsive to AI. The Boeing Aerospace Company, after pooling knowledge with AI laboratories at several universities for a number of years, opened its own AI center in 1983 to conduct research in the fields of robotics, experts systems, natural languages, and voice simulation. The company claims to have at least 12 expert systems under development including a helicopter repair service, a deep–space station design, a system to diagnose problems in airplane engines, and a program that advises division managers on database management system purchases.[7]

Another interesting expert system was developed at Lockheed. Lockheed Expert System 1 (LES) has been used for fault diagnosis on the Baseband Distribution Subsystem (BDS) which consists of 16 equipment cabinets, a terminal, and a line printer. It performs corrective maintenance on the BDS and in a very short time narrows the search area to the faulty chassis-mounted component initially causing the system failure. The search area includes some 3000 printed circuit boards, 1000 cables, and other auxiliary devices. Each component of the BDS's complex structure is framed and these frames are stored in memory. This allows for fast retrieval and an instant listing of all the related components and their locations. However, not all components in the search area are represented, only those that may possibly be faulty. Possibility information is outputted by diagnostic messages. In this way the number of components needing representation is reduced from nearly 4000 to around 100. In November 1983 Lockheed acquired a DEC VAX 11/750 computer. LES was used to identify faults that occurred in the VAX from the date of its installation. During the six months of monitoring, the expert system correctly identified 41 out of 43 faults that had occurred. The two missed cases were attributed to missing rules in the knowledge-based expert system. The amount of time spent trouble shooting was drastically reduced by 80 percent.[8]

---

[6] P. Harmon and D. King, *Expert Systems* (New York: John Wiley & Sons, 1985), p. 5.

[7] E. Horwitt, "Exploring Expert Systems," *Business Computer Systems,* March 1985, p. 48.

[8] John F. Gilmore and Kurt Gingher, "A Survey of Diagnostic Expert Systems," *Applications of Artificial Intelligence V,* ed. John F. Gilmore, Proc. SPIE 786, 1987, pp. 2–11.

Natural Languages

Natural language refers to conversational language, which, for our purposes, means English although it could be any other language. To use a computer efficiently, we have to learn some computer language or the precise language of the specific application package we are using. As a result, our efforts at interfacing with the computer are generally strained. If we were able to communicate with the computer in our native language, we could eliminate many of the problems we now encounter. Natural language processing is especially useful for just this type of interfacing with other software packages. It provides, as it were, a "front end" to the other software. Many database packages are extremely complex and hard to use because of their unique sets of commands. Managers and executives in particular simply do not have the time or the patience to learn these complicated commands. A natural language front end overcomes these barriers. With a front-end natural language, users can now concentrate on getting the job done and not on the process of getting it done.

One program that utilizes conversational English, termed INTELLECT, was developed by the Artificial Intelligence Corporation. It is one of the oldest and the most widely used of the natural languages. INTELLECT is a query program that allows users to type into the computer in conversational dialogue questions that they want answered. The program does not contain any rigid rules of grammar or syntax: the questions can be phrased in any manner. This is made possible by use of a *lexicon* — a special dictionary containing an alphabetical arrangement of the words in a language, along with a definition of each word. Besides the common English words, the lexicon contains words and phrases peculiar to any business or application environment. If used from a database, INTELLECT can perform a number of data-processing functions, including the calculation of totals and ratios, of correlations and other statistical functions, the making of comparisons, the drawing of bar graphs, and the outputting of reports.

One application of INTELLECT by the Reynolds Metals Company is to access information in a 9,000-personnel-payroll database. The company notes that its human resource department averages 100 queries per week. The type of queries answered depends on the database, but the following are representative questions:

- How many company employees, single or divorced, live in _____ and earn more than $40,000 per year?
- Give me the 10 top salespeople in terms of quota achievement.
- Compare estimated production to actual production for all departments.
- List the stores in the eastern region that had gross sales over $500,000 last year.

Of course, there are other natural language processing programs besides INTELLECT, and the number is constantly growing. Among those described in the literature are Themis, Clout, Savvy, Natural Link and others. The latest avenue of development is building a natural language into specific application programs like database managers. Examples in the personal computer realm are Q&A and Paradox.[9]

Vision

Another common AI application is machine vision. Machine vision can be regarded as the use of computer vision hardware and software for manufacturing processes. Often machine vision, when used in manufacturing, has as its object the replacement of human operators in boring or repetitive tasks. When doing tasks like these, human beings tend to become bored or tired, often missing fine details when inspecting parts, or otherwise bungling the job. Machines never get bored or tired or make mistakes like their human counterparts.

Machine vision thus can be used to automate the manufacturing process. By feeding certain information to the computer, concerning a description of the materials it will be handling or of the parts it will be making, one can not only achieve a remarkable degree of automation in the manufacturing process but also enhanced productivity, quality control, and a lowering of overall production costs and time. This is all made possible because the machine vision system operates in a highly controlled environment — with cameras and lights positioned for optimal object identification. A template or silhouette of the master object is first made by the camera and stored in a frame which is then used for template matching with the manufactured parts.

Besides its use in the manufacturing process, machine vision has found a niche for itself in quality-control applications. The camera looks at a particular object and then compares it with the template stored in its digitized memory. It can thus determine whether or not the part is up to quality specifications. The auto industry in particular has used machine vision systems for inspecting a variety of parts. Volkswagen, for instance, has used machine vision systems for inspecting the rear wheel brake assemblies for their Golf automobile.[10]

Robotics

The term *robotics* is attributed to the prolific science fiction writer, Isaac Asimov, while the "father of robotics" was none other than Joseph Enselberger, an avid reader of Asimov's literary output. He defines a robot as "a reprogrammable multifunctioning manipulator designed to move material, parts, tools, or specialized devices through variable programmed motions for the performance of a variety of tasks."[11]

---

[9] The material on INTELLECT is adapted from Artificial Intelligence Corporation documents.

[10] "Knowledge Based Vision Software for the Car Industry," in *Sensor Review* 4, no. 4 (October 1984).

[11] Anne Cardoza and Suzuee Vlk, *Robotics* (Blue Ridge Summit, Penn.: TAB Books, 1985), p. 2.

In recent years robotics has come into its own as one of the important industrial technologies of the age. The development of advanced robotic systems can, and doubtlessly will, affect nearly all segments of society. Whether completely autonomous or requiring some human-machine interface, robots are well suited for performing tasks that humans find difficult, boring, dangerous, or requiring excessive strength. Unfortunately only a few of the robotic systems currently in use employ sensory feedback to a large degree. The rest that do not are quite inflexible and need to operate in a highly structured environment. Sensor-driven robots can be more flexible and adept, more reliable and less expensive.

Robots were first developed for energy related applications. Oak Ridge and Argonne National Laboratories were the research and development sites of remote manipulators for handling radioactive materials. Since then, their applications in other related fields has grown extensively. Robots are currently used to spray paint, to drill, to weld, to grind, to inspect or to assemble parts, to detect mines, to defuse bombs, to dispose of nuclear waste, etc. Recently, the Department of Energy (DOE) launched a research program for using robots to enhance personnel safety and reduce costs of nuclear power plant construction, maintenance, operation and decommissioning.[12]

Robots are used as material handlers by a number of firms. General Electric, for example, uses an optic-sensor-equipped robot in its appliance division to "see" refrigerator compressors, pick them up, and move them from one conveyor to another. A French research group used a robotic-armed submarine to retrieve items from the sunken Titanic. Surgeons use robotic arms to guide their hands to precise locations of the body in operations. Robots are in present use as fruit pickers in some orchards. The 1979 Three Mile Island nuclear accident in Pennsylvania is still in the cleanup process with the use of a one-armed robot.

Most of the present robots are controlled by human beings and are not autonomous in the sense that they can react to the specific situations. Needless to say, many of the present-day applications will appear primitive in the next decade or so. Robots that learn are only a function of time and technology. The main ingredient of the futuristic robot is artificial intelligence. We should not be too surprised that with the advent of the thousands of expert systems in use today that the development of more useful robots is close behind.

Education

Some AI authors make a distinction between education and training. The distinction is similar to one made by industry practitioners regarding management development versus training, or, knowledge versus skills. Education, it is claimed, is more concerned with general principles or theories while training is aimed at

---

[12]M. M. Trivedi, R. C. Gonzales, and M. A. Abidi, "Developing Sensor-Driven Robots for Hazardous Environments," in *Applications of Artificial Intelligence V,* ed. John F. Gilmore, Proc. SPIE 786, 1987, pp. 185–88.

improving performance in a specific area. Although most college curricula are supposedly education oriented, many of the professional colleges such as business, law, and engineering obviously aim at developing skills. This dichotomy has existed for decades, even in the sciences. Physicists make a distinction between pure physics and applied physics; mathematicians between pure and applied math, and so on. Although this distinction is useful for some purposes, for AI it is not. It can be said that even those software programs directed toward imparting conceptual knowledge have as their implicit goal the improvement of conceptual skills.

Most AI programs are developed to solve a particular problem and, as such, would fall under the rubric of training. One such system is DELTA/CATS-1 (Diesel-Electronic Locomotive Troubleshooting Aid), an expert system developed by General Electric, that helps mechanics diagnose and solve problems in diesel locomotives. DELTA will respond to interrogation and is supported by CAD graphics and videodiscs that explain how to locate and remedy the problems. Its 1200 troubleshooting rules enable it at present to diagnose 80 percent of all the problems. After diagnosis, DELTA will show a step-by-step procedure for fixing the problem if requested. Work is still progressing on this expert system to eventually enable it to solve 100 percent of the problems with GE's diesel electric locomotives. A number of similar programs are operational in industry.

At the present time AI "education" applications are generally limited to simulations and these have been around for at least three decades, albeit they are now much more sophisticated. We can predict however, with a high degree of confidence, that many more AI education and training systems are in the offing ready to spring forth with ever-advancing computer technology. Perhaps in the not too distant future computer-aided instruction (CAI) will really come into its own with help from AI.

SUMMARY

Artificial intelligence is a term for computer-based systems that perform such functions as reasoning, learning, and making inferences in problem solving. A number of authors identify six major areas of AI applications: general problem solving, expert systems, natural language processing, vision, robotics, and education.

General problem solving is an area where AI has been of little value; most AI applications are directed to the solution of rather specific problems.

Expert systems is clearly the principal employment of AI. Expert systems are computer programs that use knowledge and inference procedures to solve problems. Knowledge consists of facts and heuristics (rules of reasoning). The inference engine (a series of commands that channels information in complex loops) manipulates the knowledge to answer questions or to draw conclusions.

Natural language means conversational language, which — for our purposes — means English. The major reason for the development of a natural language is ease of interacting with the computer. At the present time there are a number of natural languages available. Most of these employ a query program that allows users to type questions that they want answered into the computer in conversational dialogue.

Vision applications of AI have proved to be quite beneficial to manufacturing processes. Typically the specifications of manufacturing reside in a software program, and machines perform various operations on parts (machine, grind, polish, etc.) until these specifications are met. Machine vision is also used for inspection of parts.

Robotics means the development of robots programmed to perform certain functions in a structured environment. While their initial use was for safety applications, they are now utilized in factories to perform many activities such as painting, welding, etc.

Educational uses of AI have been confined to traditional simulations that have been with us for many years. However, as in all the above applications, rapid strides are being made to extend the uses of AI to evergrowing diverse fields.

KEY WORDS

Artificial Intelligence	Natural Language
Expert System	Inference Engine
Heuristic	PROLOG
Knowledge Base	Robotics
LISP	Rule

REVIEW QUESTIONS

1. How can one make the argument that artifical intelligence is an extension of cybernetics?
2. What are the essential differences between a Decision Support System (DSS) and an expert system?
3. The younger a discipline the more difficult it is to develop an expert system. Discuss.
4. What is the purpose of the inference engine in expert systems?
5. The use of natural languages will probably decrease in the future rather than increase. Discuss.
6. Speculate on the potential of robotics in future space explorations.

Additional References

Asimov, I., and K. A. Frenkel. *Robots, Machines in Man's Image*. New York: Harmony Books, 1985.

Berk, A. A. *LISP: The Language of Artificial Intelligence*. London: Collins, 1985.

Brady, M. "Artificial Intelligence and Robotics." *Artificial Intelligence* 26 (1985), pp. 79–121.

Cardoza, Anne, and Suzuee Vlk. *Robotics* (Blue Ridge Summit, Pa.: TAB Books, 1985), p. 2.

Charmiak, Eugene; Christopher K. Riebeck; and Drew V. McDermott. *Artificial Intelligence Programming*. 2nd ed. Hillsdale, N.J.: L. Erlbaum Associates, 1987.

Coxhead, Peter. *Starting LISP for AI*. Boston, Mass.: Scientific Publications, 1987.

Fu, K. S.; R. C. Gonzales; and C. S. G. Lee. *Robotics: Control, Sensing, Vision and Intelligence*. New York: McGraw-Hill, 1987.

Gilmore, John F., and Kurt Gingher. "A Survey of Diagnostic Expert Systems." In *Applications of Artificial Intelligence V*, ed. John F. Gilmore, Proc. SPIE 786 1987, pp. 2–11.

Harmon P., and D. King. *Expert Systems* (New York: John Wiley & Sons, 1985).

Holtz, Frederick. *LISP: The Language of Artificial Intelligence*. Blue Ridge Summit, Pa.: TAB Books, 1983.

Holzbock, Werner G. *Robotic Technology: Principles and Practices*. New York: Van Nostrand Reinhold, 1986.

Horwitt, E. "Exploring Expert Systems." *Business Computer Systems,* March 1985.

Jorgensen, C.; W. R. Hamel; and C. Weisbin. "Autonomous Robot Navigation." *Byte,* January 1986.

"Knowledge Based Vision Software for the Car Industry," *Sensor Review* 4, no. 4 (October 1984).

McCurduck, Pamela. *Machines Who Think*. W. H. Freeman, 1979.

Meieran, H. B. and F. E. Gelhaus. "Mobile Robots Designed for Hazardous Environments." *Robotics Engineering,* March 1986, pp. 10–24.

Rich, Elaine. *Artificial Intelligence*. New York: McGraw-Hill, 1983.

Tonimoto, Steven. *The Elements of Artificial Intelligence: An Introduction to LISP*. Rockville, Md.: Computer Science Press, 1987.

Touretzky, David S. *LISP: A Gentle Introduction to Symbolic Computation*. New York: Harper & Row, 1983.

Trivedi, M. M.; R. C. Gonzales; and M. A. Abidi. "Developing Sensor-Driven Robots for Hazardous Environments." In *Applications of Artificial Intelligence V,* ed. John F. Gilmore. Proc. SPIE 786 1987, pp. 185–88.

Wagman, Morton. *Computer Psychotherapy Systems: Theory and Research Foundations*. New York: Gordon and Breach, 1988.

White, J. R.; H. W. Harvey; and P. E. Satterlee. "Surveillance Robots for Nuclear Power Plants." *Proceedings of the American Nuclear Society Meetings,* San Francisco, Calif., November 1985.

Williamson, Mickey. *Artificial Intelligence for Microcomputers: The Guide for Business Decision Makers.* New York: Brady Communications Company, 1986.

CHAPTER 15

Environmental Scanning for Strategic Management and Planning

For knowing afar off (which is only given a prudent man to do) the evils that are brewing, they are easily cured. But when, for want of such knowledge, they are allowed to grow until everyone can recognize them, there is no longer any remedy to be found.

Niccolo Machiavelli

CHAPTER OUTLINE

MANAGEMENT USE OF EN-SCAN

APPLICATIONS OF EN-SCAN
  The Options
  Application One: Non-Computerized Paper and Pencil Version
  Application Two: Survey-Type Scanning Version
  Application Three: The PLEXSYS Version

SUMMARY

KEY WORDS

REVIEW QUESTIONS

ADDITIONAL REFERENCES

INTRODUCTION

Fifteen years ago when the first edition of this book appeared, the authors decided that they would be happy if students and managers of organizations were exposed to the philosophy of open/organic systems theory. The feeling was prevalent that if the text managed to change the manager's viewpoint away from the conventional "closed systems-parts-analysis" preoccupation and towards an "open systems-wholes-synthesis" viewpoint, it would have accomplished its mission.

Today there is evidence that the transformation did indeed take place. A very small portion of managers still adhere to the old model of organizations as closed systems. Most managers and administrators would admit that today's corporate planner requires a detailed knowledge of the external world—with all its variety and complexity—and a holistic approach to mesh the organization with a fast-changing and largely unknowable future.

The time has come to "put it all together" into a practical, pragmatic, and functional framework. In other words, the reader who desires to become a systemic manager should have a system overview that will enable him or her to become knowledgeable about the external world and the largely unknowable future.

Of all the methods available to the contemporary manager for estimating the future consequences of current decisions, the authors have chosen environmental scanning as the most practical, realistic, effective, and easiest method to implement.

We begin with a quick sketch of the overall picture: an information-based band of specialists. They resemble an orchestra, coordinated by a manager who, not being a specialist, has the luxury of being able to "see and hear" everyone's contribution to the harmonious result: the satisfaction of the customer, the audience. The coordinator's most important task is to provide for the long-range survival and prosperity of the organization. The tool for doing this is the strategic plan. The starting point and, in our view, the most important part of this plan, is the assessment of the external business environment.[1] It is the executive's evaluation of the constraints, threats, and opportunities visible in the external environment that will determine the organization's assignment of resources to its strengths and away from its weaknesses. In this chapter we will focus exclusively on the constraints, threats, and opportunities of the strategic plan. The chapter ends with a complete environmental scanning system called EN-SCAN.

---

[1] A. G. Kefalas, "Linking Environmental Scanning to the Strategic Plan," *Strategic Planning Management* 5, no. 4 (April 1987), pp. 25–33; H. E. Klein and Robert E. Linneman, "Environmental Assessment: An International Study of Corporate Practices," *The Journal of Business Strategy* 5 (1984), pp. 66–84; H. E. Klein and W. Newman, "How to use SPIRE: A Systematic Procedure for Identifying Relevant Environments for Strategic Planning," *Journal of Business Strategy* 1 (1980), pp. 32–46.

THE NEW ORGANIZATION AND THE NEW MANAGEMENT

Management literature has been filled with prognostications of management crises, but while some of these crises never materialized, others did come about. In December of 1958, for instance, Harvard Business Review published an article by Harold J. Leavitt and Thomas L. Whisler under the title "Management in the 1980s." Speculating on the long-term effects of the use of information technology on organizational structure, strategy, and decision making, the authors predicted that the new technology would have the following effects:

1. Middle management: The new technology will thin it, simplify it, program it, and separate a large part of it more rigorously from the top.

2. Top management: Top management will focus more intensively on "horizon" problems, on problems of innovation and change . . . become more abstract, more search-and-research oriented and correspondingly less directly involved in the making of routine decisions . . . spend a good deal of money and time playing games, trying to simulate its own behavior in hypothetical future environments.

3. Decision making: Information technology should make decentralization possible. It may also obviate other reasons for decentralization. For example speed and flexibility will be possible despite the large size, and top executives will be less dependent on subordinates because there will be fewer "experience" and "judgment" areas in which the junior men have more working knowledge. In addition, more efficient information-processing techniques can be expected to shorten radically the feedback loop that tests the accuracy of original observations and decisions . . . fewer people will do more work . . . the top will control the middle just as Taylorism allowed the middle to control the bottom.[2]

Our brief excursion into the world of Executive Support Systems in Chapter 13 should have convinced the reader that, even though the revolution predicted by Leavitt and Whisler is far from being completed, there are enough forces set in motion and enough momentum gained to transform today's organizations into the ones predicted by the authors. Information has indeed become the most valuable ingredient in today's organizations. Information technology has become the organizer, handler, and mover of this valuable commodity. The organization's structure does indeed resemble the one predicted: middle management has shrunk as top management's span of control has increased considerably. The top does control the middle without physical interference. The top's main preoccupation has become the non-routine, creative, and external affairs activities. Strategic planning is now top management's main responsibility.

Peter Drucker envisages the business organization of the 90s similarly. He believes that the organizations of the future will be more like "the hospital, the university, and the symphony orchestra."[3] In *Fortune's* interpretation of Drucker's

---

[2] Harold J. Leavitt and Thomas L. Whisler, "Management in the 1980s" in *Management Systems* ed. P. P. Schoderbek (New York: John Wiley & Sons, 1967), pp. 79–93.

[3] Peter F. Drucker, "The Coming of the New Organization," *Harvard Business Review,* January–February 1988, pp. 45–53.

vision, "employees in the new information-based company will know that they have to do without a flock of vice presidents feeding them information and orders. One conductor — the chief executive — will be enough to keep the oboes and cellos on the same beat. These organizations must move with startling swiftness and offer new rewards to more demanding workers. Middle managers could be in deeper trouble." In addition to these requirements of speed, better motivational schemes for workers, and leaner middle management, companies of the future will be organized along product lines; have a flatter hierarchy with a larger span of communication made possible by their integrated executive information systems; will enter into strategic alliances with competitors; and become global in thinking and operations.[4]

What type of executive will it take to run this type of organization? What traits must one have for this?

The editors of the *U.S. News & World Report* interviewed scores of executives, management consultants, and business school professors. They concluded, after an in-depth study of the managerial styles of some of the most successful U.S. executives, e.g., GE's Jack Weltch and Merck's Roy Vangelos, that to succeed in the 21st century, America's chief executives must overhaul their thinking as well as their factories. The researchers identified the following four "paramount traits":

Trait One: Global strategist (global interdependence)

Trait Two: Master of technology (computer literacy)

Trait Three: Politician par excellence (global scanner)

Trait Four: Leader/motivator (moxie and charisma)

Consultants at Arthur D. Little predict that the future belongs to the megacorporations, or global federations of multibillion-dollar operations groups, whose senior managers will concern themselves chiefly with balancing the firm's economic interests with those of the local culture. *U.S. News & World Report* believes that "perhaps the man who comes closest to fitting the megamanager mold today is American Express CEO James Robinson. He moves easily between the worlds of finance and politics, overseeing new business development that has spurred a robust 306 percent growth in the company's net income since 1977, while also finding time to jawbone Washington officials on free trade."[5]

The key ingredient of a system that will allow the top manager to keep in touch with its external environment system is knowledge — knowledge that is derived from the acquisition and evaluation of external information. It is this knowledge gained from studying and evaluating the external environment that

---

[4] "Managing for the 1990s: Smart Companies Are Getting Ready for Slower-Growing, Low-Inflation Economy, Faster, More Flexible Competitors, Computers That Give Bosses More Power, Richer, Older, Shrewder Consumers, More Demanding and Diverse Workers," *Fortune,* September 26, 1988, pp. 44–96.

[5] "Twenty-First Century Executive," *U.S. News and World Report,* March 7, 1988, pp. 48–51.

allows the management to prepare a sound long-range strategic plan. The top manager's job is to scan the external business environment for emerging issues that develop in the environment representing either constraints to be met or opportunities to be exploited. As the CEO of Thermo Electron Corp., a Waltham, Massachusetts-based, $400-million plus company told INC's editors:

Finding out what these emerging problems are is my job. I have to be out there all the time, talking with people, taking part in public-policy discussions, anything that will enable me to see broad changes before the other guy. If you do a little studying, reading the newspapers, and following things, you will begin to see what problems are coming up.

I spend a lot of my time figuring out what makes the economy tick. How can we fill some of the holes that have been created? It requires a broad understanding of society; and also of what drives our society. The more you understand the overall environment — and really, it's the whole world — the better able you'll be to find the opportunities. If you find needs that aren't being addressed, you can make a much greater contribution, and you will have a greater chance of success.

INC.:

The role of the CEO that you've defined requires much more long-term thinking than most executives practice.

Hatsopoulos:

Yes, and companies like Honda and Mazda have proven, I think, that it pays to think this way. It took them 15 years to break into the American market. If you don't think this way, you're forced into being reactive, into focusing on short-term problems, because other companies have learned to think longer term and therefore have defined the market for you. The stakes are too high to let that happen.

INC.:

So your outside activities are integral to your concept of a CEO's role, which is first and foremost defining the future.

Hatsopoulos:

Yes, this is how my outside activities come around to help the company.

INC.:

Keeping an eye out for emerging trends is important, but what's wrong with focusing on markets that are already established?

Hatsopoulos:

Nothing's wrong with it. But it's going to limit your growth.[6]

STRATEGIC PLANNING AND MANAGEMENT

Most management scholars and practitioners would agree that it is the top management's job to design the strategy that will guarantee the organization's long-term survival and prosperity. The word *strategy* has been grossly misused,

---

[6]"The Thinking Man's CEO," *INC,* November 1988, pp. 29–30.

overused, and misunderstood. Here the word will be used to describe management's task of handling the relationship between the organization and its environment. In the words of Hiroyuki Itami:

> The essence of a successful strategy lies in what I call a dynamic strategic fit, the match over time between the factors that are external to a company (for example, customer's preferences) and the internal factors (for example, the firm's reputation for good services) and the content of the strategy itself.[7]

Most traditional management books include strategic planning as the main task of strategic management. In other words, strategic management is but the implementation of the strategic plan of a firm. Strategic planning, in turn, can be defined as the planning which exhibits the following characteristics:

1. Involves only the top management level of an organization (locus).
2. Is long-term (time).
3. Focuses primarily on external factors (focus).

This tripartite model of strategic planning might be adequate for a small company and/or a company which is predominantly domestically oriented. In today's organizations with a high degree of complexity the hierarchical structure (locus), the planning horizon (time) and the types of factors involved (focus) are difficult to pinpoint. For this reason, *strategic planning and management* will be defined in this chapter by the following additional characteristics:

a. The contemplated actions involve the *integrated allocations of significant portion of organizational resources;*

b. The contemplated actions involve large *uncertainties* about possible outcomes; and

c. The contemplated actions cannot be *reversed,* except at great expense of money and time.[8]

Planning can be defined in general terms as thinking before action. More specifically, R. L. Ackoff has defined planning as, "the design of a desired future and of effective ways of bringing it about" or as "anticipatory decision making."[9] Planning is done to decide what to do and how to do it prior to taking action.

Designing a future of an organization is, of course, tantamount to predicting the organization's ability to survive in its future environment. Thus process of devising a strategic plan involves (1) thinking about and analyzing the relationship between the organization and its external environment sometime in the future,

---

[7] Hiroyuki Itami with Thomas W. Roehl, *Mobilizing Invisible Assets* (Cambridge, Mass.: Harvard University Press, 1987), 1.

[8] Roy Amara and A. J. Lipinski, *Business Planning For An Uncertain Future: Scenarios and Strategies* (New York: Pergamon Press, 1983).

[9] R. L. Ackoff, *A Concept of Corporate Planning* (New York: John Wiley & Sons, 1970), ch. 1.

and (2) finding effective ways of setting and accomplishing organizational goals. In short, organizational goals are set after carefully estimating the environment's future state and after balancing the goals against the organization's strengths and weaknesses.

Figure 15–1A lists the key questions one must ask when involved in strategic planning, and Figure 15–1B represents a corporate planning framework. The lower portion of the diagram (area III) depicts the three main elements of corporate planning: the environment, the corporation, and the stakeholders. The middle portion (area II) depicts management as thinking/analyzing the three main elements and acting on/implementing them. The upper portion (area I) symbolizes management's modeling process. The element *environment* is dealt by management via the "Environmental Model," *corporation* via the "Corporate Model," and *stakeholders* via the "Value Model." Management's overall task is to give an answer to the question posed at the top of the diagram: "WHAT IF"------------"SO WHAT."

In other words, top management must estimate the future state of the environment (create scenarios), assess the corporate strengths and weaknesses (estimate the consequences), and then tell the stakeholders how their goals will be achieved (set objectives).

THE EXTERNAL BUSINESS ENVIRONMENT: A RECONCEPTUALIZATION

As pointed out in an earlier chapter, our working definition of the environment is that *it consists of all those events, happenings, or factors with a present or future influence upon the organization that at the same time lie outside the immediate control of the organization.* This means that things which influence the organization but are under its control are not part of the environment. By the same token, neither are those things that are outside the direct control of the organization but have no influence on the organization's function, structure, or evolution. As was stressed already in the very first chapter, the two important elements in the authors' definition of (external) environment are its relevancy to the system (its influence) and the degree of control that the system has with respect to it.[10]

Table 15–1 provides a summary of various categorizations that have been developed by academicians and practitioners. These sets of categories are not rigid and inflexible but represent ways in which people feel comfortable in talking about the external business environment.

---

[10]See also A. G. Kefalas, "Defining the External Business Environment," *Human Systems Management 1,* pp. 253–60; A. G. Kefalas, "Analyzing Changes in the External Business Environment," *Planning Review* 1 (July 1981); A. G. Kefalas, "On Human Organizations: An Overview," *Human Systems Management 1* (1980), pp. 79–84.

FIGURE 15–1   Strategic Management

A. Key Questions and Descriptors

Question	Descriptors
What are our objectives?	Objectives
What principal factors and changes may influence achieving these objectives?	Scenarios
What choices do we have?	Options
How do we value these choices in light of perceived consequences vis-a-vis our objectives?	Consequences
How do we best allocate our resources?	Decisions
What do we monitor to improve the process by iteration?	Results

B. Representation of Corporate Planning Framework

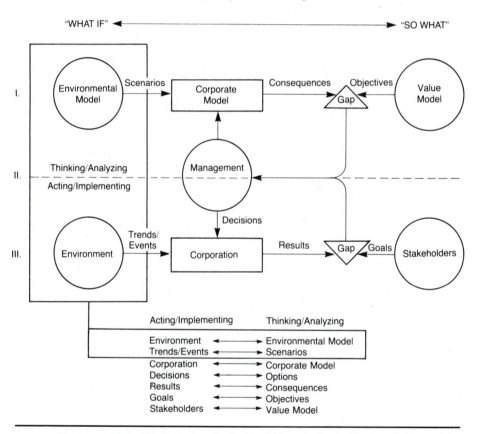

SOURCE: Roy Amara & A. J. Lipinski, *Business Planning For An Uncertain Future: Scenarios and Strategies* (New York: Pengamon Press, 1983).

TABLE 15–1   Defining the External Business Environment

Criteria	Categories	Description	Literature
(A) Nature or meaning	Constraint (C) Threat (T) Opportunity (O)	Discrepancy between goal, actual performance Acute problem, and little time for solution Missed market, etc.; could become a C or T	Academic perception; Lawrence and Lorsch, Thompson, Duncan, etc.
(B) Volatility (S,D,T)	Stable (S) Dynamic (D) Turbulent (T)	Small infrequent changes Large frequent changes Unpredictable changes; unmanageable	Academic perception: Aquilar, Lawrence and Lorsch, Thompson, Rice, Kefalas, Burns and Stalker, etc.
(C) Proximity (I,IE,GE)	Internal (I) Immediate external (IE) General external (GE)	Employee perception of firm Long-standing relationship Weak, irregular, ill-defined relationship	Management/policy literature: Terreberry and others
(D) Sector (M,G,P,T,W)	Market (M) Government (G) Public (P) Technology (T) World (W)	Buyers and sellers, competitors Administrative, legislative bodies Mr. and Mrs. Average Person Scientific developments World affairs	Practitioner literature: *Business Week, Fortune,* etc.
(E) Events- issues- trends (S,T,E,P,E)	Social (S) Technological (T) Ecological (E) Political (P) Economic (E)	What people are talking about: people, nature, technology, etc. Societal happenings Scientific developments Natural-habitat happenings Changes in political system Employment growth	Futurologists: e.g., *Futurist, Business Tomorrow, Forethought.* Public-issue, public-relations practitioners: Coates, Molitor, Palutsek, Wilson, SRI, Naisbitt, etc.

The Meaning Framework

The first way of looking at the environment is to ask the question, What do all these events, happenings, or factors mean to the organization? This framework focuses on three distinct features of the environment and their meaning to the organization:

1. Constraint (C) – a discrepancy between a desired state (a goal or a standard) and an actual state (the actual situation or performance).
2. Threat (T) – a problem for which the organization (a) has either the resources required for the solution but has neither the time nor the know-how to organize them; (b) has neither the resources nor the time and know-how to deal with them.
3. Opportunity (O) – a situation where the discrepancy between the goal and the actual performance is in favor of the latter.

The Volatility Framework

The second way of looking at the EBE is to ask the question, How volatile is the EBE? The framework that uses volatility as a criterion divides the EBE into three categories:

1. Stable (S). Very few changes occur in this environment, and the magnitude of these changes is small. Consequently, the organization has both the resources and the know-how to deal with these changes.
2. Dynamic (D). Changes in the EBE become more frequent and significantly larger. Still, the organization possesses the necessary resources and know-how to deal with the variety created by these changes.
3. Turbulent (T). Both the frequency and the magnitude of the changes that take place in the organization's EBE become unpredictable. In addition, or perhaps because of this unpredictability, the organization's ability to deal with the EBE becomes questionable.

The Proximity Framework

A large number of writers who use proximity to the organization as a criterion employ three main categories:

1. Internal (I) – employee perceptions of the organization's working climate.
2. Immediate external (IE) – long-standing relationships with other organizations (e.g., suppliers, bankers, consultants).
3. General external (GE) – ill-defined, unstructured, irregular relationships between the organization and the external world at large.

The Sector Framework

A great part of the management literature deals with the environment as a set of sectors. Five such sectors are commonly distinguished:

1. Market (M) — buyer and seller as well as competitors.
2. Government (G) — administrative, legislative, and judicial bodies that act as both regulators and supporters of the organization.
3. Public (P) — feelings, sentiments, opinions, and actions of ordinary citizens, both as individuals and as groups, that have an effect on the organization's goals and objectives.
4. Technology (T) — implementation of commercial application of scientific discoveries that impact on the organization's products, processes, or services.
5. World (W) — international developments in the above four sectors that impact on the organization's functioning.

The Events-Issues-Trends Framework

This way of looking at the environment as a set of events, issues, or trends goes under various names, such as environmental scanning, environmental analysis, trend tracking, and issue management. Whatever the name, in this view the events, issues, and trends of the environment are categorized as social, technological, ecological, political, and economic. Since this is the framework used in this book, a separate section is devoted to its explanation.

ENVIRONMENTAL SCANNING: A SURVEY OF ACTUAL APPROACHES

In Chapter 12 we saw that there are numerous methods for dealing with the future. A generic name for these methods used by futurists is futures research. Futures research ranges from Herman Kahn's genius forecasting and the Delphi method (*a fortiori* reasoning) to the Club of Rome's systems or global dynamics (computer-aided reasoning). Environmental scanning is based upon the simple dictum that *understanding the external business environment and the direction of some key trends is a prerequisite for shaping the future rather than reacting to it.*[11]

---

[11] W. L. Renfro, "Managing the Issues of the 1980's," *The Futurist,* August 1982, pp. 61–66; Roy Amara, "Managing in the 1980s," *Technology Review,* April 1981, pp. 77–82; W. L. Renfro and J. L. Morrison, "Merging Two Futures Concepts: Issues Management and Policy Impact Analysis," *The Futurist,* October 1982, pp. 54–56; John Naisbitt, *Megatrends,* (New York: Warner Books, 1982); Arnold Mitchell, *The Nine American Lifestyles* (New York: Macmillan, 1983); Frank Feather, ed. *Through the 80's: Thinking Globally, Acting Locally* (Washington, D.C.: World Future Society, 1980); A. G. Kefalas, "Managerial Futurity: Relating Today's Actions to Their Future Consequences," *Journal of Contemporary Business* 9, no. 3 (1980), pp. 123–32; A. G. Kefalas, "Emerging Issues Focus on the Future," *Athens Banner-Herald Daily News,* February 27, 1983.

Environmental Scanning in the Business World

Environmental scanning is a systematic way of looking at the world outside the organization as a set of *emerging issues* — things that society talks about, and on which there has been no definite agreement. As the Conference Board puts it, "An emerging issue may be defined as a trend or condition, internal or external, that, if continued, would have a significant effect on how a company is operated over the period of its business plan."[12]

In this portion of the text we will present general examples of corporate environmental scanning systems from a study conducted by the Conference Board entitled, "This Business of Issues: Coping with the Company's Environments." (See Exhibits 15–1, 15–2, and 15–3.) The study was motivated by the observation that attention to the external business environment was becoming a *sine qua non* of corporate survival.

Environmental Scanning in Institutions of Higher Education

Environmental scanning, which was pioneered by progressive business firms like GE, Sears, Sperry Rand, AT&T and others, has also spread into the public sector. Thus the U.S. Congress initiated a program called Trend Evaluation and Monitoring in the early seventies. In 1978 the Conference Board established an Emerging Issues System within its Management Research Division. In the field of institutions for higher education, the Georgia Center for Continuing Education at the University of Georgia has perhaps the most complete environmental scanning system in operation. The system goes beyond the conventional scanning system insofar as it attempts to feed its results into the strategic planning executive committee's task.[13] Figure 15–2 presents a sketch of the Georgia Center for Continuing Education Environmental Scanning Project.

THE EN-SCAN SYSTEM

The environmental scanning systems briefly described in the previous section were the precursors/forerunners of what eventually evolved into a full-fledged industry of providing external information for strategic decision making. As corporate offices came under stakeholder scrutiny, executives found the usual

---

[12] J.K. Brown, *This Business of Issues: Coping with the Company's Environments* (New York: The Conference Board, 1979).

[13] J. L. Morrison, E. Simpson, and D. McGinty, "Establishing an Environmental Scanning Program at the Georgia Center for Continuing Education," paper presented at the 1986 annual meeting of the American Association for Higher Education, Washington, D.C., March 1986; J. L. Morrison, "Establishing an Environmental Scanning/Forecasting System to Augment College and University Planning," *Planning for Higher Education* 15 no. 1 (1987), pp. 7–22; J. L. Morrison, W. L. Renfro, and W. I. Boucher, *Futures Research and the Strategic Planning Process: Implications for Higher Education* (Washington, D.C.: Association for Higher Education, 1984).

EXHIBIT 15–1   General Electric Company

Since 1967 a business environment research group has been at work identifying and analyzing long-term social, political, and economic trends and their potential impact on business. Originally established as a function of Personnel and Industrial Relations, this component has operated since January, 1974, as part of the Environmental Analysis Staff in Corporate Strategic Planning.

The basic objective of Business Environment Research is to promote the use of a "four-sided framework" in strategic planning, incorporating the results of social and political trend analysis, as well as economic and technological forecasting. One specific aim is to analyze changing societal expectations of corporate performance in sufficient time to enable the corporation to develop constructive, anticipatory responses to these expectations. Analysis of the past 20 years has shown that, in the absence of such a response, simple social expectations have developed into political issues, then legislated requirements and, finally litigation, progressively narrowing business' range of options at each state.

The Evolutionary Sequence of
Societal Expectations in Business Performance

Sequence	*Business' Position*	*Example*
Without proper response from business . . .	Options	Minority demands for civil rights
Societal expectations of today become . . .	Semiautonomous	Little Rock—late 1950s
Public and political issues tomorrow . . .	Defensive	1960 platforms (both parties)
Then legislation . . . and finally, *litigation and penalties*	Compliance Pay Penalties (ordered or negotiated) Loss of business	1964 Civil Rights Act 1973 backlog: 60,000 EEOC cases

SOURCE: Ian Wilson, "Corporate Environments of the Future: Planning for Major Change," (the Presidents Association, 1976).

"Nobody can predict the future" a very poor excuse. Executives demanded that staff provide them with some scenarios about the external environment. Very few executives believe in or have the time to take Mr. Hatsopoulos's approach: do-it-yourself scanning.

There are essentially two approaches to the establishment of an environmental scanning system. The first approach is to develop an in-house ES system. The other is to employ the services of the numerous external firms that specialize in ES.[14] Both approaches have their merits. Here we recom-

---

[14] For a list of some of the most well-known firms see: Myron Magnet, "Who Needs a Trend-Spotter?" *Fortune,* December 9, 1988, pp. 51–56.

EXHIBIT 15–2   Two Comprehensive Scanning and Monitoring Programs—Sears, Roebuck and Company

Sears has a corporate monitoring unit, consisting of two full-time and one half-time members. The monitors follow about 100 publications and research reports of outside services and industry groups for the purpose of tracking scores of external trends and public issues. The unit oversees as well an internal monitoring-forecasting activity covering 11 broad categories of trends (e.g., sales, income, capital expenditures, facilities), divided into a great many more subcategories.

Another component of the monitor's work—the most important one, according to the head of the unit—is face-to-face contacts with outsiders: regular visits to principals of each of the research firms, the services of which the company subscribes to or otherwise supports, and to futurists at a number of universities. The purpose of these visits is to get at the whys behind their respective forecasts and to understand the mind sets—the unstated premises and, unkindly, the biases—of the forecasters. The monitors also consult with specialists in the company who can shed light on both internal and external trends.

The monitoring unit is represented on a senior management committee of about a score of senior headquarters executives. In the meetings of this committee the findings of the monitoring staff are brought to the attention of decision makers, and the staff is informed of the agenda and priorities of the decision makers.

This obviously is an ambitious effort, which could become overwhelming were it not for adherence to several guidelines and constant attention to several caveats.

- Monitors do not—and should not—forecast; instead they rely on forecasts made by others, explaining, integrating, and, as necessary, reconciling forecasts. Monitors do not have time to forecast, and their credibility as monitors could well be compromised were they to forecast.

- Monitors have to take care not to get involved in extraneous committee work either outside or inside the company.

- They have to guard against devoting too much time to any single subject, however it may interest them. Yet they must also be able to detect "weak signals," the first blips on the radar of the scanning system.

- For the most part, they only scan, they do not study carefully, any of the publications and other documents they review. But they examine more thoroughly writings pertaining to priority trends and issues.

- In recording what is monitored, unnecessary detail is omitted. What should be set down are broad highlights, stated very briefly. Each external issue has a file folder, the key document in which is a summary of past and future trends.

SOURCE: J. K. Brown, "This Business of Issues: Coping with the Company's Environments," (New York: The Conference Board, 1979), p. 46.

mend the first approach, the development of an in-house ES system. The reason is that environmental awareness is "everybody's business." If an organization is to provide for itself a true picture of the exceedingly complex external environment, it must enlist the assistance of a rather substantial number of its managers.

EXHIBIT 15–3   External Trends Followed by Sears, Roebuck and Company

Demographics	*Demographics (Cont'd)*	*Resources*
Population	Income	Energy Supply &
Size & characteristics	Distribution by	Demand
Size	Region	Coal
Growth rate	Age	Electric/Power
Sex	Earners per family	Backlog of appropria-
Age	Education	tions
Marital status	Median income	Natural gas
Singles	Household	Petroleum
Marriages/divorces	Family	
Remarriages	Individual	Mineral & Chemical
Births	Personal income compo-	Supply
Birth expectations	nents	Imported vs. domestic
Birth/fertility rate	Disposable Personal In-	Metals
Location/mobility	come (DPI)	
Regional		Agriculture
Metro/nonmetro	Spending	Food
Farm	Personal Consumption	Fertilizer
Central cities	Expenditures (PCE)	Agribusiness
Congressional dis-	Consumer price index	
tricts	Consumer credit	Water Availability
Households/families		Supply, surface &
Age of head	Housing	underground
Average size	Existing housing	Delivery problem areas
One-person house-	Units	Drought areas
holds	Type	
Minorities	Region	Strategic Depletion
Illegal aliens	Housing costs & sales	Shortage
Spanish Americans	Housing starts	Industrial capacity
	Incomes of purchases	
Employment		Land
Civilian labor force	*Values/Lifestyles*	*Technology*
Growth rate	Values	Expenditures for R&D
Size & characteris-	Work & leisure	Federal
tics	Entitlement	Total R&D expendi-
Full time/part	Consumption vs. conser-	tures
time	vation	Defense
Sex	Consumer assertiveness	HHS
Working wives	Lifestyles	Alternate Energy Sources
Occupation	Marriage/family struc-	Nuclear
Regional distribution	ture	Solar
Labor union member-	Homes/mobility	Hydro, geothermal &
ship	Shopping habits	photochemical,
Hours	Aging/retirement	other
Benefits	Singles	Plastics

EXHIBIT 15–3 *(continued)*

Technology (Cont'd)	Government (Cont'd)	Government (Cont'd)
Electric Communications	National economic plan-	Postal Service
Computers	ning	Health Care
Personal/small busi-	Government Regulations	Taxes
ness	Agencies	Personal
Network systems	Cost/criticism	Corporate
Entertainment and	Reform	Social security
games	Corporate crimes	
Satellite communica-	Business lobby	*International*
tions	Legislation	World Population
Transportation	Antitrust	Resources
Materials	Consumerism	Food
Automobile	Employment	Energy
Electric auto	Physical conditions	Raw materials
Manufacturing Techniques	Equal opportunity	Trade
Durables/non-durables	Benefits/security	Trade/payments balance
Product Development	Economic Controls	Exports/imports
	Reporting/disclosure	Protectionism
Public Attitudes	Environment	Tariffs
Consumer Confidence	Land use	Cartels
Buying plans index	Air, water, noise	Developing Nations
Public Attitudes toward	Waster disposal	OPEC
Government regulation	Consumer Credit Physical	LDCs
Large companies	Distribution	Technology Transfer
Corporate social respon-	Transportation	Economic Indicators
sibility	Warehousing	
Industries & products	Privacy	*Economics*
Energy situation	Consumer	Gross national product
Environment	Employee	Inflation rate
Public Interest Groups	Products	Interest rate
Consumerism	Safety	Unemployment rate
	Quality/life cycle	Productivity
Government	Communication with	AAA bond interest
Operations	Customers	rate
Government purchases	Advertising/selling prac-	Capital investment re-
Government expendi-	tices	quirements
tures	Complaint procedures/	Wage levels
Social welfare	redress	Benefit-cost levels
Social security cost	Warranties	Economic forecasts
Veterans benefits	Service	Corporate profits and
Employees	Repair quality/standards/	cash flow
Public debt	licensing	Capital formation
		needs

SOURCE: J. K. Brown, "This Business of Issues: Coping with the Company's Environments" (New York: The Conference Board, 1979), p. 11.

FIGURE 15–2   Georgia Center Environmental Scanning Project

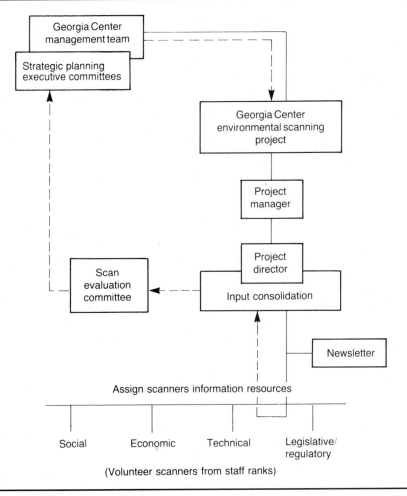

SOURCE: James L. Morrison, "Environmental Scanning Program," p. 8.

The EN-SCAN system described here is a modest beginning. The usefulness of the system lies in its simplicity and versatility. In addition, the system can take advantage of the latest developments of inexpensive and convenient sources of external information which are commercially available.[15]

Issue Development and Evolution

As can be seen from the few examples presented, there is no single theory to explain the way issues develop that is accepted by most people who deal with issue

---

[15] See for example, *The Source Information Network,* 1616 Anderson Road, McLean, Va. 22102.

identification and development. Generally, most would agree, however, that issues do not develop overnight. Rather, they follow a more or less standard process that takes time. Graham Molitor, once General Mills's director of government relations and now an independent consultant, believes that "a period of ten (10) years elapses before substantial interest in an issue develops, prompting action to be taken." In general, issue development follows the process depicted in Figure 15–3, Issue Development and Evolution.

The issue development and evolution process follows a six-step sequence beginning with emergence and ending with legislation.[16] On the average, it takes 6 to 10 years for an issue to run its course. As the issue moves through the "emergence-legislation" process, attention to the issue increases; it reaches its zenith with the passage of the law. This attention is manifested in a number of ways, such as the number of people who become involved in the issue, the amount of time and money they spend, the degree of organization and formalization which results, and so on.

Corporate response to issue development follows approximately the same pattern. The difference is that attention begins to mount later in the issue-development process but then grows at a much faster rate. In other words, once the process leaves the ignorance stage (I) and moves into the other stages, attention by the corporate world increases exponentially. (See Figure 15–4.)

The basic assumption underlying EN-SCAN is that the fate of any organization—be it a business enterprise, a hospital, a university, a city, a state, or a nation—will be influenced more by the changes in people's habits and attitudes and society's general ways of living than by deliberate and rational managerial planning. As the Conference Board Study puts it:

> Ten years ago, a prominent consultant stated, 80 percent of planning was concerned with what management wanted, 20 percent with how the world affected the company; now the figures are reversed—or at least ought to be. Hyperbole perhaps; but the trend is unmistakable, and doubtless irreversible.[17]

EN-SCAN® is a system that enables the manager to learn as much as possible about the environment by using the method of issue management. Specifically, EN-SCAN arranges the environment into a set of emerging issues. To help the manager identify and evaluate these issues, EN-SCAN employs a five-category framework called STEPE: social, technological, ecological, political, and economic issues. The logic for this arrangement in the STEPE framework is that issues first develop in the social sphere and represent the maturation and expression of visions and ideals of gifted individuals who are blessed with exceptional capabilities. Social issues will sooner or later become technological

---

[16] Hazel Henderson, "How to Cope with Organizational Future Shock," *Management Review,* July 1976, pp. 19–28; Graham T. T. Molitor, "How to Anticipate Public Policy Changes," *S.A.M. Advanced Management Journal,* Summer 1977, pp. 4–13.

[17] Ibid., p. 2. See also J. K. Brown, This Business of Issues: Coping with the Company's Environments (New York: The Conference Board, 1979), p. 2.

FIGURE 15–3   Issue Development and Evolution

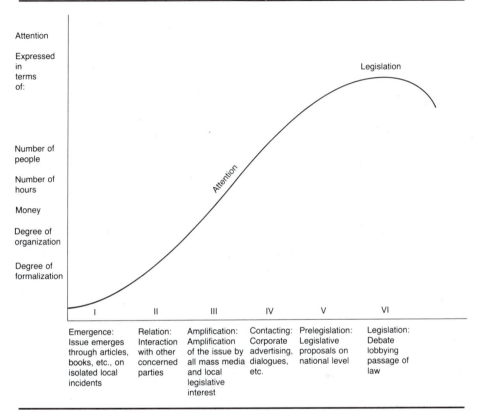

I	II	III	IV	V	VI
Emergence: Issue emerges through articles, books, etc., on isolated local incidents	Relation: Interaction with other concerned parties	Amplification: Amplification of the issue by all mass media and local legislative interest	Contacting: Corporate advertising, dialogues, etc.	Prelegislation: Legislative proposals on national level	Legislation: Debate lobbying passage of law

Attention Expressed in terms of:

Number of people

Number of hours

Money

Degree of organization

Degree of formalization

Legislation

Attention

SOURCE: Hazel Henderson, "How to Cope with Organizational Future Shock," *Management Review,* July 1976, pp. 19–28; Graham T. T. Molitor, "How to Anticipate Public Policy Changes," *S.A.M. Advanced Management Journal,* Summer 1977, pp. 4–13.

issues as ideas become realities. The use of technology will create ecological issues. Political issues are, in turn, created when society demands that the government take certain actions to control and eventually resolve these issues. Finally, all resolutions of social, technological, and ecological issues will cost money—i.e., they will become economic issues.

Neither the five categories of issues nor the sequence of issue development suggested is a concept that is well accepted and all-encompassing. STEPE is just a classification framework aimed at facilitating the organization and systemization of the scanning process. Like all classification frameworks, STEPE is an arbitrary arrangement, but the present authors find it useful. As Figure 15–5 shows, EN-SCAN provides a three-way evaluation of the environment. At top levels busy executives can concentrate on its five main categories. Questions regarding unsatisfactory or unacceptable evaluations of any of the five categories of issues can be passed on to the second and third levels for more detailed scanning. (See detailed explanation below.)

FIGURE 15–4 Corporate Response to Issue Development

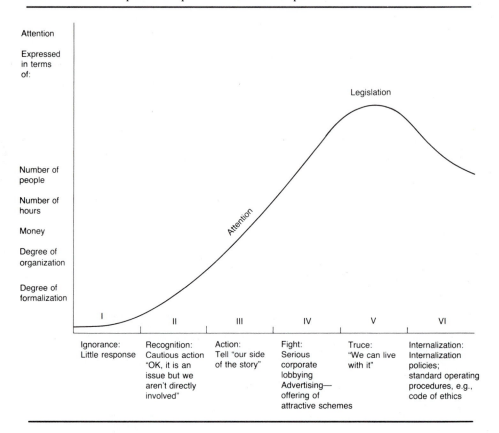

Attention

Expressed in terms of:

Number of people

Number of hours

Money

Degree of organization

Degree of formalization

I	II	III	IV	V	VI
Ignorance: Little response	Recognition: Cautious action "OK, it is an issue but we aren't directly involved"	Action: Tell "our side of the story"	Fight: Serious corporate lobbying Advertising— offering of attractive schemes	Truce: "We can live with it"	Internalization: Internalization policies; standard operating procedures, e.g., code of ethics

The EN-SCAN Software Package[18]

This package represents a true application of the systems approach to managing the organization's external relations. The package presented in these closing pages of *Management Systems: Conceptual Considerations* represents an example of the ease of application of the systems approach. The package incorporates all three steps involved in the application of the systems approach as explained in Chapter 11.

Conceptually, the package treats the organization as an open system and focuses on the external environment. The measurement/quantification step uses the "scaling method" to measure the environment. Finally, computerization is performed on a personal computer using BASIC.

---

[18] EN-SCAN is the property of Dr. A. G. Kefalas and has been registered as EN-SCAN™.

FIGURE 15–5 The Environmental Scanning System (EN-SCAN)

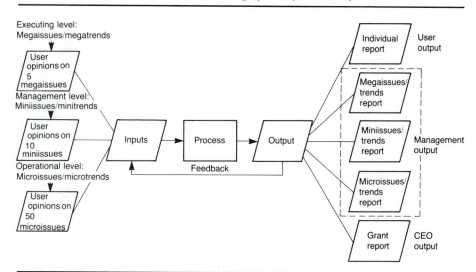

Inputs

The inputs to the EN-SCAN system are executive/management opinions on issues which characterize the environment. As can be seen from Figure 15–5 and Figure 15–6 the user has the opportunity to be as detailed about the environment as he or she desires. At the most abstract level, the user may wish to input into the system his or her understanding or degree of awareness about the environment by focusing on the five major trends which are called megaissues/megatrends. At the intermediate level, management level, the user can focus on the issues/trends which led to the development of the 5 megaissues/megatrends—i.e., the 10 miniissues or minitrends.

Finally, staff personnel in an organization can focus on the daily events and statistics which are reported by the news media or fill the government reports. The EN-SCAN system allows for 50 such microissues or microtrends. User inputs on these issues or trends are responses to questions regarding:

1. *Magnitude,* i.e., the degree of currency or attention of the issue in terms of media coverage, the number of people who talk about it, and so on. (See vertical axes of Figures 15–3 and 15–4.)

   Measurement Scale: 1   2   3   4   5
                           Low   High

2. *Importance,* i.e., what is (will be) the impact of the issues on the organization's people, money, resources, products, etc.

   Measurement Scale: 1   2   3   4   5
                           Low   High

FIGURE 15–6   The Environmental Scanning System: The STEPE Framework

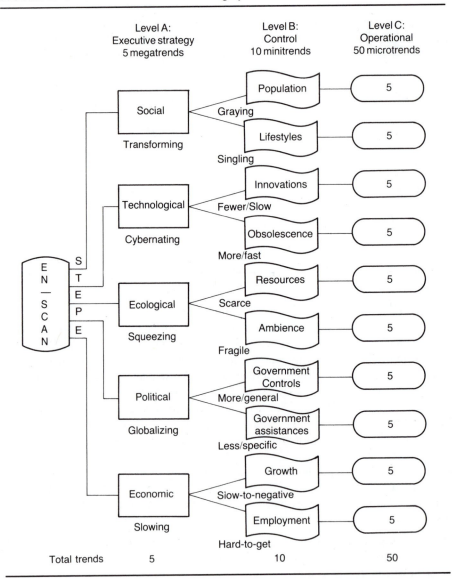

3. *Meaning,* i.e., what does the specific issue (with certain magnitude and importance) mean? Here the user is asked to state his or her opinion as to whether the issue represents

Measurement: *a.* constraint
               b. threat
               c. opportunity

Process

The user of EN-SCAN interacts with the computer by responding to questions or statements which flash on the TV screen (CRT). Improper responses are rejected by the flashing message "invalid entry . . . please reenter."

The EN-SCAN package contains everything that a complete novice needs to understand the entire environmental scanning processes. Everything appears on the TV screen in a "user friendly" form. The entire process takes no more than 30 minutes.

Outputs

There are three main types of output:

1. The individual report.
2. Management reports.
   a. Microtrends.
   b. Minitrends.
   c. Megatrends.
3. The CEO report (megatrends).

Examples of these reports are provided below in Figures 15–7, 15–8, and 15–9.

Feedback

Feedback was defined as the reintroduction of a portion of the output of a system. This portion represents a difference between (1) the expected output and

FIGURE 15–7 EN-SCAN Output: Individual Report

HERE ARE THE RESULTS OF YOUR EN-SCAN

THIS IS AN*******INDIVIDUAL REPORT*******

ISSUES	S C A N	*MAGNITUDE* 1 2 3 4 5	*IMPORTANCE* 1 2 3 4 5	*MEANING* C	T	O
SOCIAL		1.66	1.42	X		
TECHNOLOGICAL		2.22	1.9		X	
ECOLOGICAL		2.77	2.38			X
POLITICAL		2.22	1.9		X	
ECONOMIC		1.11	2.38	X		
TOTAL/AVERAGE		1.99	1.99	2	2	1

FIGURE 15–8 EN-SCAN Output: Management Reports

HERE ARE THE RESULTS OF YOUR EN-SCAN

THIS IS ###MANAGEMENT REPORT#ONE:****MEGATRENDS****

	S C A N	MAGNITUDE					IMPORTANCE					MEANING		
ISSUES		1	2	3	4	5	1	2	3	4	5	C	T	O
SOCIAL				1.33					2.35			X		
TECHNOLOGICAL				2.66					2.94					X
ECOLOGICAL				.66					.58			X		
POLITICAL				2					1.76				X	
ECONOMIC				3.33					2.35				X	
TOTAL/AVERAGE				1.99					1.99			2	2	1

HERE ARE THE RESULTS OF YOUR EN-SCAN

THIS IS ###MANAGEMENT REPORT#TWO:****MINITRENDS****

	S C A N	MAGNITUDE					IMPORTANCE					MEANING		
ISSUES		1	2	3	4	5	1	2	3	4	5	C	T	O
SOCIAL				1.87					1.33			X		
TECHNOLOGICAL				2.5					3.33				X	
ECOLOGICAL				3.12					.66					X
POLITICAL				.62					3.33					X
ECONOMIC				1.87					1.33			X		

HERE ARE THE RESULTS OF YOUR EN-SCAN

THIS IS ###MANAGEMENT REPORT # THREE:****MICROTRENDS****

	S C A N	MAGNITUDE					IMPORTANCE					MEANING		
ISSUES		1	2	3	4	5	1	2	3	4	5	C	T	O
S				0					0					
T				0					0					
E				0					0					
P				0					0					
E				0					0					

(2) the actual output. Each report in the EN-SCAN is designed to facilitate assessment of both the individual's degree of environmental awareness, as well as the collective organizational understanding of changes in the environment.

MANAGEMENT USES OF EN-SCAN

A frequently asked question in connection with environmental scanning is, What have been the payoffs from all that? In other words, is the expense worth the real or potential benefits?

The Conference Board Study contains numerous statements of corporate executives attesting to the benefits of a program of environmental analysis, scanning, and issue management. The following example is indicative of corporate leaders' opinions regarding the benefits of involvement in raising organizational, executive, and management awareness of changes in the environment.

In a talk delivered to the Corporate Section of the Public Relations Society of America meeting late in 1977, David L. Shanks, director, corporate public relations and advertising at Rexnord, had this to say:[19]

> There are many benefits to be derived in properly managing issues. I would like to list six of them. They are both protective and opportunistic.
>
> 1. Allows management to select issues that will have the greatest impact on the corporation.
> 2. Allows "management of" versus "reaction to" issues.
> 3. Inserts relevant issues into the strategic planning process.
> 4. Issue management gives the company ability to act in tune with society.
> 5. Provides opportunities for leadership roles.
> 6. Protects the credibility of business in the public mind.
>
> The fourth point may be the most important of all. Many people realize that being in tune with society may be a matter of survival. At the current rate of change, we may soon lose a great socioeconomic system unless many of us get firmly involved now.

The greatest benefit of the EN-SCAN computer package is its excellent diagnostic ability. For example, busy CEOs who desire to take a quick look at the organization's environmental awareness could examine the GRAND Report and check the correspondence between their own perceptions of the environment and that of their subordinates, and their own and subordinates' perceptions of the magnitude, importance, and meaning of environment issues and trends.

The output of this diagnosis can be used to prescribe remedial actions. If the CEO finds that most marketing senior executives assign *low* values of magnitude or importance to social issues which they see only as problems in a rapidly changing society, he or she can conclude that marketing executives are in need of some "enlightenment." Similar deliberations can be made regarding the rest of the GRAND Report output.

---

[19] Brown, "This Business of Issues," p. 27.

FIGURE 15–9 EN-SCAN Output: CEO's Grand Report

Trends Issues (Biogr. Data)	Department					Rank			Age		Education			
	1	2	3	4	5	1 Upper	2 Mid.	3 Low.	1 >40	2 <40	1 Hi	2 B.A.	3 M.A.	4 Ph.D.
Social														
Magnitude*														
Importance*														
Constraint +														
Threat +														
Opportunity +														
Technological														
Magnitude														
Importance														
Constraint														
Threat														
Opportunity														
Ecological														
Magnitude														
Importance														

Constraint													
Threat													
Opportunity													
Political													
Magnitude													
Importance													
Constraint													
Threat													
Opportunity													
Economic													
Magnitude													
Importance													
Constraint													
Threat													
Opportunity													
$\Sigma_1\ E = (M + I) / 5$**													
Σ_2 Get High + + C or T or O													

* Average of scale 1 2 3 4 5
+ Number of people
** Overall E-factor = Environmental awareness = M + I
+ + Prints C, T, or O depending on highest number of check marks

APPLICATIONS OF EN-SCAN

The Options

The EN-SCAN system can take the following forms:

1. Structured: Using the STEPE Framework described in the previous section.
2. Unstructured: Using any list of issues from any source.

Applications of the systems can be:

Category A: Noncomputerized
   1. Paper and Pencil Small Group Version.
   2. Survey-Type Version.
Category B: Computerized
   1. Standalone Computer Package (in QuickBASIC).
   2. Used in Connection with PLEXSYS.
   3. Used in Connection with any LAN or Teleconferencing.
   4. Used in Connection with any Spreadsheet software.

In the next few pages a few applications of EN-SCAN are presented using (a) the noncomputerized paper and pencil version, (b) the survey-type version, and (c) the PLEXSYS version.

Application One: Noncomputerized Paper and Pencil Version

This is the simplest version and is designed for small group presentations. The example depicted in Exhibits 15–5 and 15–6 was used with executives participating in the Georgia Executive Program. As can be seen, in a matter of a few minutes the group leader can get a sense of the group's environmental awareness as well as their optimism or pessimism. Scoring is very simple and is done by the individual. The group leader usually looks for the top three issues in the "magnitude" column and compares them with the top three issues in the "importance" column. Deviations are usually great topics for discussion.

Application Two: Survey-Type Scanning Version

To illustrate the application of EN-SCAN, the results of a recent survey undertaken by one of the authors are presented briefly in Table 15–2. The survey was sent to some 150 members of the Issue Association drawn randomly from the membership list. The same survey appeared in the November/December Issue of LINC (Linking Issues Networks for Cooperation), published by the Issue Action Publications, Inc., 105 Old Long Ridge Road, Stamford, CT 06903.

Step 1: Issue Identification.   The 20 issues that were included in this survey were identified by the NEISNET (North East Issue Network) group, in March/April 1986 by some 200 public affairs professionals in the New York area.

EXHIBIT 15–4   Instructions for Paper and Pencil ES

EN-SCAN is a method of estimating the manager's sensitivity to issues (events, occurrences) in the external business environment that may impact the organization's future survival, success, development, and/or performance. Once the issues of greatest significance have been identified, the organization can integrate (incorporate) this information into the strategic planning process.

Definitions

Issue: An event, happening, or factor in the contemporary environment over which the organization has no direct control, but which influences the organization's future. Criteria such as magnitude, importance, and meaning describe issues.

Magnitude: The amount of concern or attention to the issue in terms of media coverage, the number of people who talk about it, etc.

Importance: The impact of the issue on the organization's personnel, money, resources, products, etc.

Meaning: The nature, implication, significance, or urgency of the issue to the organization. Does the issue represent a constraint, threat, or opportunity?

Constraint: Limits, restricts, or regulates the organization's capacity for goal attainment.

Threat: Indicates a severe problem which endangers the existence of the organization.

Opportunity: Offers an open avenue, benefit, or advantage that enhances the organization's capacity for goal attainment.

Directions: For each issue listed check the number which describes the magnitude and importance of the issue to your organization. (1 = low, while 5 = high). Check the column which describes your assessment of the issue as a constraint, threat, or opportunity.

a. Count the number of issues checked in each column and place in the appropriate box.

b. Multiply the order of magnitude or importance (1, 2, 3, 4 or 5) by the number of checks in each box to obtain a score.

c. Add each of the five column scores under MAGNITUDE and divide by the total number of issues to obtain the Average M. Using the IMPORTANCE columns, follow the same procedure to obtain an Average I.

d. (1) To determine scanning ability, add the Average M and the Average I and divide by 2.

Very Poor   Poor   Good   Very Good   Excellent

(2) To determine outlook percentage, divide the total number of checks for C, T, O obtained in (a) by the number of issues, multiply by 100, and place the percentage in the box.

Outlook: Pessimistic if C = over 60%; Hysterical if T = over 60%; Optimistic if O = over 60%

ENVIRONMENTAL SCANNING: EN-SCAN

EN-SCAN is a method of estimating the manager's sensitivity to issues (events, occurrences) in the external business environment that may impact the organization's future survival, success, development, and/or performance. Once the issues of greatest significance have been identified, the organization can integrate (incorporate) this information into the strategic planning process.

EXHIBIT 15–5 Example of GS: Computer Firm

Issues Identification	Issues Evaluation										Meaning		
	Magnitude					Importance 1 (low) – 5 (high)					C	T	O
	1	2	3	4	5	1	2	3	4	5			
1. China as a world economic power				✓						✓			✓
2. Controlling health care costs					✓			✓			✓		✓
3. Campus unrest			✓			✓					✓		
4. Birth defects				✓				✓			✓		
5. Saving Africa					✓		✓					✓	
6. Distribution of wealth created by machines (taxing robots)									✓		✓		
7. Individual lifestyle vs. community rights			✓				✓				✓		
8. Land use (food growth vs. tobacco growth)				✓			✓						✓
9. Role of foreign capital in the U.S.				✓				✓			✓		
10. Regulation of information transfer (personal data security)				✓					✓		✓		
11. Role of the computer education			✓							✓			✓

12. Climate changes (ozone, acid rain)
13. Right to quiet (noise pollution)
14. Fringe benefits for part-time employees
15. Human language/computer language (excessive emphasis on computer literacy will impair language skills)
16. International regulation of financial markets

a. Number checked
b. Score
c. Total
d. (1) Scanning ability
   (2) Outlook

Average M = 3.56

Average I = 3.44

$$\frac{\text{Average M} + \text{Average I}}{2} = 3.5$$

a. Count the number of issues checked in each column and place in the appropriate box.

b. Multiply the order of magnitude or importance (1, 2, 3, 4 or 5) by the number of checks in each box to obtain a score.

c. Add each of the five column scores under MAGNITUDE and divide by the total number of checks to obtain an Average M. Using the IMPORTANCE columns, follow the same procedure to obtain an Average I.

d. (1) To determine SCANNING ABILITY, add the Average M and the Average I and divide by 2.

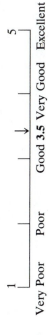

1				5
Very Poor	Poor	Good **3.5** Very Good	Excellent	

(2) To determine outlook percentage, divide the total number of checks obtained in (a) by the number of issues, multiply by 100, and place the percentage in the box.

Outlook: Pessimistic if ©= over 60%; Hysterical if T = over 60%; Optimistic if O = over 60%

EXHIBIT 15–6   Example of GS: Computer Firm

ENVIRONMENTAL SCANNING: EN-SCAN

EN-SCAN is a method of estimating the manager's sensitivity to issues (events, occurrences) in the external business environment that may impact the organization's future survival, success, development, and/or performance. Once the issues of greatest significance have been identified, the organization can integrate (incorporate) this information into the strategic planning process.

Issues Identification	Issues Evaluation Magnitude 1	2	3	4	5	Importance 1 (low) 1	2	3	4	5 (high)	Meaning C	T	O
1. China as a world economic power			✓				✓					X	
2. Controlling health care costs		✓				✓						X	
3. Campus unrest		✓					✓				X		
4. Birth defects		✓				✓						X	
5. Saving Africa			✓			✓						X	
6. Distribution of wealth created by machines (taxing robots)		✓				✓						X	
7. Individual lifestyle vs. community rights		✓				✓						X	
8. Land use (food growth vs. tobacco growth)		✓					✓				X		
9. Role of foreign capital in the U.S.			✓				✓				X		
10. Regulation of information transfer (personal data security)			✓			✓						X	
11. Role of the computer education			✓			✓							X

Issues	Magnitude (1–5)					Importance (1–5)					Outlook		
12. Climate changes (ozone, acid rain)			✓									✓	
13. Right to quiet (noise pollution)		✓					✓					X	X
14. Fringe benefits for part-time employees		✓						✓					X
15. Human language/computer language (excessive emphasis on computer literacy will impair language skills)		✓						✓				X	X
16. International regulation of financial markets		✓							✓				X
a. Number checked	0	9	7	0	0	0	9	6	1	0	5	10	1
b. Score	0	18	21	0	0	9	12	6	1	3			
c. Total	Average M = 2.44					Average I = 2.18					31	63	6

d. (1) Scanning ability $\dfrac{\text{Average M} + \text{Average I}}{2} = 2.31$

(2) Outlook

$$\underset{\text{Very Poor}}{1} \quad \underset{\text{Poor } \mathbf{2.3}}{\longrightarrow} \quad \underset{\text{Good}}{ } \quad \underset{\text{Very Good}}{ } \quad \underset{\text{Excellent}}{5}$$

a. Count the number of issues checked in each column and place in the appropriate box.

b. Multiply the order of magnitude or importance (1, 2, 3, 4 or 5) by the number of checks in each box to obtain a score.

c. Add each of the five column scores under MAGNITUDE and divide by the total number of issues to obtain the Average M. Using the IMPORTANCE columns, follow the same procedure to obtain an Average I.

d. (1) To determine SCANNING ABILITY, add the Average M and the Average I and divide by 2.

(2) To determine outlook percentage, divide the total number of checks obtained in (a) by the number of issues, multiply by 100, and place the percentage in the box.

Outlook: Pessimistic if C = over 60%; Hysterical if Ⓣ = over 60%; Optimistic if O = over 60%

TABLE 15–2   Issue Evaluation of the Top 20 Issues Identified by NEISNET Issue Evaluation

Issues	Mean Magnitude	Mean Importance	Cross Mean	M-I
1. Liability insurance/tort legislation	3.62	3.69	3.66	−0.07
2. Tax reform	4.19	3.75	3.97	0.44
3. Superfund reauthor/toxic waste	3.25	2.94	3.10	0.31
4. Trade legislation	3.06	2.69	2.88	0.37
5. South African divestiture/apartheid	3.62	2.37	3.00	1.25
6. U.S. dollar/U.S. economy	3.37	3.62	3.50	−0.25
7. Oil import fee	2.62	3.00	2.81	−0.38
8. Gramm-Rudman/deficit	3.37	3.13	3.250	0.25
9. Deregulation/privatization	2.50	3.62	3.06	−1.12
10. FCC constraints	1.69	2.00	1.85	−0.31
11. Shifting responsibility– federal and local to state	2.69	3.31	3.00	−0.62
12. Health care costs	3.56	3.56	3.56	0.00
13. Trans border data flow	2.25	2.56	2.41	−0.31
14. Nutritional labeling	2.50	1.95	2.22	0.56
15. Irradiation of food	2.00	2.19	2.10	−0.19
16. Crisis management	2.06	2.81	2.44	−0.75
17. Environmental effects of chemicals	3.56	3.29	3.38	0.37
18. Emissions control	3.06	2.94	3.00	0.12
19. Process patents	2.12	2.56	2.34	−0.44
20. Child care	2.81	3.06	2.94	−0.25
Overall Average	2.90	2.94	2.92	−0.05

Step 2: Issue Evaluation.   These 20 issues were evaluated by the LINC members by using the system explained in the first portion of this part.

Table 15–2 shows the issue evaluation results for 50 of the LINC respondents. The columns give data on: (a) the magnitude; (b) the importance; (c) cross mean (i.e., M + I); and (d) the directionality (M − I) of these issues. Table 15–3 provides data on the top five issues only.

These tables help illustrate the usefulness of the EN-SCAN framework. Thus, by forcing respondents to not only list the most "important" issues, as is usually done by the popular pollsters, but also to assess the "importance of the impact that the issue will have on the organization," respondents go beyond impulsive naming.

Put differently, through the use of EN-SCAN, the individuals who do environmental scanning for the organization focus only on issues that are truly relevant to the organizational participants. This helps insure that issues that make

TABLE 15–3   Top Five Issues

Top Five Issues by Magnitude (M)	
1. Tax reform	4.19
2. Liability insurance/tort legislation	3.62
3. South Africa divestiture/apartheid	3.62
4. Environmental effects of chemicals	3.56
5. Health care costs	3.56
Top Five Issues by Importance (I)	
1. Tax reform	3.75
2. Liability insurance/tort legislation	3.69
3. U.S. dollar/U.S. economy	3.62
4. Regulation/privatization	3.62
5. Health care costs	3.56
Top Five Issues by Cross Mean (M + I)/2	
1. Tax reform	3.97
2. Liability insurance/tort legislation	3.66
3. Health care costs	3.56
4. U.S. dollar/U.S. economy	3.50
5. Environmental effects of chemicals	3.38
Top Five Issues by M > I (Greater Magnitude)	
1. South Africa divestiture/apartheid	1.25
2. Nutritional labeling	0.56
3. Tax reform	0.44
4. Trade legislation	0.37
5. Environmental effects of chemicals	0.37
Top Five Issues by M < I (Greater Importance)	
1. Deregulation/privatization of utilities	1.12
2. Crisis management	0.75
3. Shifting responsibility—federal and local to state	0.62
4. Process patents	0.44
5. Oil import fee	0.38

it to the next step do not waste the time of the professional planners just because they are on the popular list of some issue identification organization, but that they will help address the truly significant challenges that will face the organization in the future.

Consider, for example, the issue that appears at the top of both EN-SCAN's magnitude and importance lists: tax reform. One immediately notes that even though the respondents rated the issue 4.19 on magnitude, their rating of the issue's importance was considerably lower, i.e., 3.75—a difference of 0.44. This suggests that while the public media assign substantial importance to the issue, the actual impact on the individual respondent's organization is perceived as only intermediate.

FIGURE 15–10 Environmental Scanning (EN-SCAN): An Assistant Expert System

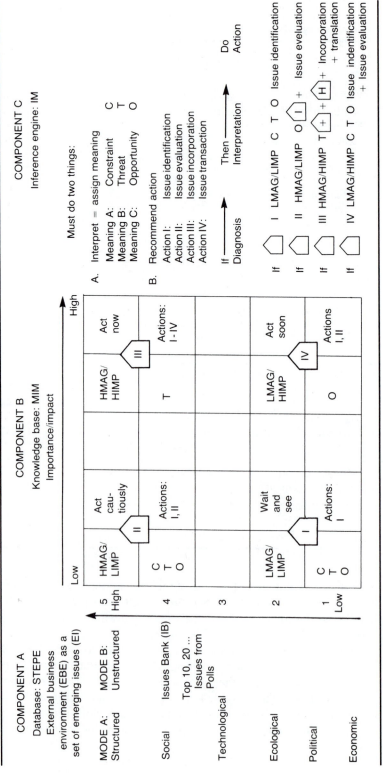

SOURCE: A. G. Kefalas.

Step 3: Issue Incorporation.   The purpose of this step is to incorporate the findings of the evaluation process into the organization's strategic plan. EN-SCAN incorporates its evaluation of the issues into an organization's strategic plan by going through a rather elaborate process of "scenario writing."

Scenario Writing.   The purpose of scenario writing is to help planning managers present key external developments in a language familiar to both the planning staff and line executives.

The scenario writing process begins with the design of the Magnitude/Awareness and Importance/Impact matrix (Figure 15–10). Each of the five issues is positioned in the MAG/IMP matrix. The position in the MAG/IMP matrix will automatically dictate certain actions. In the above example, the top issue (tax reform) is positioned in the high magnitude quartile of the matrix. As a result of this positioning, the issue is judged as a threat. Accordingly, it is recommended that the legal and tax staff work with the strategic planners and adjust the next year's cash flows to reflect the higher taxes incorporated in the new law. The same is true with the second issue, liability insurance/tort legislation. This issue is also a threat.

The next issue, South Africa divestment/apartheid, which received a 3.62 in the magnitude scale, was judged as rather unimportant, since it only received a 2.37 on the importance scale. Thus, this issue will be positioned as a high magnitude/low importance issue and accordingly EN-SCAN recommends that no action be taken.

After the process of positioning the five top issues on the MAG/IMP matrix, it is now time to actually write the scenarios. Such narrative descriptions of the environment could be developed assuming that the environment consisted of only the five top issues in any of the five columns of the table. However, some writers may find it useful to describe the environment by writing three alternative scenarios (a pessimistic, an optimistic, and a most likely scenario) on the five issues using the magnitude scale. For other purposes and/or other bosses, the top five issues on the importance scale might be more appropriate. Using the last set of top issues where importance is greater than magnitude, one could write a scenario that runs as follows: Deregulation of public utilities seems to be imminent. Sources in Washington suggest that the U.S. Congress is determined to pass some type of deregulation legislation. Influences on our cash flow streams is great. In addition, the staff estimates that hopes for future rate increases already included in next year's strategic plans must be abandoned.

Step 4: Issue Translation.   This step seeks to assign quantitative measurements to the consequences of the impact of each scenario on the various organizational strategies that make up the strategic plan. Basically, this procedure adjusts each major estimate of human, technological, and monetary resource utilization in the organization's strategic plan for the events described in each scenario. This demonstrates the impact these adjustments have on the results projected in the plan.

Unfortunately, space does not permit a more detailed description of this step. The interested reader is referred to Amara and Lipinski's text, *Business Planning*

TABLE 15–4   PLEXSYS Report System

TABULATION OF VOTES -- NUMBER OF TIMES IN POSITIONS

	1	2	3	4	5	6	7	8	9	10	11	12	13	14	15	16	17	RANK
AIDS	9	2	1	2	1	2	–	–	1	–	–	–	–	–	–	–	–	274
Budget a	2	2	3	2	3	2	–	–	1	–	–	–	1	2	–	–	–	224
EDUCATIO	3	–	4	2	1	2	–	–	1	1	–	1	2	–	1	–	–	212
NUCLEAR	2	1	3	2	–	–	1	2	4	–	2	–	1	–	–	–	–	209
NATIONAL	–	5	–	2	1	–	1	1	1	3	2	–	1	–	–	1	–	196
MEDICAL	–	2	2	1	1	2	–	3	–	1	1	3	–	1	–	1	–	182
Ecologic	–	2	–	1	–	3	–	1	2	4	1	1	–	–	2	–	1	162
Racism a	1	1	–	1	1	2	3	–	1	1	1	1	1	–	1	3	–	161
HOMELESS	–	–	3	1	1	–	1	2	2	1	2	–	1	2	1	1	–	161
Trade Ba	1	1	–	2	2	–	2	–	2	–	3	1	–	–	1	1	2	161
Glasnost	–	–	1	1	3	2	1	1	–	–	2	1	3	1	–	1	1	155
DRUG TES	–	–	1	1	–	1	3	1	1	3	1	2	–	–	–	2	2	142
Central	–	–	–	–	2	1	3	–	–	1	–	4	2	3	–	1	1	128
Mideast	–	1	–	–	1	1	1	2	–	–	–	1	2	4	4	1	–	118
Space Pr	–	1	–	–	–	–	2	–	–	2	1	2	1	1	3	1	4	97
Aparthei	–	–	–	–	1	–	–	1	2	1	–	–	2	3	3	2	3	87
Social S	–	–	–	–	–	–	–	4	–	–	2	1	1	1	2	3	4	85

TABULATION OF VOTES – HISTOGRAM

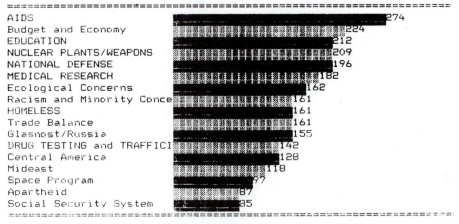

AIDS	274
Budget and Economy	224
EDUCATION	212
NUCLEAR PLANTS/WEAPONS	209
NATIONAL DEFENSE	196
MEDICAL RESEARCH	182
Ecological Concerns	162
Racism and Minority Conce	161
HOMELESS	161
Trade Balance	161
Glasnost/Russia	155
DRUG TESTING and TRAFFICI	142
Central America	128
Mideast	118
Space Program	97
Apartheid	87
Social Security System	85

Rank = the number of votes for AIDS, for example, is multiplied by the rank number and summed (e.g., $(9 \times 17) + (2 \times 16) + (1 \times 9) = 274$)

for an Uncertain Future, for the techniques available for performing such calculations. In general, planning staff will revise the strategic plan using their best interpretation of the scenarios and advise operating personnel to do the same in their own operational plans. Finally, all functional areas will be notified of the appropriate changes.

Application Three: The PLEXSYS Version

In Chapter 13 we provided a brief description of the University of Arizona's package called PLEXSYS. Here we will illustrate the use of this package by students of the MAN 905 course, The Systems Approach, a graduate course at the University of Georgia, College of Business Administration. The students who used this text experimented with the entire process of environmental scanning and PLEXSYS.

At the beginning of the quarter students were asked to scan a number of news articles both printed and audiovisual, for a period of one month. They were asked to develop a matrix with the news they scanned and the main emerging issues they covered. Upon completion of the scanning period, the students were asked to gather at the PLEXSYS lab. Each student sat at a microcomputer, followed the instructions on the screen, and entered the issues that he or she had identified. When all the issues were entered—some 175 of them—the process of "cleaning up" the issues began. After about 30 minutes of deleting duplicate issues and entering or renaming a few, the 17 issues were identified (the limit is set by PLEXSYS). Subsequently the 17 issues were arranged in the order depicted in Table 15–4.

In the next session students were asked to write scenarios using PLEXSYS's Policy Formulation Tool. Subsequently, scenarios were voted on using PLEXSYS's Voting Tool.

SUMMARY

Few managers would ever dare submit a plan for longer than a few months without making a concerted effort to assess the external business environment. However, equally few managers would regard that task as either trivial or easy. Practicing managers discover sooner or later that their performance appraisals will be determined not so much by what they control (i.e., the strategies) but rather by what they do not control (i.e., the environment).

EN-SCAN is an easy and inexpensive way of getting a large number of concerned managers involved in the exceedingly difficult task of becoming aware of the world surrounding the organization, and not concentrating on the way the organization impacts the world. This task is too complex and too important to be left to specialists within or without the organization.

The convenience offered by the microcomputer is one of the greatest strengths of EN-SCAN. Specifically, each designated scanner has a data diskette onto which he or she stores identified issues and evaluates them. Then the scanner turns over the diskette to another individual, usually the facilitator, who copies all responses into the main memory, runs the program, and prints out the results. If an organization has some type of office network, everything could be on a server to which each individual user will have access.

This chapter presented the logic and justification for the organizational design and implementation of an environmental scanning system. After presenting in summary form a reconceptualization of the environment, the chapter focused on industry-tried approaches to environmental scanning, analysis, or issue management. Finally, the last section of the chapter briefly described the environmental scanning system called EN-SCAN. The system can run on any microcomputer (personal computer) such as IBM PC, TRS-80, etc.

With the design of EN-SCAN, the authors of *Management Systems: Conceptual Considerations* have shown that the systems approach is indeed an idea whose time has come and, in particular, that environmental scanning, the central theme of this book, has finally arrived.

KEY WORDS

The New Organization	PLEXSYS
The 21st Century Executive	Strategic Planning
Strategic Management	Planning
Issue Management	Strategy
Issue Development	Environment
EN-SCAN	Constraint
STEPE	Threats
Megatrends	Opportunities

REVIEW QUESTIONS

1. Defend or refute the Machiavelli quotation at the opening of this chapter.

2. Explain the difference between a constraint and a threat. Use some examples from a real business experience.

3. Cite some firms or industries which in your opinion belong to a (1) stable, (2) dynamic, or (3) turbulent environment.

4. How are the proximity and sector frameworks related, and how are they different?

5. How are the sector and the events-issues-trends frameworks related, and how are they different?

6. Cite some emerging issues for the food industry.

7. Cite some emerging issues for the auto and steel industries.

8. Develop a scanning or monitoring system for the IBM company using any of the examples given in the chapter.

9. Develop a critical issues or trends analysis for your own state governor's office.

10. Create a personal computer critical trends tracking system for your university's administration.

Additional References

Ackoff, Russell L. *Creating the Corporate Future.* New York: John Wiley & Sons, 1981.

Boe, Archie R. "Fitting the Corporation to the Future." *Public Relations Quarterly 24,* no. 4 (Winter 1979), pp. 4–6.

Brooke, Michael. "Multinational Corporate Structures: The Next Stage." F86 Futures 11, no. 2 (April 1979), pp. 111–21.

Chase, W. H. *Issue Management.* Stamford, Conn.: IAP, 1984.

Coates, J. F. *Issue Management.* Mt. Airy, Md.: Lomand Publishing, 1986.

Coates, T. F. "The Potential Impacts of Robotics." *The Futurist* 17, no. 1 (February 1983), pp. 28–32.

"The Global 2000 Report: Entering the Twenty-First Century." *World Future Society Bulletin* 14, no. 5 (September–October 1980), pp. 115–21.

Handy, Charles. "Through the Organizational Looking Glass." *Harvard Business Review* 58, no. 1 (January–February 1980), pp. 115–21.

Harman, Willis W. *An Incomplete Guide to the Future.* Palo Alto, Calif.: Stanford Alumni Association, 1976.

Jones, Barry. *Sleepers, Wake! Technology and the Future of Work.* Brighton, England: Harvester Press, 1982.

Kahn, Herman. *The Next 200 Years.* New York: William Morrow, 1976.

Kefalas, Asterios G. "On Managerial Philosophy: Competing Future-Views." *World Future Society Bulletin* 14, no. 1 (January–February 1980), pp. 17–26.

Laszlo, Ervin. *A Strategy for the Future.* New York: George Braziller, 1974.

Magorah, Maruyama. "Beyond Management: The Shifting Focus of Our Economic and Business Rethinking." *Futures* 5, no. 3 (1981), pp. 213–46.

Morf, Martin. "Eight Scenarios for Work in the Future." *The Futurist* 17, no. 3 (June 1983), pp. 24–29.

Nanus, B. "The Future-Oriented Corporation." *Business Horizons* 18, no. 1 (February 1975), pp. 5–12.

Pitt, Douglas, and Simon Booth. "Paradigms Lost? Reflections on the Coming Organizational Revolution." *Futures* 15, no. 3 (June 1983), pp. 193–204.

Sheppard, C. Steward, and Donald C. Carroll, eds. *Working in the Twenty-First Century.* New York: John Wiley & Sons, 1980.

Solomon, Lewis D. *Multinational Corporations and the Emerging World Order.* Kent, England: Bailey and Swinfen, 1979.

Toffler, Alvin. *The Third Wave.* New York: William Morrow, 1980.

Williams, Trevor A. *Learning to Manage Our Futures.* New York: Wiley Interscience, 1982.

Comprehensive Project on the Systems Approach*

GUIDELINES FOR THE CLASS PROJECT

Introduction

This course is designed to facilitate the development of the *conceptual skills* necessary for the creation of a vision which is demanded by today's complexity in the business environment. Conceptual skills are defined by Katz as the manager's ability to (1) see the organization as a whole, (2) focus on the relationships among the parts of that whole, and (3) take a long term view of the organization's future.

In this course an effort is made to provide a conceptual framework which will allow the student to see the organization as a whole and focus on the relationships among the parts of that whole as they are affected by and affect each other's goal-directedness. This framework is drawn from the systems field and specifically General Systems Theory and Cybernetics.

General Systems Theory emphasizes the openness of the system as a prerequisite to its survival and growth. Cybernetics, as *The Science of Control and Communication in The Animal and The Machine* (Norbert Wiener, 1948), introduced the concept of "implicit control" as a universal phenomenon in all exceedingly complex and probabilistic systems. Management Cybernetics, as the hybrid field of the application of General Systems Theory and Cybernetics to the study of large scale systems, conceives of organizations as open systems with a closed loop (self-regulation).

Thus, the subject of inquiry in this class is organizations that are complex systems whose behavior can only be described in probabilistic terms and are, however, self-regulating.

A basic philosophy implicit in this conceptual framework is that this type of organization depends for its evolution not as much on past/historical behavior but rather its existence is closely correlated with the ability to (1) accurately assess the external environment, and (2) timely develop the appropriate plans to deal with problems and opportunities associated with that environment.

Operationally, the framework conceives of the organization as a control system which feeds on information. Both the organization's assessment of the external environment and the design of the appropriate "systems" in its structure and strategy to deal with the environment depend on its ability to (1) gather, (2) organize, (3) process, and (4) disseminate the right information to the right decision points (persons) at the right time for minimum cost.

Practical applications of this framework have thus far been scarce. Systems

---

*This is an example of a class assignment that was used by A. G. Kefalas at the University of Georgia.

people of the GST and Cybernetics camps have assumed that somehow making managers aware of the conceptual issues would take care of the applications. Management Information Systems (MIS) people, on the other hand, have also assumed that once the manager has all the information on the internal aspects of the business the so-called environmental issues will take care of themselves.

The Linkage: Artificial Intelligence

The systems approach, as it is operationalized in this book, attempts to bridge this gap. In other words, the systems approach attempts to link the conceptual systems emphasis on the environment with the MIS emphasis on the internal aspect of the organization.

Recently some new developments have taken place which point to the direction of the Systems Approach. These developments have different names. For the lack of a better name, we will discuss them here under one name, artificial intelligence.

It is becoming exceedingly obvious that the human thinking process is not as incomprehensible and unpredictable as we had believed. At the same time, recent developments in the so-called cognitive sciences, a metadiscipline of information theory, logic, psychology, computer sciences, and mathematics, enabled humans to study and replicate their own thought processes. Applications of these developments via information technology are known as expert systems or knowledge systems. The basic idea behind it is that it is now possible to preserve and duplicate the heuristics of some exceptional individuals with the help of computer systems. These systems can then be used by ordinary individuals and act as "knowledge amplifiers" of these individuals just like a power motor acts as a "muscle amplifier."

The Subject: The Executive

Our subject here is the so-called executive. This person is defined as the ultimate decision maker in an organization. Structurally, the executive is positioned at the top of the hierarchically organized system called the organization. This executive is a "knowledge executive." In other words, the "stuff" he or she must deal with is not things or objects but knowledge about things or objects. The executive must have knowledge about both the internal and the external worlds of the organization. Knowledge is a function of information which in turn depends on the existence of data which in turn depends on the existence of a system which gathers these data.

For all practical purposes, it is assumed that the existing MIS structure is adequate in serving the knowledge executive as far as the internal world is concerned. By the same token, it is assumed that the MIS structure is *in*adequate in servicing the executive's knowledge needs about the external environment.

Environmental scanning is assumed to be the appropriate tool for designing an information system that will supply the executive the knowledge about the external environment which he or she needs to manage the long-term survival and evolution of the organization.

The Project

The task is to design an executive information system (EIS) using the systems framework which will supply the executive with the necessary information to create the knowledge needed for the management of a complex system in an open and very turbulent world.

The Approach

You will work as a team. Team building is the latest buzzword in management. Four students will form a consulting company that specializes in the design of executive information systems with an emphasis on external information. The firm uses environmental scanning as the shell for the development of these systems. Six weeks will be allowed for this project.

It is desirable that the project will go through all three steps of the systems approach: Conceptualization, measurement, and computerization.

Methodology

The organization, like all organic systems, is a goal-directed system. Management is the agent who sets the organizational goals. The overall mission of the organization is its profitable survival and growth/prosperity.

Goals in systems nomenclature are jointly set by the organization and its environment. The organization/management makes an offer to the environment that it wishes to pursue a certain goal. The environment, in turn, accepts or rejects this offer. Once a congruence has been reached between the organization's wishes/goals and the environment's tolerances/acceptances, management formalizes these goals in its strategic plan.

Procedure

Start with setting goals for the organization that you have chosen for the project. Work backwards towards the outputs of the system, then to the inputs, and then back to the processes.

Step 1: Set/Determine goals/preferable . . . what is desired . . .

Step 2: Determine the environmental tolerances/probable . . . may happen . . .

Step 3: Determine outputs/possible . . . can do . . .

Step 4: Create the management vision

Step   5:   Determine informational needs

Step   6:   Design the system's architecture

Step   7:   Draw up the system's policies

Step   8:   Hook your system to the strategic plan

Step   9:   Test the system's effectiveness

Step 10:   Prepare for the final presentation

The Basic Question: What is an executive?
What kinds of information does the executive need to set the organizational goals, choose the appropriate strategies and evaluate the degree of accomplishment of these goals?
What are the basic sources of this information?
What is the most appropriate mode of transmission?

Details

You may wish to use a real/actual organization or may wish to invent one. In either case, a complete description of the organization must be provided. Financial and other information about organizations can be found in the library, or can be electronically retrieved via the computer system at the library.

GOOD LUCK!

Glossary

accompanying attributes Those characteristics of an object whose presence or absence would not affect the designation or definition of the object under study.

adaptability of a system The ability of a system to learn and to alter its internal operations in response to changes in the environment.

algorithm A terminating finite set of instructions organized and designed so that its mechanical processing will result in the solution of a specific problem.

analysis The breaking up of study subjects into smaller and more manageable components for individual examination and evaluation (compare synthesis).

artificial intelligence A subfield of computer science concerned with the concepts and methods of symbolic inference by a computer and the symbolic representation of the knowledge to be used in making inferences. A field aimed at pursuing the possibility that a computer can be made to behave in ways that humans recognize as "intelligent" behavior in each other.

attributes Properties of objects or relationships that manifest the way something is known, observed, or introduced in a process.

black box A component of a system that is considered only in terms of its inputs and outputs and whose internal mechanisms are unknown or unknowable.

block diagram A basic schematic tool for illustrating functionally the components of a control system. The four basic symbols employed are the arrow, the block, the transfer function, and the circle.

boundaries of a system The line forming a closed circle around selected variables, where there is less interchange of energy and/or information across the line of the circle than within the delimiting circle.

bounded rationality The idea that all managerial decisions involve limited, imperfect knowledge. Because human beings, for various reasons, cannot explore all possible alternatives, the decisions made are considered to be good enough (satisficing), not the very best possible (optimal).

Classical Organization Theory An early theory based on two major themes: the use of people as adjuncts to machines in the performance of routine production tasks, and the formal structure of the organization.

closed loop   A system in which part of the output is fed back to the input in such a way that the system's output can affect its input or some of the system's operating characteristics.

closed system   A system that does not take in or give out anything to its environment (compare open system).

collaborative decision support systems (CDSS)   Systems that are designed to facilitate group work via electronic brainstorming and issue management.

complexity   That property of a system resulting from the interaction of four main determinants: the number of system elements, their attributes, the number of interactions among the elements, and the degree of organization of the elements.

cones of resolution   Hierarchically arranged levels of conceptualization with the most abstract at the top and the most concrete at the bottom. Each distinguishable feature at one level may represent a wealth of detail when examined on a larger scale at a lower level.

cost-benefit analysis   A method of analyzing alternatives in which the costs and benefits of each alternative are determined and compared.

cybernetic system   A system characterized by extreme complexity, probabilism, and self-regulation.

cybernetics   The science of control and communication in the animal and in the machine.

data   Material of little or no value to an individual in a specific situation. Also, material that does not reduce the amount of ignorance or the range of uncertainty in the mind of the decision maker (compare information).

decision system   A subsystem of a goal-seeking system, serving as a channel through which the goal-seeking system acts on the environment.

defining attributes   Those characteristics without which an entity would not be designated or defined as it is.

demography   The field of human population analysis, with special reference to the size, density, etc., of the population; also applied to the study of other animal populations.

deterministic   Having a specific outcome, or a situation in which outcomes must follow from specific courses of action or inputs in a perfectly predictable way (compare stochastic).

domain   An area or sphere of interest; a field of one's expertise.

ecosystem   A natural unit of living and nonliving elements which interact to produce a stable system.

enterprise-environment interaction system   An open system that functions by importing the necessary resources and/or information from the environment and by exporting the product of the combinations of these resources into the environment.

entropy   A measure of the degree of disorder in a closed system; also, the natural tendency of objects to fall into a state of disorder.

environment   That which not only lies outside the system's control but which also determines in some way how the system performs.

environment-organization interactive system   In systems terminology, the superordinate system: the whole.

environmental factors   A set of measurable properties of the environment perceived directly or indirectly by the organization operating in the environment.

environmental scanning   The process of acquiring information about the organization's environment.

epistemological   Concerning the theory of the nature, origin, content, and validity of knowledge.

equifinality   A characteristic of systems in which final results may be achieved with different starting conditions and with varying inputs and in different ways.

executive support systems   On-line systems connecting PCs to mainframes which allow the executive to obtain status reports and drill-down.

expected value of perfect information   The difference between the expected value under certain prediction and the expected value of the optimal strategy under uncertain prediction.

expert system   A computer-based system composed of a user interface, an inference engine, and a knowledge base derived from one or more experts in a particular field.

feedback   The return of some of the output of a system as input (see closed loop).

feedback control system   A system that tends to maintain a prescribed relationship between one system variable and another by comparing functions of these variables and using the difference as a means of control; also, the transmission of a signal from a later to an earlier stage.

first-order feedback system   A simple automatic goal-maintenance system.

flow diagram   A graphical representation of the sequence of data transformations needed to produce an output data structure from an input data structure.

focal system   The system or subsystem that is the object of study.

game theory   A mathematical approach to idealized problems of games with conflict or competition among the units.

General Systems Theory   The theory of open organic systems that possess certain characteristics such as organization, dynamic equilibrium, self-regulation, and teleology. Its main domain, the growth and evolution of general systems, evolved from Bertalanffy's concept of organismic evolution.

global modeling   The application of the systems approach to world problems.

goal-seeking systems   Systems that can so modify their output through a goal-feedback mechanism that they tend toward a preset state or goal.

group decision support systems (GDSS)   Systems which are designed to facilitate group decision making.

groupware   Another name of CDSS. Systems that are designed to facilitate group work and team building.

heuristics   Rules of thumb or educated guesses learned from experience that experts use in solving problems in their domain of specialty. An independent discovery method for learning.

hierarchical system   A system composed of interrelated subsystems, all of which are ranked or ordered such that each is subordinate to the one above it, until the lowest elementary subsystem level is reached.

holistic   Concerned with wholes or complete systems rather than with parts of the system. (Sometimes erroneously spelled wholistic.)

homeostasis   The maintenance of static or dynamic equilibrium between the different and interdependent elements of an organism, irrespective of external effects.

ideal-seeking system   A system which upon attainment of any of its goals or objectives, seeks another goal or objective that more closely approximates its ideal.

industrial dynamics   The study of the information-feedback characteristics of industrial activity to show how organizational structure, amplification (in policies), and time delays (in decisions and actions) interact to influence the success of the enterprise. Also known as system dynamics.

inference engine   That part of an expert system that enables one to generate new facts from existing facts by applying already acquired procedures to new situations. That part of a knowledge based system that contains the procedures for reaching a conclusion.

information   Evaluated data for specific individuals, working on a particular problem at a specific time and for achieving a specific goal. Also, selected data for reducing the amount of ignorance or the range of uncertainty in the mind of the decision maker. The selection of a given message from a set of possible messages.

input   A start-up force or signal that provides the system with its operating necessities. Or it can be a stimulus or excitation applied to a system from an external source eliciting a response from the system. Or the importation of data and instructions which translate the external form into a set of symbols that can be read and interpreted by the computer's electronic circuitry.

internal organization system   A subsystem of a goal-seeking system, that acts as an evaluator of the scanning system.

isomorphic systems   Two systems whose elements exist in a one-to-one correspondence with each other. There is also a correspondence between the systems' operational characteristics.

knowledge base   That body of facts, rules, principles, laws and heuristics that forms the basis of a knowledge system.

knowledge-based system   A decision support system containing a knowledge base and an inference engine.

knowledge engineering   The acquisition, organizing, and designing of a knowledge base consisting of facts, rules, laws, principles, and heuristics.

law of requisite variety   There must be as much variety in the control mechanism as there is in the system being controlled. Only variety can destroy variety (Ashby's Law).

LISP   A largely procedural programming language that represents information in the form of lists. It is widely used in artificial intelligence research. Stands for LISt Processing.

matrix organization   The combination of project organization and functional organization in which a pool of specialists is assigned to a particular project for the duration of the project. These personnel are subject to the horizontal authority of the project manager as well as to the vertical, functional authority.

metatechnology   The conceptualization, design, and implementation of ways of organizing man and machine into systems for the collection, storage, processing, dissemination, and use of information.

mismatch signal   The signal that conveys to the organizing system feedback information and changes in the external environment of a system, with one known intended state and one known actual state.

model   A simplified representation of something to be made or already existing. Models may be physical, schematic, or mathematical.

Modern Organization Theory   Theory characterized by three parallel developments: the continuation of earlier classical and neoclassical theory, the emergence of behavioral science research, and the emergence of operations research.

modern systems approach   The view that organizations are systems which are in constant interaction with their environment (see open systems).

negative feedback   The return of some of the outputs of a system as inputs in such a way that they are deviation-counteracting systems. Such mechanisms are control-maintaining devices.

natural language   Natural language refers to conversational language, which, for our purposes, means English although it could be any other language.

Neoclassical Organization Theory   The theory that accepts the basic postulates of the classical theory but modifies them by superimposing changes in operating methods and structure evoked by individual behavior and the influence of the informal group.

nesting of a system   The division and subdivison of a system into subsystems and subsubsystems depending on the particular resolution level desired.

noise   Any disturbance which does not represent any part of a message from a specified source. Usually refers to random disturbances.

objects The inputs, processes, outputs, and feedback control of a system.

open-loop system A system in which the output of the system is not coupled to the input for measurement.

open system A system that interacts with its environment (compare closed system).

operations research A systems approach to problem solving, using a set of mathematical techniques for the management of organizations. Also, the application of scientific methods, of mathematical and statistical techniques, and of other tools to problems involving the operations of systems in order to provide those in control of the operations with optimum solutions to problems.

organizational effectiveness The degree to which the organization achieves its goals or objectives. Also, the ability of the organization to exploit its environment in the acquisition of scarce and valued resources.

organized complexity Interactions within and among subsystems, that, viewed as a hierarchy, can be described as a series of feedback loops arranged in an ascending order of complexity.

output The result of the process, or alternatively, the purpose for which the system exists.

paradigm A pattern or example.

parameters Elements outside a designated system which have an effect on one or more variables of the designated system (compare variables).

positive feedback The return of some of the outputs of a system as inputs in such a way that they are deviation-amplifying systems. Such mechanisms are growth-promoting devices.

process The manner of combining the inputs so that the system will achieve a certain result. Or the process that transforms the input into an output.

PROLOG A non-procedural symbolic logic programming language now widely used in artificial intelligence research. Originally designed to evaluate theorems to see if they are true. The language adopted by the Japanese for the fifth generation computers that they are designing. Stands for PROgramming in LOGic.

random input Outputs from previous systems that are potential inputs to the focal system.

recursion A computer programming technique whereby a function calls on itself to solve a problem. The execution of each new operation, however, unlike that in iteration, is dependent upon the result of the previous one.

redundant relationship Those relationships that duplicate other relationships. They increase the probability that a system will operate all of the time instead of just some of the time.

relationships The bonds that link objects together.

resources All the means available to the system for the execution of the activities necessary for goal realization.

robotics   A reprogrammable multifunctioning manipulator designed to move material, parts, tools, or specialized devices through variable programmed motions for the performance of a variety of tasks.

satisficing behavior   The idea that people strive for accomplishments that they consider good enough rather than for the very best possible.

scanning   The process whereby the organization acquires information for decision making.

scanning system   A subsystem of a goal-seeking system, serving as a channel through which the goal-seeking system receives information about its environment.

search   A mode of scanning which aims at finding a particular piece of information for solving a specific problem.

second-order feedback system   A system with a memory unit that is able to initiate alternative courses of action in response to changed environmental conditions, and is also able to choose the best alternative for the particular set of conditions. It is an automatic goal-changing system.

serial input   The result of a previous system with which the focal system is serially or directly related.

simulation   Technique involving the use of a mathematical model to determine how the real system would behave under changed conditions by observing the behavior of the system's model under changed values of its variables and of the environment's parameters.

state of a system   The set of relevant processes in a system at a given moment of time, determined by the accumulation or integration of the past rates or flows.

steady state   A situation where inputs and outputs are constant.

stochastic   Having a probabilistic outcome. Or one where the next event is but randomly related to previous events.

strategic problem of open systems   The maintenance and regulation of flow of information between the system and its environment.

suboptimization   Decisions by subunits or individuals that are desirable or optimal for them but which are harmful to and less than optimal for the larger organization of which they are a part.

surveillance   A mode of scanning which aims at finding some general knowledge for the information seeker.

symbiotic relationship   One without which the connected systems cannot continue to function.

synergistic relationship   One in which the cooperative action of semi-dependent subsystems taken together produces a total output greater than or superior to the sum of their outputs taken independently.

synergy   The system's output in which the total effect is greater than or superior to the effects obtained through the parts functioning independently. Often represented by $2 + 2 = 5$.

synthesis   The combining of disparate elements into a whole.

system variety   The possible state of affairs (complexity) in a system which is equal to the number of states raised to the power of the number of relationships.

systems   A set of objects, together with relationships between the objects and between their attributes, connected or related to each other and to their environment in such a way as to form a whole.

systems analysis   The organized step-by-step study of the detailed procedures for the collection, manipulation, and evaluation of data about an organization for the purpose of determining not only what must be done but also to ascertain the best way to improve the functioning of the system.

systems approach   A philosophy that conceives of an enterprise as a set of objects with a given set of relationships between the objects and their attributes, connected or related to each other and to their environment in such a way as to form a whole. It views a problem as a whole.

system's behavior   A series of change in one or more structural properties of the system or of its environment.

systems chart   Chart that focuses on the inputs and the outputs of the system; it identifies programs, procedures, and data structures by name.

systems dynamics   The science of feedback behavior in multiple-loop nonlinear social systems. (See also industrial dynamics.)

systems thinking   A way of conceptualizing with the objective of reversing the subdivision of the sciences into smaller and more highly specialized disciplines through an interdisciplinary synthesis of existing scientific knowledge.

teleological behavior   Goal-directed behavior of an organism or machine controlled by an error-correcting mechanism (negative feedback).

third-order feedback system   A system with a memory, able to initiate alternative courses of action in response to changed environmental conditions and to choose the best alternative for the particular set of conditions, coupled with the ability to reflect upon its past decision making. It is a reflective goal-changing system.

uncertainty   Inability to predict the final state of the system given its initial state.

variables   Elements within a designated system.

Name Index

Subject Index